Post-2020 Climate Action

Shinichiro Fujimori • Mikiko Kainuma
Toshihiko Masui
Editors

Post-2020 Climate Action

Global and Asian Perspectives

Springer

Editors
Shinichiro Fujimori
Center for Social and Environmental Systems Research
National Institute for Environmental Studies (NIES)
Tsukuba, Ibaraki, Japan

International Institute for Applied Systems Analysis
Laxenburg, Austria

Mikiko Kainuma
Institute for Global Environmental Strategies (IGES)
Hayama, Kanagawa, Japan

Center for Social and Environmental Systems Research
National Institute for Environmental Studies (NIES)
Tsukuba, Ibaraki, Japan

Toshihiko Masui
Center for Social and Environmental Systems Research
National Institute for Environmental Studies (NIES)
Tsukuba, Ibaraki, Japan

ISBN 978-981-10-3868-6 ISBN 978-981-10-3869-3 (eBook)
DOI 10.1007/978-981-10-3869-3

Library of Congress Control Number: 2017936199

Printed on acid-free paper

This Springer imprint is published by Springer Nature
The registered company is Springer Nature Singapore Pte Ltd.
The registered company address is: 152 Beach Road, #21-01/04 Gateway East, Singapore 189721, Singapore

Preface

In 2015, the twenty-first session of the Conference of the Parties (COP 21) to the United Nations Framework Convention on Climate Change adopted the Paris Agreement. The Paris Agreement provides a framework for global action to address climate change in the period after 2020. The objective of the agreement is to limit the increase in global temperatures to well below 2 °C above preindustrial levels while also considering a limit of 1.5 °C. In preparation for COP 21, each country submitted an emissions reduction target as their intended nationally determined contribution (INDC). Although the individual countries' pledges do not legally bind their emissions, this global policy trend is intended to change the landscape of the climate change discussion.

My initial reaction to observing this news was that it would help to make the world more sustainable. However, several aspects of the effects of the Paris Agreement must be clarified. Here, I would like to introduce three of these concerns. The first is related to the temporal scope of the agreement. We must better understand the short-, medium-, and long-term implications of the Paris Agreement and determine whether the short- and long-term objectives are consistent. The second is related to its spatial scope. Because individual countries declare their targets from their own perspective, the intensity of the reduction target varies among countries, which could induce political conflicts. The third concern is related to economic development and the potential negative impacts of climate mitigation actions on development. In particular, such impacts would have greater consequences for low-income countries. Asian region should be highlighted in the sense that they are expected to grow economically coming decades.

In this book, we have attempted to approach these fundamental questions using an integrated assessment model (IAM), the Asia-Pacific Integrated Model/Computable General Equilibrium (AIM/CGE) model. I believe that we have succeeded in addressing most of the greatest concerns regarding the Paris Agreement in this book, although several questions remain for future studies. Moreover, because details of the model structure are included in the final two chapters, this book could be used as a CGE model textbook in the context of climate policy assessments.

The AIM modeling team has more than 25 years of experience in assessing climate mitigation policies and climate change impact. The majority of this activity has been focused on developing a computer simulation model that allows us to better understand the implications of climate mitigation policy and then applying this model. In addition, the AIM modeling team has a unique characteristic in the IAM community in that it has a strong long-standing relationship within Asian researchers from countries such as Cambodia, China, Japan, India, Indonesia, Korea, Malaysia, Nepal, Thailand, and Vietnam. Needless to say, this book is one of the fruits of the cooperation and trust among these researchers. I greatly appreciate the contributions of my colleagues to this book and beyond of this activity. I hope that our success over the last several decades will continue in the future. Moreover, I hope that the issues surrounding climate change are resolved along with the broader sense of sustainable development.

I am deeply indebted to the initial AIM founders. Foremost I wish to thank Prof. Yuzuru Matsuoka, who raised me from an amateur to a full-fledged researcher. I am still on the way to find the philosophy of model development and its policy assessment, but it turns out that the injected gene became a great foundation for me. I have seen that Prof. Tsuneyuki Morita's spirit is still alive in the AIM team. It is a regret that I could not directly see and learn from him since he passed away in 2003, which was more than 7 years before I started to work at NIES. I touch upon his great influence from his old friends. I believe that this book would have never been realized without insights of these two professional pioneers. I also greatly appreciate Dr. Mikiko Kainuma and Dr. Hideo Harasawa who are still working hard together now and support the AIM activity. In addition to the chapter authors, there are a number of team members in NIES who either directly or indirectly contribute to the research activities related to this book. I would like to express deep thanks to all of them.

The editors are grateful to the financial support from the Global Environmental Research Fund (2–1402, S-12-2) of the Ministry of the Environment, as well as JSPS KAKENHI (Grant Number JP16K18177) of the Ministry of Education, Culture, Sports, Science and Technology in Japan.

Finally, my special thanks to my wife, Tomoko Hasegawa, who shares her life with me as a great partner and a reliable colleague.

On behalf of the editors.

Tsukuba, Ibaraki
Japan

Shinichiro Fujimori

Contents

Chapter 1
Introduction: Overview and Key Messages

Mikiko Kainuma, Shinichiro Fujimori, and Toshihiko Masui

Abstract This book analyzes the roles of technologies and their prevalence in implementing intended nationally determined contributions (INDCs), estimates the economic impacts and co-benefits of INDCs, clarifies the gaps between the current INDCs and the long-term target of the Paris Agreement to stay well below 2 °C, and investigates measures to narrow these gaps. Analyses of the INDCs of six Asian countries with the Asia-Pacific Integrated Modeling/Computable General Equilibrium (AIM/CGE) model are introduced. The book is also intended to support climate policy analyses by introducing methodologies to analyze the economic impacts of climate policies using AIM/CGE. These analyses show the feasibility of INDCs, their implications for the long-term climate goal, and the challenges to increase the levels of ambition of INDCs.

Keywords Climate change • Intended nationally determined contribution (INDC) • Paris agreement • Mitigation • CGE model

M. Kainuma
Institute for Global Environmental Strategies (IGES), 2108–11 Kamiyamaguchi, Hayama, Kanagawa 240–0115, Japan

Center for Social and Environmental Systems Research, National Institute for Environmental Studies (NIES), 16–2 Onogawa, Tsukuba, Ibaraki 305–8506, Japan
e-mail: kainuma@iges.or.jp

S. Fujimori (✉)
Center for Social and Environmental Systems Research, National Institute for Environmental Studies (NIES), 16–2 Onogawa, Tsukuba, Ibaraki 305–8506, Japan

International Institute for Applied Systems Analysis, Schlossplatz-1, Laxenburg 2361, Austria
e-mail: fujimori.shinichiro@nies.go.jp

T. Masui
Center for Social and Environmental Systems Research, National Institute for Environmental Studies (NIES), 16–2 Onogawa, Tsukuba, Ibaraki 305–8506, Japan

S. Fujimori et al. (eds.), *Post-2020 Climate Action*,
DOI 10.1007/978-981-10-3869-3_1

1.1 Introduction

At the 21st Conference of the Parties (COP21) to the United Nations Framework Convention on Climate Change (UNFCCC) held in Paris in December 2015, world leaders agreed to reduce greenhouse gas (GHG) emissions from human activities to zero during the latter half of the twenty-first century, moving toward the target of limiting the average global temperature increase to no more than 2 °C, and to drive efforts to limit the temperature increase even further to 1.5 °C above preindustrial levels (UNFCCC 2015).

Before COP21, the COP invited all parties to communicate their intended nationally determined contributions (INDCs) to the secretariat well in advance of COP21 in a manner that facilitates the clarity, transparency, and understanding of the INDCs (UNFCCC 2014). As of April 4, 2016, 161 out of 189 countries had submitted INDCs to UNFCCC (UNFCCC 2016a). Many of the INDCs are national in scope, and several contain quantified emissions reduction targets, which take a variety of forms such as absolute targets, relative targets for reducing emissions below the business-as-usual (BAU) level, and intensity targets. The Paris Agreement asked that each party prepare, communicate, and maintain successive nationally determined contributions (NDCs) that it intends to achieve. If a party had communicated an INDC before joining the agreement, the INDC is considered to be a NDC.

The Paris Agreement entered into force on 4 November 2016, 30 days after the date on which at least 55 parties agreed to the convention and their share of the global greenhouse gas emissions exceeds 55% of the global GHG emissions. The parties were also asked to formulate long-term low GHG emission strategies, considering their common and different responsibilities and respective capabilities, in light of different national circumstances (UNFCCC 2015).

Although implementation of the communicated INDCs would lead to sizably lower aggregate global GHG levels compared with the pre-INDC trajectories, the GHG emission levels of the INDCs are expected to be higher than the emission levels under the least-cost 2 °C scenarios (UNFCCC 2016b). The Paris Agreement adopted a mechanism to increase the GHG reduction target by facilitating dialogue in 2018 to consider the collective efforts of parties relative to progress toward the long-term goal. By 2020, parties are urged to communicate a new NDC and to do so every 5 years thereafter (UNFCCC 2015).

There have been many analyses of INDCs (UNFCCC 2016b; Climate Action Tracker 2015; World Resource Institute 2015; MILES Project Consortium 2015; UNEP 2016; Amponin and Evans 2016; Pandey et al. 2016). These reports have analyzed the INDCs with 2025–2030 target years, and most of them have explored the feasibility of staying well below 2 °C in the long run. They are also intended to support the UNFCCC dialogue in 2018, which facilitates dialogue to consider the collective efforts of parties relative to progress toward the long-term goal, and other global stocktakes that will take place every 5 years.

This book analyzes the roles of technologies and their prevalence in implementing INDCs, estimates the economic impacts and co-benefits of INDCs, clarifies the gaps between the current INDCs and the long-term target of the Paris Agreement to stay well below 2 °C, and investigates measures to narrow these gaps. It introduces analyses of the INDCs of several countries using the Asia-Pacific Integrated Modeling/Computable General Equilibrium (AIM/CGE) model. It also is intended to support climate policy analyses by introducing methodologies to analyze the economic impacts of climate policies with AIM/CGE.

1.2 A Modeling Approach for Tackling Climate Change

Many projects have been developed to analyze long-term scenarios and INDCs, including the Modelling and Informing Low Emission Strategies (MILES) project (MILES Project Consortium 2015), Deep Decarbonization Pathways Project (DDPP) (DDPP Project 2015), and CD-LINKS (CD-LINKS 2016). The MILES project analyzes the INDCs of six countries and the European Union (EU). In the DDPP, 16 modeling teams will develop scenarios through 2050 that explore the possibilities of deep decarbonization. CD-LINK is a consortium that is investigating the transformation to low-carbon societies. The results of these scenarios have been stored in the IIASA database (Schleussner et al. 2016; Rogelj et al. 2016).

The AIM team has participated in these projects and has analyzed low-carbon policies. The AIM team has also supported the capacity to conduct integrated assessment modeling in several countries in Asia for more than 20 years (AIM 2016).

Figure 1.1 shows the framework of AIM (Masui 2015), which consists of a family of analytical models, each of which is developed to analyze a specific set of sectors and policy issues. These models are interlinked to provide an integrated system for assessing emission-climate-impact/adaptation.

AIM is a tool designed to provide pathways to a low-carbon society, while ensuring consistency among sectors and policies. It provides comprehensive assessments of various policies, including macroeconomic policies and technological measures. It can analyze trade-offs between rapid economic growth in Asia and its environmental impact and assess sustainable development policies.

The integrated system has several core emission models and element models. The core emission models are the Computable General Equilibrium (CGE) model, Enduse model, Extended Snapshot (ExSS) tool, and Agriculture, Forestry and Other Land Use (AFOLU) model. Low-carbon society scenarios can be developed by linking these models.

The 2003 AIM book focused on the AIM/Enduse model (Kainuma et al. 2003). This book focuses on AIM/CGE and presents several analyses of INDCs using AIM/CGE. A detailed description of AIM/CGE is also presented to facilitate the development of national models with AIM/CGE.

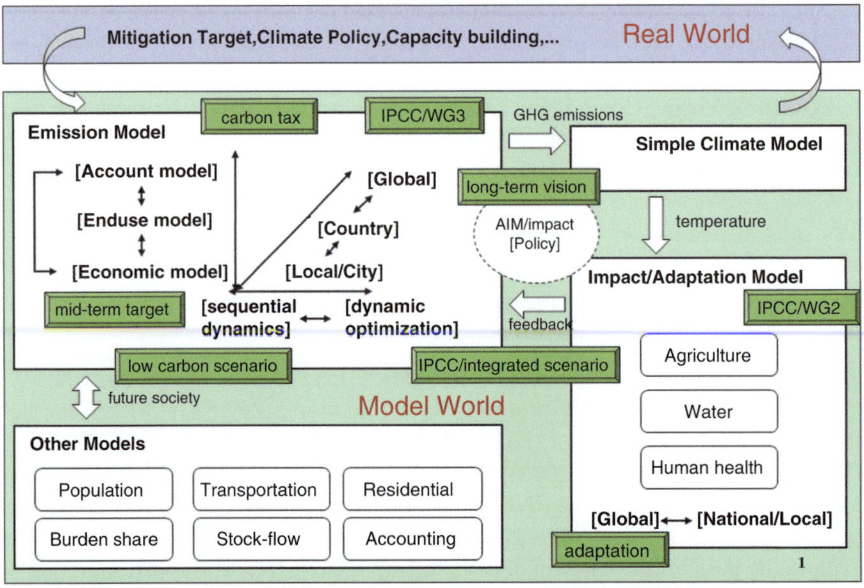

Fig. 1.1 Integrated framework of the AIM used for analyzing climate policies

1.3 Book Structure

The scope of this book is threefold. First, it examines long-term scenarios and INDC scenarios. Second, it examines the impact of climate change under different scenarios. Finally, it examines how the AIM/CGE is used to analyze the INDC of individual countries.

The book consists of three parts. Part I (Chaps. 2, 3, 4 and 5) presents climate policy analyses from global and regional perspectives. Part II (Chaps. 6, 7, 8, 9, 10 and 11) presents country analyses of INDCs. Part III (Chaps. 12 and 13) introduces AIM/CGE and its manual.

Chapter 2 investigates the implications of the near-term emissions targets (for around the year 2030) in the context of the long-term climate mitigation goal using the Asia-Pacific Integrated Model (AIM) framework. In the midterm (from 2030 to 2050), it may be necessary to consider rapid changes to the existing energy or socioeconomic systems, while long-term measures (after 2050) will rely on the substantial use of bioenergy with carbon capture and storage (BECCS) technology or afforestation. This chapter suggests that if the emissions in 2030 are as indicated in the INDCs, the emissions reductions in the period from 2030 to 2050 need to be drastic, and large negative emissions in the latter half of century will be required to meet the 2 °C goal.

Chapter 3 provides an overview of the relevant scientific knowledge about the climate-related risks corresponding to the different levels of mitigation effort, including the two long-term goals in the Paris Agreement, as well as the expected consequences of extending the INDCs into the future without strengthening them. While the trajectory of climate change may not differ significantly among the different scenarios used to attain the same long-term temperature goal, the risks associated with climate change could differ significantly. As a specific example, with the increased dependency on BECCS technology in the latter half of this century, following a weaker near-term mitigation effort, competition for land and water between food production and bioenergy production is expected to increase, resulting in an increased risk of food deficits.

Chapter 4 evaluates the intergenerational and interregional equity of the INDCs by comparing scenarios with INDC emissions targets in 2030 and with an immediate emissions reduction associated with a uniform global carbon price. It shows that compared with an immediate emissions reduction scenario, the intergenerational equity status is not favorable for the INDC scenario and the future generation suffers more from delayed mitigation. In comparison, the INDC scenario has better interregional equity in the early part of the century, while the interregional equity worsens later in the century.

Chapter 5 analyzes the effectiveness of the International Emissions Trading (ET) scheme. The INDC without ET resulted in a 0.4% welfare loss in 2030 globally, while with the ET scenario, the loss was 0.1%. The OECD countries tend to have larger losses in the scenario without ET, whereas their losses are reduced substantially in the scenario with ET.

Chapter 6 investigates the effectiveness of different carbon reduction countermeasures at achieving the INDC target for China and the 2 °C target and assesses the economic impacts of achieving the carbon reduction targets. It shows that China should not only rely on optimizing industry structure and restricting its industry output, but more importantly it should rely on low-carbon technologies in the power and end-use sectors, as well as low-carbon style consumption.

Chapter 7 assesses three scenarios for India up to 2030: the Reference, INDC, and 2 °C scenarios. The results show that even after full implementation of the INDCs, the emissions would still be 25 GtCO2 eq. higher than the 2 °C stabilization target over the period from 2016 to 2030. Enhanced actions like the penetration of renewable technologies, end-use demand management, and improved energy efficiency could help to close this emission gap.

Chapter 8 assesses the INDCs of Indonesia using an AIM/CGE coupled with an AFOLU model. The model shows that the emissions reduction target of the INDCs can be achieved at low economic cost [less than 1% of the gross domestic product (GDP)] and that the mitigation actions required would not harm economic development. The energy sector is expected to grow rapidly and become more important in the future. Therefore, climate-related policymakers should focus equally on land use and the energy sector in the future.

In Chap. 9, the NDC of Japan is assessed using the technology selection model (AIM/Enduse) and the Asia-Pacific Integrated Modeling/Computable General Equilibrium (AIM/CGE) model. In addition, the consistency between the NDC in 2030 and the 80% reduction target in 2050 is assessed using AIM/Enduse. The analysis by 2030 suggests that the NDC can be achieved with improved energy efficiency in the end-use sectors and the deployment of low-carbon energies on the supply side. The 2050 target to reduce GHG by 80% is also technically feasible, although rapid, large-scale transformation of the energy system is required between 2030 and 2050, including electrification in the end-use sectors.

Chapter 10 assesses the economic impacts of the Thai INDC. Four scenarios are analyzed with the GHG emission constraints and the renewable power generation target. The results show that under the power development plan of 2015, the INDC target is achievable. The macroeconomic loss is low with the modest reduction target but high with the stricter reduction target. In addition, the latter scenario needs a greater push toward renewable energy to realize stringent climate policy. Therefore, the availability of land for deploying renewable solar, wind, and biomass energy technologies needs to be evaluated to meet the higher GHG emissions reduction.

Chapter 11 analyzes the INDC of Vietnam from the perspectives of the energy production system and the economy and assesses the role of renewables in achieving the INDC. Seven scenarios are assessed, each with the same socioeconomic development, but with different shares of renewables in power generation. The results show that the GDP and welfare losses could be reduced through the double deployment of wind and solar photovoltaic power generation systems.

Chapter 12 describes the technical aspects of the AIM/CGE model and explains the social accounting matrix (SAM) used in the model, the model structure, and the mathematical statements. The formula shown in this chapter is applicable to both global and national scales.

Chapter 13 documents the main model structure and how to implement future scenario assumptions for the analysis of long-term climate mitigation with AIM/CGE. In this document, higher priority is set for showing how the scenario outcomes of this model are generated, over just explaining either the model structure or the scenario assumptions. Six topics are explained to understand the behaviors of the model in conjunction with future scenarios: (1) macroeconomics, labor, and population; (2) energy supply; (3) energy demand; (4) agriculture and land use; (5) GHG reduction measures other than changing the energy system; and (6) how to implement new production sectors or goods (not accounted for in the base year).

1.4 Key Messages

This book focuses on the analyses of country INDCs and their implications for global climate change mainly using AIM/CGE, which is a powerful tool for analyzing the economic impacts of climate policies. The main messages of this book are as follows:

1. Interactions between short- and midterm policies and long-term climate goals:
    (a) The current emissions reduction targets indicated by INDCs are insufficient to achieve the long-term goal of staying well below 2 °C. Each country must increase the emissions reduction target in the INDCs. One of the keys to stabilizing the climate is how effectively the global stocktake will work.
    (b) The current INDCs are unfavorable from the perspective of intergenerational equity, as the future generation suffers more from delayed mitigation. However, the INDC scenario has better interregional equity in the early part of the century than does the immediate emissions reduction scenario, in which we assume a global carbon price up to 2030. However, as immediate reduction after 2030 is required to achieve the long-term climate goal, it is necessary to consider the economic development of low-income countries.
    (c) The Paris Agreement seeks to achieve a balance between anthropogenic emissions from sources and the removal of GHGs by sinks in the second half of this century. Delayed mitigation efforts could increase the adverse effects on land use and ecosystems, as well as the uncertainties of long-term climate impacts.
2. Relation between the peak period and long-term goals. To achieve the long-term temperature goal, it is important to reach the global peak in GHGs as soon as possible. If the peak is delayed, greater reduction efforts will be required subsequently. The scenario that is in line with both the Copenhagen commitment and INDC of China indicates that the emissions in China will peak in 2030. The Chinese case study also shows that it is feasible to achieve an emissions peak in 2020 that meets the stringent target of 2 °C. Under the INDC scenario, the most important measures after decarbonizing power are adjusting industry structure and improving end-use energy efficiency, followed by carbon capture and storage (CCS) technology, fuel switching, and lowering the activity level of energy-intensive industries. In the stringent scenario, end-use energy efficiency is by far the most important measure, followed by CCS technology.
3. Climate impacts and mitigation goals. If the global temperature increase can be limited to below 1.5 °C above the preindustrial level, this would reduce the risks from climate change as evaluated from several perspectives. However, some of the risks (such as the effects on unique and threatened systems and the risks from extreme events) would still be considerable at 1–2 °C above the preindustrial level. Taking definitive action and devising ways of dealing with climate uncertainty are of greater importance than the specific choice of a goal (i.e., 1.5, 2.0, or 2.5 °C) in terms of the global risks, although a more comprehensive assessment could alter this finding.
4. Innovation: Renewables, CCS, and energy efficiency improvement:
    (a) The share of renewables in power generation must increase considerably due to the policy push aimed at achieving the decarbonization target pledged in the INDCs. As renewables are one of the most important options for achieving the long-term target, more innovation is necessary to obtain renewables with increased reliability, better quality, and decreased costs.

(b) CCS is not a major option in INDCs, but its share will be increased in the long-term 2 °C/1.5 °C scenarios, as most long-term scenarios assume the overshoot of GHG emissions and negative CO_2 emissions can be realized by BECCS or afforestation.
(c) Improved energy efficiency also plays a key role for achieving INDCs and the global emissions goal. Research and development is required to increase energy efficiency. Both demand-side push, such as setting building codes and energy efficiency requirements for products, and demand-side pull, such as encouraging consumers and manufacturers to purchase energy-efficient products, are effective for increasing energy efficiency.

5. Carbon pricing and emissions trading. Increasing focus must be placed on putting a price on carbon as a means of reducing CO_2 emissions by increasing investment in cleaner options, as well as pushing consumers to buy energy-efficient technologies. The international carbon emissions trading system is a cost-effective way to reduce the total global mitigation cost. Although there are some difficulties with implementing the market mechanism, such as establishing a monitoring, reporting, and verification (MRV) system and preparing market infrastructure, ET will become more important as the emissions reduction targets become more stringent. In terms of the financial flow associated with ET, money was transferred from Organization for Economic Cooperation and Development (OECD) countries to non-OECD countries because OECD countries imported emissions from non-OECD countries. These imports helped to reduce the mitigation cost in OECD countries and to increase mitigation efforts in non-OECD countries.
6. International cooperation. Although there is room to achieve deeper emissions reductions in some developing countries like Indonesia, support from the developed countries could enhance the emission cuts.

References

AIM (2016) AIM International Workshop. http://www-iam.nies.go.jp/aim/index.html

Amponin JA, Evans JW (2016) Assessing the intended nationally determined contributions of ADB developing members, Asian Development Bank, No. 44, August 2016. https://www.adb.org/sites/default/files/publication/189882/sdwp-044.pdf

CD-LINKS (2016) Linking Climate and Development Policies – Leveraging International Networks and Knowledge Sharing (CD-LINKS). http://www.iiasa.ac.at/web/home/research/researchPrograms/Energy/Research/CD-LINKS.html

Climate Action Tracker (2015) Climate pledges will bring 2.7°C of warming, potential for more action, Carbon Action Tracker. http://climateactiontracker.org/news/253/Climate-pledges-willbring-2.7C-of-warming-potential-for-more-action.html

DDPP Project (2015) Pathways to deep decarbonization, IDDRI & SDSN. http://www.iddri.org/Publications/Rapports-and-briefing-papers/ddpp_2015synthetisreport.pdf

Kainuma M, Morita T, Matsuoka Y (eds) (2003) Climate Policy Assessment. Springer. ISBN: 978-4-431-67979-0 (Print) 978-4-431-53985-8 (Online)

Masui T (2015) Overall activities in FY 2014–2015 and future research plan, 21st international AIM workshop, Tsukuba, Japan. http://www-iam.nies.go.jp/aim/aim_workshop/aimws_21/presentation/s01_02_masui.pdf
MILES Project Consortium (2015) Beyond the numbers: understanding the transformation induced by INDCs, IDDRI. http://www.iddri.org/Publications/Collections/Analyses/MILES%20report.pdf
Pandey R, Kainuma M, Ishikawa T, Nishioka S (2016) A primer of low carbon societies. LCS-Rnet Secretariat, IGES, Japan. http://lcs-rnet.org/pdf/publications/lcs_Primer_on_Low_Carbon_Societies_e.pdf
Rogelj J, den Elzen M, Höhne N, Fransen T, Fekete H, Winkler H, Schaeffer R, Sha F, Riahi K, Meinshausen M (2016) Paris Agreement climate proposals need a boost to keep warming well below 2°C. Nature 534:631–639. doi:10.1038/nature18307
Schleussner C, Rogelj J, Schaeffer M, Lissner T, Licker R, Fischer EM, Knutti R, Levermann A, Frieler K, Hare W (2016) Science and policy characteristics of the Paris Agreement temperature goal. Nat Clim Chang 6:827–835. doi:10.1038/nclimate3096
UNEP (2016) Emissions Gap Report 2016. http://uneplive.unep.org/media/docs/theme/13/Emissions_Gap_Report_2016.pdf
UNFCCC (2014) Report of the Conference of the Parties on its nineteenth session, held in Warsaw from 11 to 23 November 2013, FCCC/CP/2013/10/Add.1. http://unfccc.int/resource/docs/2013/cop19/eng/10a01.pdf
UNFCCC (2015) Paris Agreement http://unfccc.int/files/home/application/pdf/paris_agreement.pdf
UNFCCC (2016a) NDC registry. http://unfccc.int/focus/ndc_registry/items/9433.php
UNFCCC (2016b) Aggregate effect of the intended nationally determined contributions: an update, FCCC/CP/2016/2. http://unfccc.int/resource/docs/2016/cop22/eng/02.pdf
World Resource Institute (2015) Interpreting INDCs: assessing transparency of post-2020 greenhouse gas emissions targets for 8 top-emitting economies, Working paper, December 2015. http://www.wri.org/sites/default/files/WRI_WP_InterpretingINDCs.pdf

Chapter 2
Implications of the Paris Agreement in the Context of Long-Term Climate Mitigation Goals

Shinichiro Fujimori, Xuanming Su, Jing-Yu Liu, Tomoko Hasegawa, Kiyoshi Takahashi, Toshihiko Masui, and Maho Takimi

Abstract The Paris Agreement confirmed the global aim to achieve a long-term climate goal, in which the global increase in mean temperature is kept below 2 °C compared to the preindustrial level. We investigated the implications of the near-term emissions targets (for around the year 2030) in the context of the long-term climate mitigation goal using the Asia-Pacific Integrated Model (AIM) framework. To achieve the 2 °C goal, a large greenhouse gas (GHG) emissions reduction is required, either in the early or latter half of this century. In the midterm (from 2030 to 2050), it may be necessary to consider rapid changes to the existing energy or socioeconomic systems, while long-term measures (after 2050) will rely on the substantial use of biomass combined with carbon capture and storage (CCS) technology or afforestation, which will eventually realize so-called negative CO_2 emissions. With respect to the policy context, two suggestions are provided here. The first is the review and revision of the nationally determined contributions (NDCs) in 2020, with an additional reduction target to the current NDCs being one workable alternative. The second suggestion is a concrete and numerical midterm emissions reduction target, for example, to be met by 2040 or 2050, which could also help to achieve the long-term climate goal.

S. Fujimori (✉) • T. Hasegawa
Center for Social and Environmental Systems Research, National Institute for Environmental Studies (NIES), 16–2 Onogawa, Tsukuba, Ibaraki 305–8506, Japan

International Institute for Applied Systems Analysis, Schlossplatz-1, Laxenburg 2361, Austria
e-mail: fujimori.shinichiro@nies.go.jp

X. Su • J.-Y. Liu • K. Takahashi • T. Masui
Center for Social and Environmental Systems Research, National Institute for Environmental Studies (NIES), 16–2 Onogawa, Tsukuba, Ibaraki 305–8506, Japan

M. Takimi
Mizuho Information & Research Institute, Inc., 2–3 Kanda-Nishikicho, Chiyoda-ku, Tokyo 101–8443, Japan

S. Fujimori et al. (eds.), *Post-2020 Climate Action*,
DOI 10.1007/978-981-10-3869-3_2

Keywords AIM • Integrated assessment model • Land use • Climate change mitigation

Abbreviations

AIM	Asia-Pacific Integrated Model
INDCs	Intended nationally determined contributions
COP	Conference of the Parties
UNFCCC	United Nations Framework Convention on Climate Change
PA	Paris Agreement
NDCs	Nationally determined contributions
MAC	Marginal abatement cost
CGE	Computable general equilibrium
AIM/PLUM	Asia-Pacific Integrated Model/integration Platform for Land-Use and Environmental Modeling
MAGICC	Model for the Assessment of Greenhouse-gas Induced Climate Change
CES	Constant elasticity substitution
LES	Linear expenditure system
IPCC AR5	Fifth Assessment Report of Intergovernmental Panel on Climate Change
LPJmL	Lund-Potsdam-Jena managed Land Dynamic Global Vegetation and Water Balance Model
VISIT	Vegetation Integrative Simulator for Trace Gases
SCM4OPT	Simple Climate Model for Optimization
RCP	Representative concentration pathways
SSP	Shared socioeconomic pathways
BECCS	Biomass energy combined with CCS

2.1 Introduction

The Conference of the Parties (COP) 21 to the United Nations Framework Convention on Climate Change (UNFCCC) adopted the Paris Agreement (PA) (United Nations Framework Convention on Climate Change 2015). The PA provides a framework for global action to address climate change in the period after 2020. The objective of the agreement is to maintain the increase in global temperatures well below 2 °C above preindustrial levels.

The PA requires parties to prepare nationally determined contributions (NDCs), indicating an individual country's emissions reduction commitments, take measures to achieve their objectives, and report on progress. To raise the level of ambition over time, parties must submit updated NDCs every 5 years. Each party's new NDC must be more ambitious than its previous NDC. Over 180 parties to the

UNFCCC communicated their intended nationally determined contributions (INDCs) for 2025/2030 before COP21.

There have already been several assessments related to INDCs published either in scientific papers, websites, or reports (Meinshausen et al. 2015; Fawcett et al. 2015; Iyer et al. 2015; International Energy Agency 2015; Climate Action Tracker 2015; Lomborg 2015; Benveniste et al. 2015; Kitous and Keramidas 2015; Hof et al. 2015). Some have proposed alternative scenarios to achieve the 2 °C goal, because INDCs based on emissions have led to scenarios with a temperature rise larger than 2 °C. Some assessments have made comparisons with the recent Fifth Assessment Report of Intergovernmental Panel on Climate Change (IPCC AR5) scenario database (International Institute for Applied Systems Analysis 2015a) and made allocations based on the multiple effort-sharing schemes. The consensus among the assessments at this stage is that current INDCs are not in line with the 2 °C goal, which was also stated in the PA (United Nations Framework Convention on Climate Change 2015). To achieve the 2 °C goal, either a further emissions reduction before 2030 or more drastic and quick reductions after this date are required.

The difficulties in reducing greenhouse gas (GHG) emissions after 2030 are not obvious. Therefore, we investigated two issues with respect to the relationship between INDCs and the long-term climate mitigation goal. The first is what difficulties exist in realizing the 2 °C goal under the current INDCs. We considered difficulties that might arise in the medium and long term. Here, we defined medium and long term as from 2030 to 2050 and after 2050, respectively. The second issue is what policy suggestions can be derived from such assessments. Regarding the first issue, we also considered the modeling limitations in the current integrated assessment model.

To address these issues, we conducted a scenario assessment using the Asia-Pacific Integrated Model (AIM) framework. The remainder of this paper is organized as follows. Section 2.2 provides the overall methodology, as well as the model and scenario framework. In Sect. 2.3, the results are presented and analyzed. An interpretation of the results and limitations of the study are provided in Sect. 2.4, and the conclusions are presented in Sect. 2.5.

2.2 Methodology

Figure 2.1 shows the overall modeling framework. Several models exchanged information with each other within this framework, and the core of the simulation was a global computable general equilibrium model: AIM/computable general equilibrium (CGE). AIM/CGE produces a marginal abatement cost (MAC) curve by experimenting with specific carbon price pathway scenarios (e.g., 3% per year increase). This information was used for the calibration of the MAC function (the actual parameter adjusted by this information is the reduction control rate) in a Dynamic Integrated Climate-Economy (DICE)-type intertemporal optimization

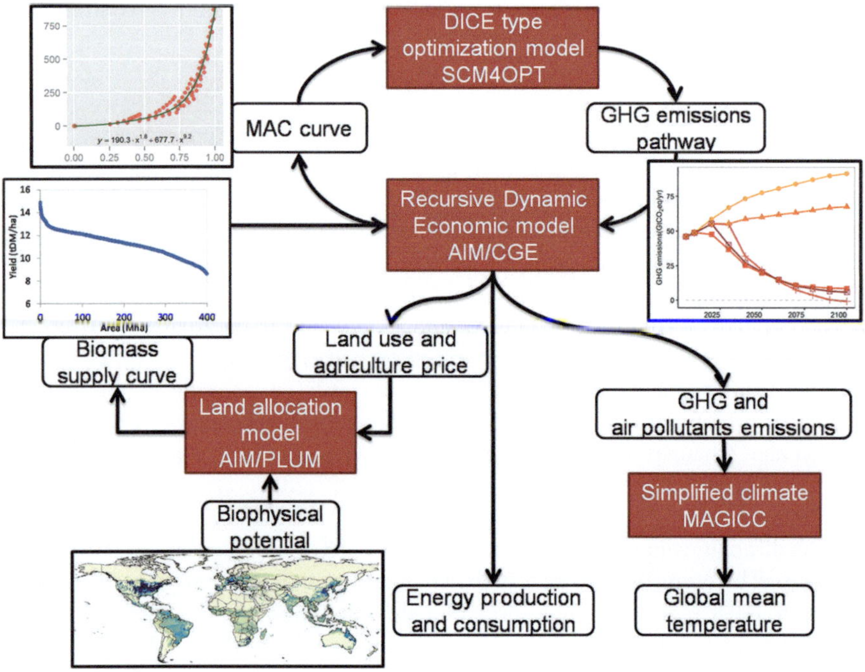

Fig. 2.1 Asia-Pacific Integrated Model (AIM) framework

model (Nordhaus and Sztorc 2013) – the Simple Climate Model for Optimization (SCM4OPT). SCM4OPT was used to produce GHG emissions pathways under particular climate constraints for AIM/CGE. Then, AIM/PLUM (integration Platform for Land-Use and Environmental Modelling) (Hasegawa et al. 2017) was used to conduct a land-use allocation of agricultural cropland and pastureland. AIM/PLUM downscales the regionally aggregated land-use information provided by AIM/CGE into gridded information. This land-use allocation allowed us to obtain biomass energy crop supply curves for each region, which represented the relationship between the bioenergy cropland area and its yield. These supply curves were fed into the AIM/CGE together with emissions constraints from SCM4OPT. Finally, the GHG and air pollutant emissions, which are computed by AIM/CGE, were input to a simplified climate model, the Model for the Assessment of Greenhouse-gas Induced Climate Change (MAGICC) version 6 (Meinshausen et al. 2011), to simulate global mean temperature and radiative forcing under the detailed emissions information produced by AIM/CGE. SCM4OPT could also be used for the final climate estimates, but here we prioritized its comparability with other integrated assessment modeling (IAM) studies, which have predominantly used MAGICC. The simulation covers the period from 2005 to 2100.

2.2.1 AIM/CGE

The CGE model used in this study is a recursive dynamic general equilibrium model that covers all regions of the world and is widely used in climate mitigation and impact studies (Hasegawa et al. 2014, 2015a, b, 2016; Mittal et al. 2016; Fujimori et al. 2014b, c, 2015). The main inputs of the model are assumptions about population growth, GDP growth, energy technology efficiency, resource reserves, and related factors.

It was important that the industrial classification of energy sectors, including power sectors, were disaggregated in detail, because energy systems and technological descriptions were crucial for the purposes of this study. Moreover, to assess bioenergy and land-use competition appropriately, agricultural sectors also need to be highly disaggregated (Fujimori et al. 2014a). The details of the model structure and mathematical formulas are given in Chap. 12.

The production sectors were assumed to maximize profits under multi-nested constant elasticity substitution (CES) functions and at each input price. Energy transformation sectors input energy and value added as a fixed coefficient, whereas energy end-use sectors have elasticities between energy and value added. They were treated in this manner to deal appropriately with energy conversion efficiency in the energy transformation sectors. Power generated from several energy sources was combined by a logit function, although a CES function is often used in other CGE models. We chose this method for the consideration of the energy balance because the CES function does not guarantee a material balance (Schumacher and Sands 2006). Household expenditure on each commodity was described by a linear expenditure system (LES) function. The saving ratio was endogenously determined to balance saving and investment, and capital formation for each item was determined using a fixed coefficient. The Armington assumption was used for trade, and the current account was assumed to be balanced.

In addition to energy-related CO_2 emissions, CO_2 from other sources, CH_4, and N_2O (including changes resulting from land-use and nonenergy-related emissions) were included as GHG emissions in this model. Global warming potentials were used when considering emissions of the six gases covered by the Kyoto protocol as specified in the fourth IPCC assessment report (IPCC 2007).

Once an emissions constraint was placed on a region, a carbon tax became a complementary variable to that constraint. Such a tax raises the price of fossil fuel goods when emissions are constrained and promotes energy savings and the substitution of fossil fuels by alternative energy sources with lower emissions. An emissions tax, called the GHG emissions price, is also an incentive to reduce nonenergy-related emissions. The revenue from this tax is assumed to go to households.

2.2.2 AIM/PLUM

The AIM/PLUM is a global land-use allocation model used to downscale the AIM/CGE's aggregated regional land-use projections into a spatial gridded land-use pattern (0.5° × 0.5°) for the interactive assessment of human activities and biophysical elements. Regional-scale land demand estimated by AIM/CGE (17 regions) was fed into the AIM/PLUM land-use allocation model and was spatially distributed into grid cells (0.5° × 0.5°). The cropland and afforestation area was allocated based on optimization (profit maximization), where a land owner was assumed to decide the mix of land uses to obtain the highest profit for a given biophysical land productivity condition (e.g., crop yield production per unit area). Because the optimization was solved for each region that had the same regional classification used in AIM/CGE, land transactions across the regions were not allowed. The allocation was conducted in 5-year steps. There were seven crop types, with or without irrigation. Land for harvested wood was excluded from the model framework.

The potential yield of crops was based on estimates from the Lund-Potsdam-Jena managed Land Dynamic Global Vegetation and Water Balance Model (LPJmL) (Bondeau et al. 2007) prepared for Inter-Sectoral Impact Model Intercomparison Project (ISI-MIP) (Rosenzweig et al. 2014). The bioenergy crop yield and forest carbon sequestration were based on estimates from the Vegetation Integrative Simulator for Trace Gases (VISIT) (Ito and Inatomi 2012). Please see Hasegawa et al. (2017) for more details.

2.2.3 SCM4OPT

SCM4OPT is an intertemporal optimization model based on DICE2013R, in which the discounted total welfare is maximized under given socioeconomic and climate conditions. The main extension from DICE consisted of two parts. The first part considered anthropogenic emissions. In the original DICE, only industrial CO_2 emissions are explicitly treated as endogenous emissions, while the rest are exogenously given as radiative forcings. In our extension, we incorporated the major GHGs (land-related CO_2, CH_4, N_2O, and F gases) and other climate forcers (SOx, NOx, CO, OC, BC, NH_3, and VOCs). For the baseline case, AIM/CGE emissions were used directly, whereas for the climate policy cases, the MAC curves, which were represented by a function of the carbon price, were applied to all gases. These MAC curves were calibrated from the AIM/CGE outcomes.

The second part of the extension was in the climate module. Because many climate forcers are incorporated, the functions from emissions to forcings were added mostly from MAGICC 6.0. Natural forcing and the products of the chemical reactions of emissions were also added (e.g., cloud albedo, stratospheric ozone,

Table 2.1 List of scenarios

Scenario name	GHG emissions reduction		
	2015–2020	2020–2030	2030–2100
Baseline	None		
INDCSamePrice	Cancun pledge	INDCs	Same carbon price in 2030
450ppmeRCP	Same as RCP2.6 emissions pathway		
450ppmeCancunP	Cancun pledge	Equivalent to cumulative emissions in 450ppmeRCP	
450ppmeINDC	Cancun pledge	INDCs	Equivalent to cumulative emissions in 450ppmeRCP

tropospheric ozone, stratospheric water vapor from CH_4 oxidization, solar and volcanic inputs).

2.2.4 *Scenarios*

We simulated five scenarios in this study, and they are summarized in Table 2.1. The first scenario was a baseline scenario, which includes no climate policies. A climate policy was defined as the implementation of emissions constraints or a carbon price. The second scenario was the INDCSamePrice scenario, which implemented the Cancun pledge by 2020, and the GHG emissions then followed INDCs targets for individual regions until 2030. The carbon price was assumed to be constant at the 2030 level for all dates afterwards. The remaining three scenarios were intended to stabilize the atmospheric CO_2 concentration at 450 ppm CO_2-equivalent after 2100. This concentration stabilization is often considered equivalent to the 2 °C goal, i.e., maintaining the temperature increase from preindustrial level below 2 °C with a probability of >66%. However, the emissions pathways differed among the three scenarios. The first scenario, 450ppmeRCP, roughly followed the Representative Concentration Pathway 2.6 (RCP2.6) (Vuuren et al. 2011), with a global uniformed carbon price applied after 2015, given the exogenous emissions pathway. The 450ppmeCancunP scenario met the 2020s Cancun pledge and then reduced emissions using a global uniformed carbon price. The 450ppmeINDC scenario followed the INDC targets and then reduced emissions with a global uniformed carbon price. The GHG emissions pathways in 450ppmeCancunP and 450ppmeINDC after 2020 and 2030 were determined using SCM4OPT, which intertemporally optimizes the emission levels for each GHG by maximizing the social welfare. While ultimate climate goals of 450ppmeCancunP and 450ppmeINDC scenarios are the same, the emissions pathways are different due to the INDC 2030s target, and that eventually differ from intermediate mitigation strategy between two scenarios.

Here, the population and GDP assumptions under Shared Socioeconomic Pathway 2 (SSP2) (International Institute for Applied Systems Analysis 2015b) were used for the drivers of basic human activity, and other technological assumptions were based on Fujimori et al. (2016).

2.3 Results

2.3.1 GHG Emissions

GHG emissions (the six gases covered by the Kyoto protocol) are shown in Fig. 2.2. The GHG emissions at baseline steadily increase during this century due to the population increase and GDP growth. The rate of increase is higher in the first part of the period than the latter part because the GDP growth rate is relatively high in the early period. In 2100, GHG emissions reach around 90 $GtCO_2eq$/year. The GHG emissions in INDCSamePrice are slightly lower than at baseline in 2020. In 2030, the emissions are 57 $GtCO_2eq$/year, while baseline emissions are 75 $GtCO_2eq$/year. After 2030, due to the implementation of the continuous carbon price, the GHG emissions in 2100 are lower than at baseline at 67 $GtCO_2eq$/year. The global mean increases in temperature from the preindustrial level in 2100 for

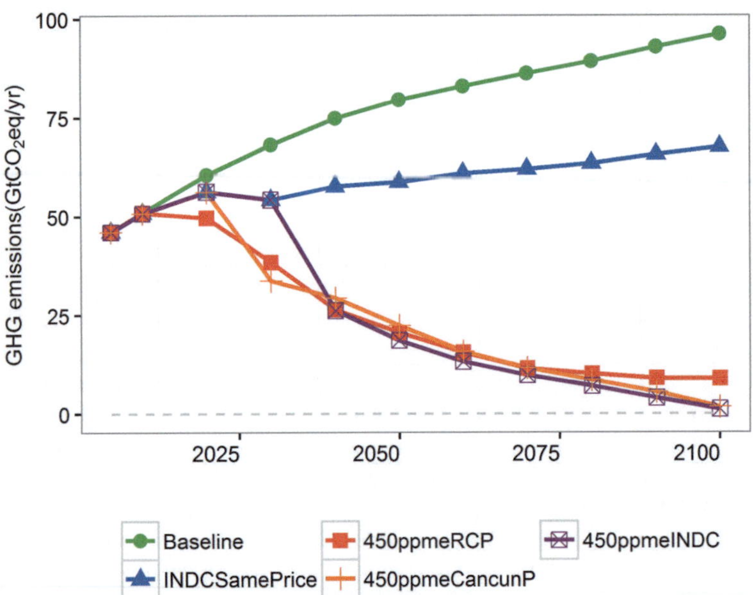

Fig. 2.2 Global total greenhouse gas (GHG) emissions

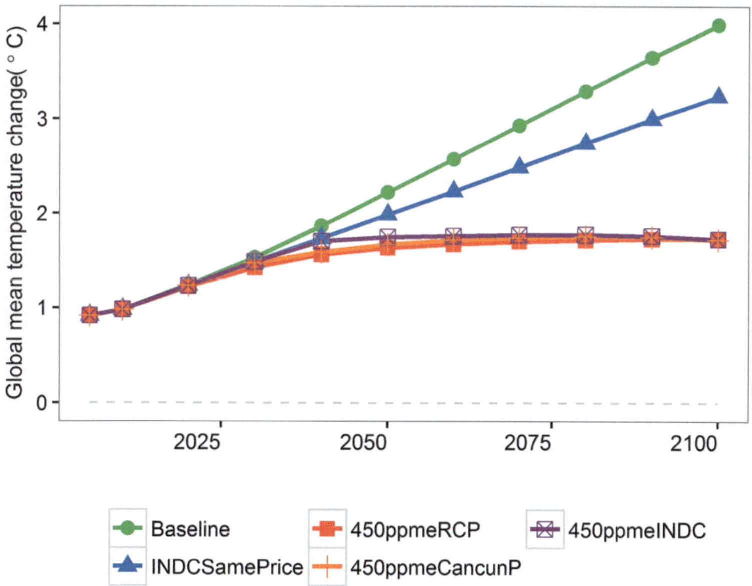

Fig. 2.3 Global mean change in temperature relative to the preindustrial level

baseline and INDCSamePrice are 4.0 °C and 3.1 °C, respectively (in Fig. 2.3). In 450ppmeRCP, emissions are reduced even in 2020, with an ongoing reduction throughout the entire century. The emissions in 2100 are 9 $GtCO_2eq$/year. In 450ppmeCancunP, the emissions reduction is delayed compared to 450ppmeRCP, with a value of 55 $GtCO_2eq$/year in 2020. From 2020, this scenario shifts to a stage of drastic reductions, and the final emissions are lower than those in 450ppmeRCP in 2100 (2 $GtCO_2eq$/year). In 450ppmeRCP and 450ppmeCancunP, the emissions are similar in the middle of the 2020s. In 450ppmeINDC the emissions are reduced by half from 2030 to 2050 and are deeper than in the other scenarios in the latter half of the century. The scenario results in almost zero emissions in 2100. This scenario does not result in large negative CO_2 emissions; however, if independent CO_2 emissions are the focus, the negative CO_2 emissions would be large (this information is shown in a later section).

In 2030, the emissions in 450ppmeINDC are around 15 $GtCO_2eq$/year larger than in 450ppmeRCP, and therefore these additional emissions until 2030 need to be compensated for during the latter period. The global change in mean temperature in the three 450 ppm CO_2-equivalent stabilization scenarios is around 1.7 °C at the end of the century (to maintain the >66% probability, the average change in temperature is 1.7 rather than 2.0 °C). From 2030 to 2050, 450ppmINDC shows a rapid emissions reduction, with the annual reduction rate reaching 5% per year. This value is quite high and would present difficulties for medium-term mitigation based on the INDCs.

2.3.2 Primary Energy Supply and Final Energy Consumption

Figure 2.4 shows the global primary energy supply for each scenario. At baseline, the total primary energy supply is the largest at around 1020 EJ/year in 2100, which is 2.3 times the current level (in 2005). In INDCSamePrice, the total primary energy supply in 2100 is around 10% less than at baseline (930 EJ/year) due to energy savings in the energy end-use sectors and the shift in power generation from fossil fuel to renewable energy sources (the transformation efficiency is high in renewable energy). In the 450 ppm stabilization scenarios (450ppmeRCP, 450ppmeCancunP, and 450ppmeINDC), the energy supply is much smaller than at baseline at 730, 700, and 630 EJ/year in 2100, respectively. This decrease is realized by the large-scale saving of energy in energy end-use sectors and the transformation of power generation sources. Among the three 450 ppm scenarios, the total energy in 2100 differs because of the timing of the drastic emissions reduction. As the timing of the reduction is delayed, the total energy supply in 2100 becomes lower. In terms of the composition of energy sources, fossil fuel remains the dominant source of energy supply in the baseline scenario. Renewable energies such as biomass, wind, and solar become the main energy sources in the 450 ppm stabilization scenarios. Carbon capture and storage (CCS) is almost fully implemented in the remaining

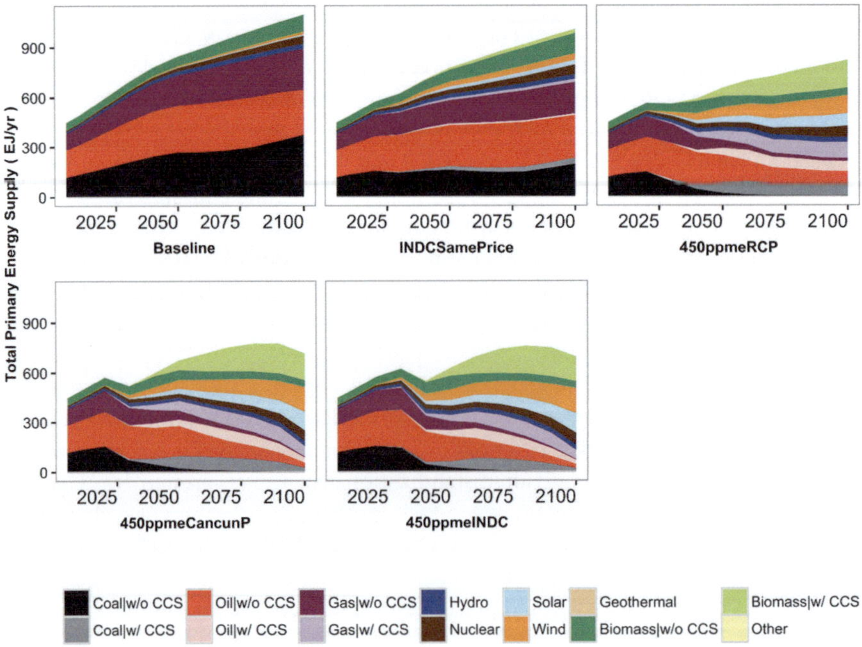

Fig. 2.4 Global primary energy supply and its composition

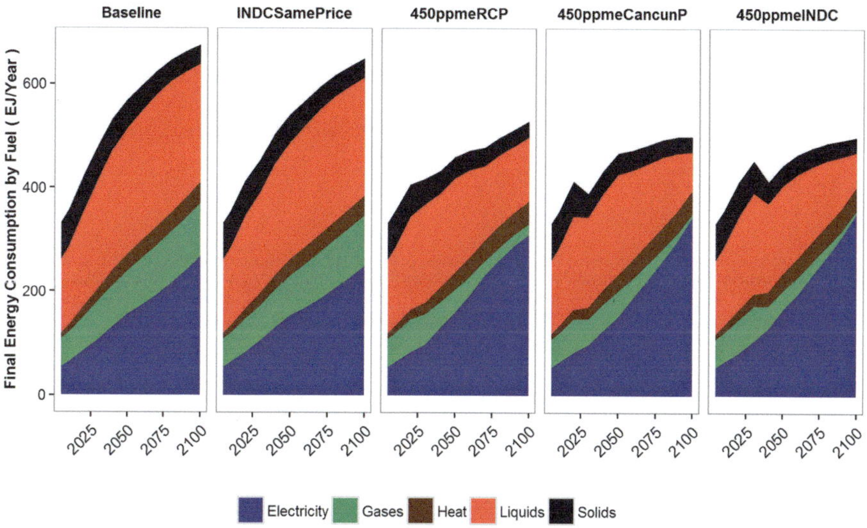

Fig. 2.5 Global gross domestic product (GDP) loss rate and carbon prices

industries with large fossil fuel consumption, such as steel and power generation. In 2030, there are primary energy supply reductions in 450ppmeCuncunP and 450ppmeINDC, but the reasons for this differ. In 450ppmeCuncunP, it is due to both the rapid emissions reduction and higher GHG emissions price (shown in Fig. 2.5, although the differences in the 2030s are not apparent). In 450ppmeINDC, it is due to the INDC commitment and the strong emissions reduction by large emitters in OECD countries, with the global total primary energy production in 2030 being low as a result.

The rapid emissions reduction in 450ppmeINDC and 450ppmeCancunP during the period from 2030 to 2050 also relies on final energy consumption-side transformation. As illustrated in Fig. 2.6, total energy consumption drops rapidly during that period. Furthermore, regarding the fuel mix, fossil fuel-oriented solid (mainly coal), liquid (mainly petroleum oil), and gases decrease, whereas electricity increases.

2.3.3 Mitigation Cost

The mitigation cost (here we use GDP loss rates as a metric of mitigation cost) in INDCSamePrice is 0.4% in 2030 and then is almost constant or slightly declines, as shown in Fig. 2.5. This is because the same carbon price is assumed after 2030, and the absolute cost of mitigation is kept almost the same, while GDP continuously grows. Comparing the three 450 ppm stabilization scenarios, 450ppmeRCP and 450ppmeCancunP, which have drastic emissions reductions by 2020, have high

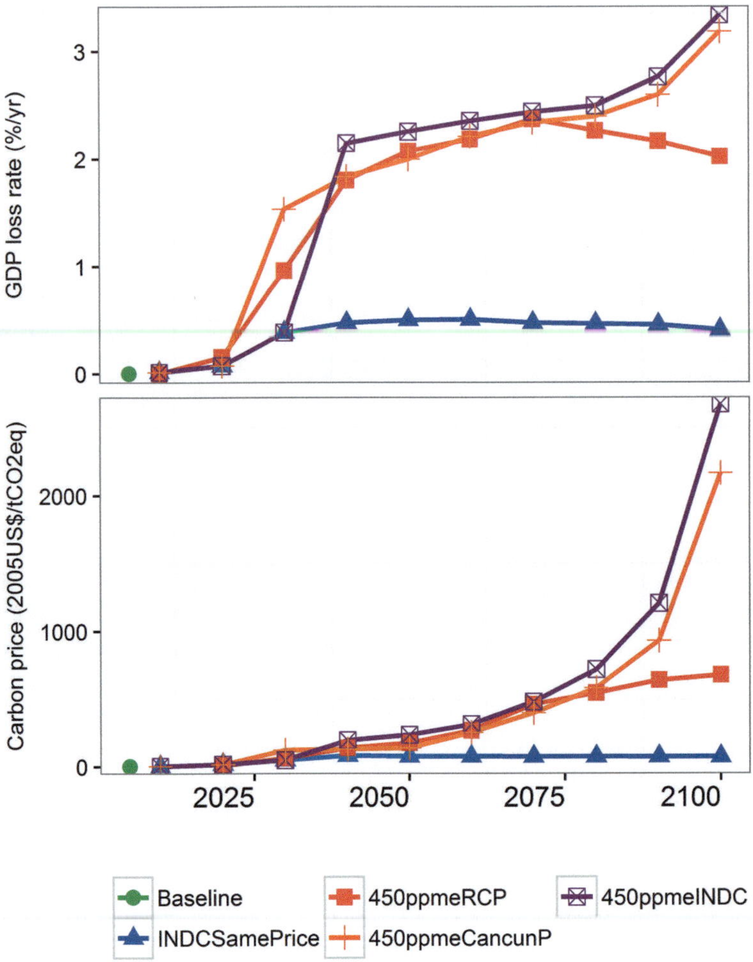

Fig. 2.6 Global final energy consumption by fuels

GDP loss rates in the early period. In contrast, 450ppmeINDC has a large GDP loss in the latter period. However, 450ppmeINDC and 450ppmeCancunP are not very different in the latter period. The GDP loss rate in 2050 is around 2.0% for 450ppmeqRCP and 450ppmeCancunP but is 2%, 3.0%, and 3.2% in 2100 for 450ppmeRCP, 450ppmeCancunP, and 450ppmeINDC, respectively. The cost in 2100 is slightly lower than in IPCC AR5 (Clarke et al. 2014) but is almost the same in 2050.

There are apparent differences in the carbon price between 450ppmeRCP and the other two 450 ppm stabilization scenarios (450ppmeINDC and 450ppmeCancunP). In 2100, 450ppmeRCP is around 650 $/tCO_2eq$, whereas 450ppmeINDC exceeds 3000 $/tCO_2eq$. In particular, because the marginal

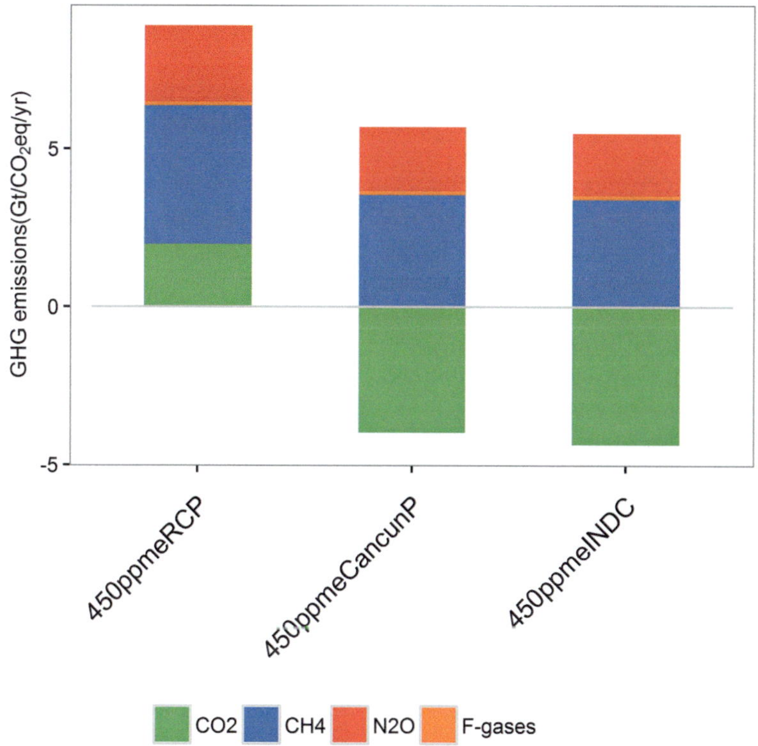

Fig. 2.7 Greenhouse gas (GHG) emissions profile in 2100 for the 450 ppm stabilization scenarios

emissions reduction space in 450ppmeINDC after 2070 is quite limited, the carbon price becomes much more sensitive to the incremental emissions reduction. Figure 2.7 shows the GHG emissions profile in 2100 for the 450 ppm stabilization scenarios. The emissions of CH_4, N_2O, and F gases are not very different among the three 450 ppm scenarios. This implies that there is limited scope for further non-CO_2 emissions reduction in the high carbon price area. Under such conditions, the additional GHG emissions reduction must be realized by a CO_2 emissions reduction. As a result, the total GHG emissions in 450ppmeINDC and 450ppmeCancunP in the latter half of the century are around zero, and it is obvious that large-scale negative CO_2 emissions are required. From the short-lived climate forcer's (SLCF) point of view, CH_4 is classified as SLCF and thought to play an important role for near-term climate policy aligning with air pollution policy. In 2030, 450ppmeRCP has a larger reduction in the near-term than other 450 ppm scenarios.

2.3.4 Land-Use and Land-Based CO_2 Emissions Reduction

A negative CO_2 emission can be realized only by biomass energy combined with CCS (BECCS) or afforestation in AIM/CGE. Although there is technology available that can directly capture CO_2 from the atmosphere and store it underground (i.e., direct air capture), AIM/CGE does not consider it. Both BECCS and afforestation are measures that are strongly related to land use. Thus, we considered the CO_2 sequestration achieved by BECCS and afforestation (Fig. 2.8). All of the 450 ppm scenarios have large negative CO_2 emissions of more than 5 $GtCO_2$eq/year in 2100. Remarkably, 450ppmeINDC and 450ppmeCancunP achieve 8 $GtCO_2$/year in 2100, with a particularly pronounced increase in afforestation (4 $GtCO_2$/year). This implies that around 300 Mha is required for afforestation, assuming a mean annual carbon sequestration of 3 tC/year/ha. Furthermore, bioenergy crops cover around 260 Mha in 2100 in 450ppmeINDC. The area used for land-based emissions reduction measures eventually accounts for around 40% of the current cropland area (1500 Mha). The differences in land-use changes among models under various mitigation scenarios have been assessed previously

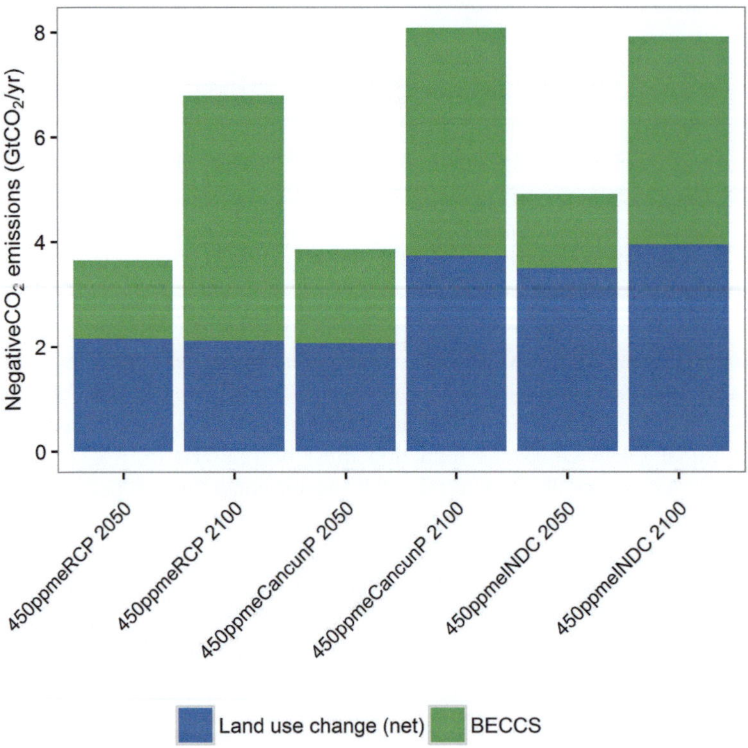

Fig. 2.8 Negative CO_2 emissions in 2050 and 2100

(Popp et al. 2014). The area of cropland used for bioenergy in these scenarios is 24–36% of the total cropland, and our results are compatible with this figure.

2.4 Discussion

Our results indicate that the 2 °C goal is achievable (in terms of a modeling exercise), even if the 2030s emissions are on track with the INDC targets. If the emissions in 2030 follow the INDCs, the emissions reduction after 2030 should be more drastic, and several important characteristics are observed. First, a substantial and rapid emissions reduction is required from 2030 to 2050. Second, large-scale negative CO_2 emissions and land-based CO_2 emissions reduction measures are required. The scale is more than that of 450ppmeRCP at the end of this century. Third, the carbon price becomes sensitive to the low GHG emissions area in the latter half of the century, and the mitigation cost also increases.

Emissions should be drastically reduced after 2030 to achieve the 2 °C goal going through the INDC targets. This is more obvious in the midterm until 2050 than in the latter period after 2050. This observation implies that the energy system would require rapid transformation. Gokul et al. (2015) reported that the additional energy system capacity could be huge with such rapid transformation. That is one of the aspects that we need to consider. Not only energy supply side but also energy demand side showed rapid changes. Furthermore, the acceptability and capability of the social and economic system could be a major discussion point. AIM/CGE numerically provides a mathematical solution, but this does not necessarily mean that the real world can easily implement it, because the primary focus of AIM/CGE is to consistently foresee the long-term interaction of economic, energy, agriculture, and land-use factors. In that sense, the model with the best or most detailed representation of energy technology and economic systems as well as energy demand-side structure would be the most appropriate to use.

With regard to land use, there seems to be a requirement to consider a wide range of environmental concerns. If bioenergy crops need irrigation, water resources are placed under additional pressure. The ecosystems and biodiversity that would be affected by land-use changes should be considered. If bioenergy crops require the application of nitrogen fertilizer, the nitrogen cycle would be affected, and nitrogen pollution might become a concern. Large-scale afforestation would be implemented using productive species in terms of carbon sequestration, but such measures may limit the number of tree species and would influence the biodiversity and ecosystem balances. The AIM framework has so far considered energy, monetary, and land-use factors in the economic system, although water bodies, ecosystems, and other biophysical interactions are not explicitly taken into account. Thus, we need to incorporate them into the integrated framework. Some IAM studies, such as one on the Global Change Assessment Model (GCAM) (Kim et al. 2016), have already considered this.

The fact that the carbon price becomes sensitive in the low GHG emissions area implies that the emissions pathways in the latter half of the century should be carefully considered. In this study, we used the outputs of a DICE-type intertemporal model, SCM4OPT, as the total GHG emissions constraint for AIM/CGE. However, we might be able to lessen the emissions reduction in the last few decades by a few gigatons of CO_2-equivalent emissions per year, to keep the carbon price and the mitigation cost relatively low and realistic, while the temperature increase would be close to 1.7 °C in 2100. Therefore, we should not interpret the results of this study as being the only emissions pathways to realize the 2 °C goal with INDCs. Rather, it would be better to be interpreted that the mitigation cost at the end of this century in INDCs is much larger than RCP2.6 to achieve 2 °C goal. Moreover, a very low level of GHG emissions is a necessary condition in the latter half of the century, which is similar to the RCP2.6 emissions pathways. More importantly, if the emissions in the first half of the century cannot be reduced as shown in this study, the carbon price would drastically increase. This also implies that numerically concrete emissions targets not only for 2030 but also for the midterm could be important to maintain the feasibility of long-term mitigation goals in the policy context relating to the PA. There is currently only a near-term, which is 2025–2030, emissions target and a long-term ultimate goal. A midterm goal could help to achieve the long-term goal.

As discussed above, there are mid- and long-term difficulties to achieving the 2 °C goal based on meeting the INDC target, including rapid emissions reduction and negative CO_2 emissions. However, this does not necessarily indicate that we should not aim to meet the 2 °C goal. First of all, the immediate policy reaction such as shown in 450ppmeRCP would reduce both mid- and long-term difficulties. Second, it is obvious that the climate mitigation would certainly reduce the climate change impact. The economics of that benefit is presented in IPCC report (Arent et al. 2014). Three degree celsius increase which roughly corresponds to INDCSamePrice scenario could cause 2–3% GDP loss, whereas 1–2 °C increase would be almost zero impact. Considering these difficulties, we need to have a better assessment model and policy decisions that consider the impact of climate change and adaptation to its consequences. The climate change issue is complex in the sense that the polluters and victims differ temporally and spatially. A simple cost-benefit analysis would not provide the best solution. The uncertainty in the impact of climate change also makes the problem complex (IPCC 2014). However, the review and revision of NDCs in 2020 is a great opportunity to consider the above mitigation difficulties.

We have already discussed the current modeling limitations that need to be improved in the future. Here, we note three additional limitations of this study. First, the range of energy technologies considered in AIM/CGE could be expanded. One important technology that has not yet been incorporated is hydrogen. BECCS can be used in association with hydrogen fuel as a decarbonizing technology. For example, in a non-large point source but hard to decarbonize sector such as the transport sector, the use of hydrogen fuel would be an interesting research area. Second, we use only single climate sensitivity 3 °C, but if the climate sensitivity differs from the assumption made in this study, the conclusion might be affected. Considering a probability in climate model may help to get better understanding (Rogelj et al. 2013). In particular, by the end of this century, the carbon price is very sensitive to the emissions reduction requirement, and slight changes in the climate response to the emissions reduction are clearly important. Third, although the global uniform carbon price is applied to all scenarios (after 2030 for 450ppmeINDC and after 2020 for 450ppmeCancunP) as is often done in many IAM studies, this is obviously an idealized modeling exercise, and real policy implementation would be difficult.

2.5 Conclusions

This study assessed long-term climate mitigation scenarios, to meet the 2 °C goal, considering INDCs. If the emissions in 2030 are as indicated in the INDCs, the emissions reduction in 2030–2050 needs to be quite drastic, and large negative emissions in the latter half of the century will be required to meet the 2 °C goal. We confirmed that there are midterm and long-term difficulties to achieve the 2 °C goal with INDCs. To avoid them, additional reduction targets are an attractive option when counties review and revise their NDCs in 2020. Moreover, the emissions reduction commitments in the following decades (i.e., 2040s or 2050s) are also important to achieve the 2 °C goal. On the other hand, the current AIM framework does not sufficiently represent the fundamental elements and needs to better incorporate land-related aspects, such as water bodies and ecosystems, as well as a better representation of energy technologies and economic systems for the midterm assessment.

References

Arent D, Tol R, Faust E, Hella J, Kumar S, Strzepek KM et al (2014) Key economic sectors and services. In: Field CB, Barros VR, Dokken DJ, Mach KJ, Mastrandrea MD, Bilir TE, Chatterjee M, Ebi KL, Estrada YO, Genova RC, Girma B, Kissel ES, Levy AN, MacCracken S, Mastrandrea PR, White LL (eds) Climate change 2014: impacts, adaptation, and vulnerability. Part A: global and sectoral aspects. Contribution of Working Group II to the fifth assessment report of the Intergovernmental Panel on Climate Change. Cambridge University Press, Cambridge/New York, pp 659–708

Benveniste H, Criqui P, Boucher O, Breon F-M, Guivarch C, Prados E et al (2015) The INDC counter, aggregation of national contributions and 2°C trajectories. http://hal.univ-grenoble-alpes.fr/hal-01245354

Bondeau A, Smith PC, Zaehle S, Schaphoff S, Lucht W, Cramer W et al (2007) Modelling the role of agriculture for the 20th century global terrestrial carbon balance. Glob Chang Biol 13 (3):679–706. doi:10.1111/j.1365-2486.2006.01305.x

Clarke L, Jiang K, Akimoto K, Babiker M, Blanford G, Fisher-Vanden K et al (2014) Assessing transformation pathways. In: Climate change 2014: mitigation of climate change. contribution of Working Group III to the fifth assessment report of the Intergovernmental Panel on Climate Change. Cambridge University Press, Cambridge/New York

Climate Action Tracker (2015) Tracking INDCs. http://climateactiontracker.org/. Accessed 2 Feb 2016

Fawcett AA, Iyer GC, Clarke LE, Edmonds JA, Hultman NE, McJeon HC et al (2015) Can Paris pledges avert severe climate change? Science 350(6265):1168–1169. doi:10.1126/science.aad5761

Fujimori S, Hasegawa T, Masui T, Takahashi K (2014c) Land use representation in a global CGE model for long-term simulation: CET vs. logit functions. Food Sec 6(5):685–699. doi:10.1007/s12571-014-0375-z

Fujimori S, Kainuma M, Masui T, Hasegawa T, Dai H (2014a) The effectiveness of energy service demand reduction: a scenario analysis of global climate change mitigation. Energ Policy 75 (0):379–391. doi:http://dx.doi.org/10.1016/j.enpol.2014.09.015

Fujimori S, Masui T, Matsuoka Y (2014b) Development of a global computable general equilibrium model coupled with detailed energy end-use technology. Appl Energy 128(0):296–306. doi:http://dx.doi.org/10.1016/j.apenergy.2014.04.074

Fujimori S, Masui T, Matsuoka Y (2015) Gains from emission trading under multiple stabilization targets and technological constraints. Energy Econ 48:306–315. doi:http://dx.doi.org/10.1016/j.eneco.2014.12.011

Fujimori S, Hasegawa T, Masui T, Takahashi K, Silva DH, Dai H et al (2017) SSP3: AIM implementation of shared socioeconomic pathways. Glob Environ Chang. 42:268–283

Hasegawa T, Fujimori S, Shin Y, Takahashi K, Masui T, Tanaka A (2014) Climate change impact and adaptation assessment on food consumption utilizing a new scenario framework. Environ Sci Technol 48(1):438–445. doi:10.1021/es4034149

Hasegawa T, Fujimori S, Shin Y, Tanaka A, Takahashi K, Masui T (2015a) Consequence of climate mitigation on the risk of hunger. Environ Sci Technol 49(12):7245–7253. doi:10.1021/es5051748

Hasegawa T, Fujimori S, Takahashi K, Masui T (2015b) Scenarios for the risk of hunger in the twenty-first century using Shared Socioeconomic Pathways. Environ Res Lett 10(1):014010

Hasegawa T, Fujimori S, Takahashi K, Yokohata T, Masui T (2016) Economic implications of climate change impacts on human health through undernourishment. Clim Chang 136(2):189–202

Hasegawa T, Fujimori S, Takahashi K, Ito A, Masui T (2017) Global land-use allocation model linked to an integrated assessment model. Science of The Total Environment 580:787–796.

Hof A, van Soest H, van den Berg M, de Boer HS, den Elzen M, Harmsen M et al (2015) Raising the ambition level of INDCs allows for a smoother energy transition. Assessment of the implications of INDCs for achieving the 2°C climate goal. PBL Netherlands Environmental Assessment Agency, Bilthoven, The Netherlands

International Energy Agency (IEA) (2015) World energy outlook Special report 2015: energy and climate change, International Energy Agency

International Institute for Applied Systems Analysis (IIASA) (2015a) IAMC AR5 scenario database. https://secure.iiasa.ac.at/web-apps/ene/AR5DB/dsd?Action=htmlpage&page=about#intro

International Institute for Applied Systems Analysis (IIASA) (2015b) Shared socioeconomic pathways (SSP) database version 0.9.3. https://secure.iiasa.ac.at/web-apps/ene/SspDb

IPCC (2007) Climate change 2007: the physical science basis. contribution of Working Group I to the fourth assessment report of the Intergovernmental Panel on Climate Change

IPCC (2014) IPCC, 2014: climate change 2014: synthesis report. contribution of Working Groups I, II and III to the fifth assessment report of the Intergovernmental Panel on Climate Change. IPCC

Ito A, Inatomi M (2012) Water-use efficiency of the terrestrial biosphere: a model analysis focusing on interactions between the global carbon and water cycles. J Hydrometeorol 13(2):681–694. doi:10.1175/JHM-D-10-05034.1

Iyer GC, Edmonds JA, Fawcett AA, Hultman NE, Alsalam J, Asrar GR et al (2015) The contribution of Paris to limit global warming to 2°C. Environ Res Lett 10(12):125002

Kim SH, Hejazi M, Liu L, Calvin K, Clarke L, Edmonds J et al (2016) Balancing global water availability and use at basin scale in an integrated assessment model. Clim Chang 136 (2):217–231. doi:10.1007/s10584-016-1604-6

Kitous A, Keramidas K (2015) Analysis of scenarios integrating the INDCs. European Commission

Lomborg B (2015) Impact of current climate proposals. Global Policy n/a-n/a. doi:10.1111/1758-5899.12295

Meinshausen M, Raper SCB, Wigley TML (2011) Emulating coupled atmosphere-ocean and carbon cycle models with a simpler model, MAGICC6 Part 1: model description and calibration. Atmos Chem Phys 11(4):1417–1456. doi:10.5194/acp-11-1417-2011

Meinshausen M, Jeffery L, Guetschow J, Robiou du Pont Y, Rogelj J, Schaeffer M et al (2015) National post-2020 greenhouse gas targets and diversity-aware leadership. Nat Clim Change 5 (12):1098–1106. doi:10.1038/nclimate2826. http://www.nature.com/nclimate/journal/v5/n12/abs/nclimate2826.html#supplementary-information

Mittal S, Dai H, Fujimori S, Masui T (2016) Bridging greenhouse gas emissions and renewable energy deployment target: comparative assessment of China and India. Appl Energy 166:301–313

Nordhaus W, Sztorc P (2013) DICE 2013R: introduction and user's manual, 2nd edn. http://www.econ.yale.edu/~nordhaus/homepage/documents/DICE_Manual_103113r2.pdf

Popp A, Rose SK, Calvin K, Van Vuuren DP, Dietrich JP, Wise M et al (2014) Land-use transition for bioenergy and climate stabilization: model comparison of drivers, impacts and interactions with other land use based mitigation options. Clim Chang 123(3):495–509. doi:10.1007/s10584-013-0926-x

Rogelj J, McCollum DL, Reisinger A, Meinshausen M, Riahi K (2013) Probabilistic cost estimates for climate change mitigation. Nature 493(7430):79–83. doi:10.1038/nature11787

Rosenzweig C, Elliott J, Deryng D, Ruane AC, Müller C, Arneth A et al (2014) Assessing agricultural risks of climate change in the 21st century in a global gridded crop model intercomparison. Proc Natl Acad Sci 111(9):3268–3273. doi:10.1073/pnas.1222463110

Schumacher K, Sands RD (2006) Innovative energy technologies and climate policy in Germany. Energ Policy 34(18):3929–3941. doi:10.1016/j.enpol.2005.09.022

United Nations Framework Convention on Climate Change (UNFCCC) (2015) Adoption of the Paris Agreement. Proposal by the President (1/CP21). http://unfccc.int/resource/docs/2015/cop21/eng/10a01.pdf. Accessed 2 Feb 2016

Vuuren DP, Stehfest E, Elzen MGJ, Kram T, Vliet J, Deetman S et al (2011) RCP2.6: exploring the possibility to keep global mean temperature increase below 2°C. Clim Chang 109 (1–2):95–116. doi:10.1007/s10584-011-0152-3

Chapter 3
Risks from Global Climate Change and the Paris Agreement

Kiyoshi Takahashi, Seita Emori, Shinichiro Fujimori, and Toshihiko Masui

Abstract The objectives of the Paris Agreement (PA) include limiting the increase in the global average temperature to well below 2 °C ("the 2 °C goal") and pursuing efforts to limit the increase to 1.5 °C ("the 1.5 °C goal"). The purpose of this chapter is to provide an overview of the relevant scientific knowledge on the risks from climate change corresponding to different levels of mitigation efforts, including the two long-term goals in the PA, as well as the expected consequences of extending the Intended Nationally Determined Contributions (INDCs) without strengthening them.

According to figures summarizing risks from climate change in the Fifth Assessment Report of the Intergovernmental Panel on Climate Change (IPCC-AR5), if the global temperature increase could be limited to below 1.5 °C above the preindustrial level, this would reduce the risks from climate change as evaluated from several perspectives. However, some of the risks (such as the effects on unique and threatened systems and the risks from extreme events) would still be considerable even at 1–2 °C above the preindustrial level.

Even with a similar degree of climate change, the levels of risk could differ under different degrees of exposure and/or vulnerability. To increase the accuracy of assessments of global climate risks for different temperature increases, explicit consideration of future changes in exposure and vulnerability will be necessary. Internationally coordinated development of new socioeconomic scenarios (Shared Socioeconomic Pathways, SSPs) is expected to promote such studies for the next round of the IPCC report.

K. Takahashi (✉) • T. Masui
Center for Social and Environmental Systems Research, National Institute for Environmental Studies (NIES), 16–2 Onogawa, Tsukuba, Ibaraki 305–8506, Japan
e-mail: ktakaha@nies.go.jp

S. Emori
Center for Global Environmental Research, National Institute for Environmental Studies (NIES), Tsukuba 305-8506, Ibaraki, Japan

S. Fujimori
Center for Social and Environmental Systems Research, National Institute for Environmental Studies (NIES), 16–2 Onogawa, Tsukuba, Ibaraki 305–8506, Japan

International Institute for Applied Systems Analysis, Schlossplatz-1, Laxenburg 2361, Austria
e-mail: fujimori.shinichiro@nies.go.jp

S. Fujimori et al. (eds.), *Post-2020 Climate Action*,
DOI 10.1007/978-981-10-3869-3_3

Keywords Climate risk • Reasons for concern • Temperature goal • Paris Agreement • Vulnerability

Key Messages to Policy Makers

- If the global temperature increase could be limited to below 1.5 °C above the preindustrial level, this would reduce the risks arising from climate change, evaluated from several perspectives. However, some of the risks from climate change would still be considerable even at 1–2 °C above the preindustrial level.
- The number of climate impact studies focusing on both changes in climate and socioeconomic factors, of which there were only a few before IPCC-AR5, is increasing. Furthermore, after the adoption of the Paris Agreement, the risks associated with a global temperature increase of 1.5 °C have been receiving more attention.

3.1 Introduction

At the 21st Conference of the Parties to the United Nations Framework Convention on Climate Change (UNFCCC) (COP21) held in Paris in December 2015, 195 countries adopted the first ever universal, legally binding global climate deal, the so-called Paris Agreement (PA). The objectives of the PA, defined in its Article 2, include limiting the increase in the global average temperature to well below 2 °C ("the 2 °C goal") and pursuing efforts to limit the increase to 1.5 °C ("the 1.5 °C goal"). This agreement on long-term temperature goals is a political decision among the parties based on understanding and perception of various factors including the anticipated climate risks, mitigation potential and cost, and adaptation potential and cost. While the agreement is value-laden, and science cannot deduce the long-term goals exclusively, the recent advances in scientific understanding of the problem, especially those reported in the Fifth Assessment Report of the Intergovernmental Panel on Climate Change (IPCC-AR5) published in 2014 (IPCC 2014), have served as a strong basis for the agreement. Without continuous advances in scientific knowledge in relation to climate risks, and suitable management measures to address these risks, the challenging task of fulfilling the PA's objectives cannot be achieved.

This book focuses on the ways in which Asian countries can achieve the mitigation target voluntarily proposed as the Intended Nationally Determined Contributions (INDCs) in the PA and discusses how the target could be increased in Chap. 2. The purpose of this chapter, as a background to later chapters, is to provide an overview of the relevant scientific knowledge on the climate-related risks corresponding to the different levels of mitigation effort, including the two long-term goals in the PA, as well as the expected consequences of extending the INDCs into the future without strengthening them.

In Sect. 3.2, both the global and Asian climate risks corresponding to different levels of global mean temperature increase are summarized based on IPCC-AR5. Next, in Sect. 3.3, the relevant scientific literature published after IPCC-AR5 is reviewed to outline the latest research trends. Finally, in Sect. 3.4, future research needs are discussed.

3.2 Global Mean Temperature Increase and Risks from Climate Change Assessed in IPCC-AR5

3.2.1 Temperature Increase Under the Greenhouse Gas (GHG) Emission Scenarios of Extending the INDCs

According to a 2016 report by the United States Agency for International Development (USAID 2016), globally, the INDCs submitted to the UNFCCC represent 98.6% of global greenhouse gas (GHG) emissions. The global implementation of the INDCs is estimated to result in aggregate global emission levels of 55 $GtCO_2eq$ in 2025 and 57 $GtCO_2eq$ in 2030. These emissions represent a decrease in the rate of increase of emissions by a third from 2010 to 2030 compared to the rate of increase from 1990 to 2010. If INDC commitments are met, global GHG emission levels would decrease by 3 $GtCO_2eq$ in 2025 and 4 $GtCO_2eq$ in 2030 compared to pre-INDC trajectories of emission levels. Aggregate global emission levels resulting from the implementation of INDCs would remain 9 $GtCO_2eq$ higher in 2025 and 15 $GtCO_2eq$ higher in 2030 than the least-cost 2 °C scenario.

The UNFCCC Secretariat report published immediately prior to COP21 (UNFCCC 2015) did not directly assess the implications for temperature change by the end of the century under the INDCs because this required information on emissions beyond 2030. It has also been claimed that there is a lack of clarity regarding the INDCs on the assumptions and approaches to be taken in the accounting of the land use, land-use change, and forestry (LULUCF) sector, including the approaches used for estimating, projecting, and accounting for GHG emissions and removals from the sector, which has resulted in the need to express the estimated aggregate emissions in 2025 and 2030 as a range (USAID 2016).

However, several independent analyses, based on a range of assumptions, methodologies, and data sources, have attempted to estimate the impact of the INDCs on the global temperature leading to a range of average values below, at, or above 3 °C (UNFCCC 2015). In other words, there is a consensus that the intended contributions represented in the INDCs are an improvement over business-as-usual trends leading to 4 °C or greater warming, but that they are not sufficiently ambitious to meet the 2 °C goal (e.g., 2.7 °C by Climate Action Tracker, 3.5 °C by Climate Interactive, and 3.2 °C by the Asia-Pacific Integrated Model (AIM) (Fujimori et al. 2016)). The background of the range of temperature estimates is covered in detail by Levin and Fransen (2015, website). Here, for simplicity, the risks from climate change under the extension of the INDCs are discussed below, assuming that the extension would lead a global mean temperature increase of 3 °C.

3.2.2 *Reasons for Concern in IPCC-AR5: Global Mean Temperature Increase and Its Risks*

The most prominent summary of the risks from climate change, assessed in IPCC-AR5 corresponding to different levels of climate change, is provided by the five "reasons for concern (RFCs)," also known as a "burning embers diagram" (Fig. 3.1; for details of individual RFCs, see Table 3.1). This has its origins in the Third Assessment Report of the IPCC (IPCC-TAR) published in 2001 and was updated in 2014 for IPCC-AR5 based on assessment of the latest available scientific literature and expert judgment by the report's lead authors. The RFCs provide a framework

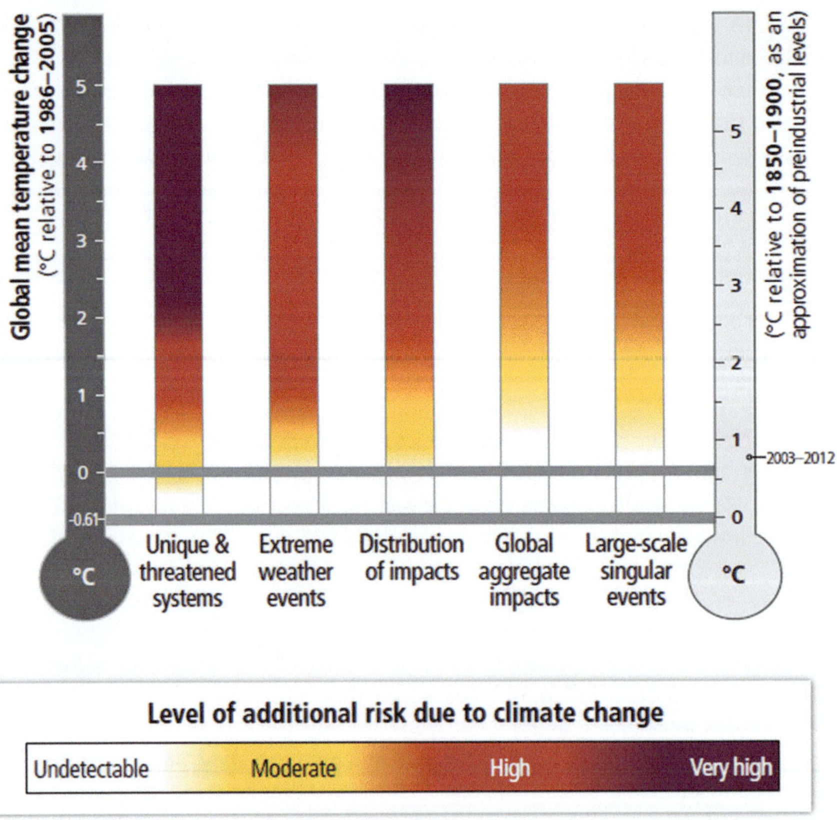

Fig. 3.1 A global perspective on climate-related risks (Adapted from IPCC (2014))

Table 3.1 Five integrative reasons for concern

Five RFCs	Assessment of risk levels
Unique and threatened systems	Some unique and threatened systems, including ecosystems and cultures, are already at risk from climate change (high confidence). The number of such systems at risk of severe consequences is higher with additional warming of around 1 °C. Many species and systems with limited adaptive capacity are subject to very high risks with additional warming of 2 °C, particularly Arctic sea ice and coral reef systems
Extreme weather events	Climate change-related risks from extreme events, such as heat waves, extreme precipitation, and coastal flooding, are already moderate (high confidence) and high with 1 °C additional warming (moderate confidence). Risks associated with some types of extreme events (e.g., extreme heat) increase further at higher temperatures (high confidence)
Distribution of impacts	Risks are unevenly distributed and are generally greater for disadvantaged people and communities in countries at all levels of development. Risks are already moderate because of regionally differentiated climate change impacts on crop production in particular (medium to high confidence). Based on projected decreases in regional crop yields and water availability, risks of unevenly distributed impacts are high for additional warming above 2 °C (medium confidence)
Global aggregate impacts	Risks of global aggregate impacts are moderate for additional warming between 1 and 2 °C, reflecting impacts on both Earth's biodiversity and the overall global economy (medium confidence). Extensive biodiversity loss with associated loss of ecosystem goods and services results in high risks around 3 °C additional warming (high confidence). Aggregate economic damages accelerate with increasing temperature (limited evidence, high agreement), but few quantitative estimates have been completed for additional warming around 3 °C or above
Large-scale singular events	With increasing warming, some physical systems or ecosystems may be at risk of abrupt and irreversible changes. Risks associated with such tipping points become moderate with 0–1 °C additional warming, due to early warning signs that both warm-water coral reef and Arctic ecosystems are already experiencing irreversible regime shifts (medium confidence). Risks increase disproportionately as temperature increases between 1 and 2 °C additional warming, and become high above 3 °C, due to the potential for a large and irreversible sea-level rise from ice sheet loss. For sustained warming greater than some threshold, near-complete loss of the Greenland ice sheet would occur over a millennium or more, contributing up to 7 m of global mean sea-level rise

All temperatures below are given as global average temperature change relative to 1986–2005 ("recent")
Adapted from IPCC (2014)

for summarizing the key risks across sectors and regions and illustrate the implications of warming and of adaptation limits for people, economies, and ecosystems. Increases in the magnitude of warming increase the likelihood of severe, pervasive, and irreversible impacts. The overall risks from climate change impacts can be reduced by limiting the rate and magnitude of climate change. The diagram has been utilized for evaluating dangerous anthropogenic interference with the climate system. Since there are a number of articles with in-depth interpretations and arguments on the concept and background of the RFCs (e.g., Schneider 2009, Smith et al. 2009), the topic is not discussed here.

Based on the two long-term goals of the PA (2 °C and 1.5 °C), the resulting climate change scenarios of the INDCs (about 3 °C), and business-as-usual (no-mitigation) emissions (4.0 °C), we can interpret the RFC figure as follows:

1. A global mean temperature increase of at least 4 °C from the preindustrial level, with no mitigation scenario, would cause high- to very high-risk levels in all of the five RFCs. These assessments are based on the projections of the severe and widespread impacts on unique and threatened systems, substantial species extinction, substantial risks to global and regional food security, and the combination of high temperature and humidity compromising normal anthropogenic activities, including growing food or working outdoors in some regions for parts of the year. The precise levels of climate change sufficient to trigger tipping points (thresholds for abrupt and irreversible change) remain uncertain, but the risk associated with crossing multiple tipping points in the Earth system or in interlinked human and natural systems increases with rising temperature (IPCC 2014).
2. With the achievement of the INDCs specified in the PA but without the more stringent mitigation efforts, if the corresponding temperature increase in this century is 3 °C, high-risk levels are still anticipated for the first three of the five RFCs. Even for the remaining two RFCs, the risk levels are moderate to high.
3. Even with the achievement of the 2 °C goal of the PA, the risk levels for the first two RFCs are still expected to be high. This implies that adaptation is necessary even with this range of strict goals and that we will also need to accept some degree of residual damage.
4. If the temperature increase could be limited to below the 1.5 °C proposed in the PA, the risk levels would be reduced to some extent for all of the five RFCs. However, some risks from climate change are considerable even at 1 or 2 °C above the preindustrial level. It should be borne in mind that higher than moderate risks would still remain for the first two RFCs even with the achievement of the very stringent goal for global temperature increase of 1.5 °C.

The risks associated with reasons for concern (RFCs) are shown for increasing levels of climate change. The colored shading indicates the additional risk due to climate change when a temperature level is reached and then sustained or exceeded. Undetectable risk (white) indicates no associated impacts are detectable and attributable to climate change. Moderate risk (yellow) indicates that associated impacts are both detectable and attributable to climate change with at least medium confidence, also accounting for the other specific criteria for key risks. High risk (red) indicates severe and widespread impacts, also accounting for the other specific criteria for key risks. Purple shows that very high risk is indicated by all specific criteria for key risks.

3.2.3 *Climate Risks in Asia and Global Mean Temperature Increase*

While comprehensive consideration of risks from climate change at the global scale is most relevant for international discussion of the goal of long-term stabilization, the risks at the regional scale are also important for assessing this goal as well as for consideration of adaptation options. IPCC-AR5 summarized three key regional risks for each of the nine regions of the world. Figure 3.2 shows the three regional risks identified for Asia, as an example. Detrimental effects from floods to human life, heat stress mortality due to extremely hot weather, and drought-related water and food shortages were identified as the three key risks for the region by the regional chapter authors. In IPCC-AR5, key risks are defined as potentially severe impacts relevant to Article 2 of the UNFCCC, which refers to "dangerous anthropogenic interference with the climate system."

The usefulness of this regional summary is that we can compare the differences in the risks anticipated for global temperature increases at the end of this century of 2 °C and 4 °C, respectively. The risks from an intermediate temperature increase of 3 °C, roughly corresponding to the extension of the INDCs, cannot be directly read from this table, but we can at least gain an insight into the regional aspects of the benefits of mitigation (expected abatement of risks) of the 2 °C goal. In addition, the risk levels for the near-term period (2030–2040) are also assessed. In the near term, projected levels of global mean temperature increase do not diverge substantially for different emission scenarios; therefore, there is only one assessment for this period. Although it does not appear to be explicitly mentioned in IPCC-AR5, the approximate expected risks from climate change might be considered for the 1.5 °C goal if we take into account the emission-indifferent rate of climate change through the period (approximately 0.2 °C/decade) and the historical trend (0.85 °C from preindustrial to present levels).

According to the IPCC-AR5 assessment, the risks of flood damage in Asia are already at the medium level, although there is some potential to reduce the risks through additional adaptive efforts such as reduction in the vulnerability of lifeline infrastructure and services. Due to changes in driving forces such as rain intensity, typhoons, and sea-level rise, risk levels at the end of the century would be very high for the 4 °C condition, while they could be partially reduced by achieving the 2 °C goal. From Fig. 3.2, it can also be seen that concurrent implementation of mitigation and adaptation would be effective for maintaining lower-risk levels. While the key risks for Asia are presented here as examples, discussion of temperature goals focusing on the regional aspects of climate change could be supported by the corresponding assessments for other regions.

Each key risk is characterized as very low to very high for three time frames: the present, near term (here; assessed for 2030–2040), and longer term (here; assessed for 2080–2100). For the longer term, risk levels are presented for two scenarios of global mean temperature increase (2 °C and 4 °C above the preindustrial level).

Key Risk	Adaptation issues & prospects	Climatic drivers
Increased riverine, coastal, and urban flooding leading to widespread damage to infrastructure, livelihoods, and settlements in Asia (medium confidence)	• Exposure reduction via structural and non-structural measures, effective land-use planning, and selective relocation • Reduction in the vulnerability of lifeline infrastructure and services (e.g., water, energy, waste management, food, biomass, mobility, local ecosystems, telecommunications) • Construction of monitoring and early warning systems; Measures to identify exposed areas, assist vulnerable areas and households, and diversify livelihoods • Economic diversification	Extreme precipitation Damaging cyclone Sea level
Increased risk of heat-related mortality (high confidence)	• Heat health warning systems • Urban planning to reduce heat islands; Improvement of the built environment; Development of sustainable cities • New work practices to avoid heat stress among outdoor workers	Extreme temperature
Increased risk of drought-related water and food shortage causing malnutrition (high confidence)	• Disaster preparedness including early-warning systems and local coping strategies • Adaptive/integrated water resource management • Water infrastructure and reservoir development • Diversification of water sources including water re-use • More efficient use of water (e.g., improved agricultural practices, irrigation management, and resilient agriculture)	Extreme temperature Drying trend

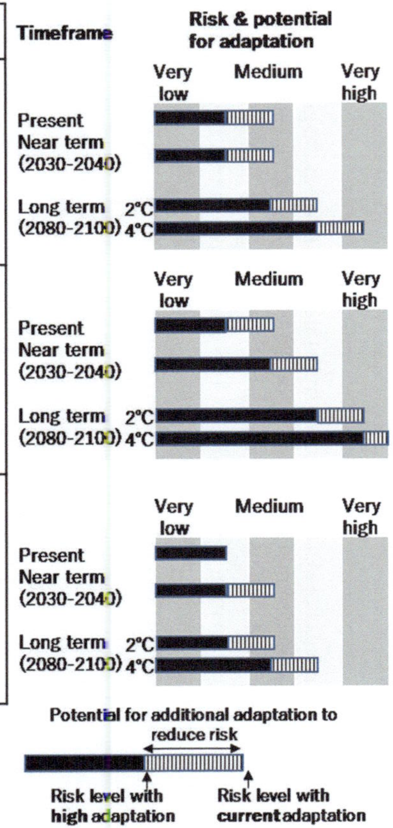

Fig. 3.2 Key regional risks from climate change and the potential for reducing risks through adaptation and mitigation (Redrawn based on IPCC (2014))

3.3 Studies Relevant to the PA Temperature Goals Published after IPCC-AR5

The PA includes a dual temperature goal: "holding the increase in the global average temperature to well below 2 °C above preindustrial levels and pursuing efforts to limit the temperature increase to 1.5 °C." Although the risks from climate change of a 2 °C increase in global temperature from the preindustrial level have been studied and reviewed extensively both before and after the adoption of the 2 °C goal recognized in COP15 in 2009 (Copenhagen Accord) and agreed in COP16 in 2010 (Cancun Agreement), there have been few studies focusing on the risks caused by a 1.5 °C increase. However, following the PA, the climate risks of a 1.5 °C have attracted significantly more attention.

Needless to say, the review and strengthening of the climate goals in the PA reflect the averseness of countries to the risks from climate change that could still exist even in a 2 °C warmer world. Both within and outside the arena of international negotiation of climate policies, small island states have repeatedly and continuously claimed that critical impacts of sea-level rise are inevitable even if the 2 °C goal is achieved. The updated scientific findings on climate risks including IPCC-AR5, clearly, also affected the inclusion of the lower-temperature increase goal in the PA. In other words, science has played a major role in the reviews of the long-term goals.

In Paris, the UNFCCC invited the IPCC to assess the latest science on the climate risks relevant to the 1.5 °C goal, and the IPCC plans to publish its special report on this topic by 2018. Thus, we can expect a large number of studies associated with the 1.5 °C goal over the next 2 years. Here, we highlight various examples of new relevant studies on the risks from global climate change published after IPCC-AR5.

Although peer-reviewed literature exclusively focusing on the risks from climate change caused by a 1.5 °C temperature increase is scarce, Schleussner et al. (2016) provided an assessment of the key impacts of climate change at warming levels of 1.5 °C and 2 °C, including extreme weather events, water availability, agricultural yields, sea-level rise, and risk of coral reef loss, and revealed substantial differences in the impacts of a 1.5 °C and those of a 2 °C warming that are highly relevant for the assessment of dangerous anthropogenic interference with the climate system. According to Schleussner et al. (2016), for the scenario of an end-of-century warming of 2 °C, virtually all tropical coral reefs would be at risk of severe degradation due to temperature-induced bleaching from 2050 onward. For the 1.5 °C scenario, this fraction would fall to about 90% in 2050 and was projected to decline to 70% by 2100. Regarding water availability, the regional reduction in median water availability for the Mediterranean would almost double from 9 to 17% between a warming of 1.5 °C and 2 °C, and the projected lengthening of regional dry spells would increase from 7 to 11%. The best estimate of sea-level rise indicated a 50 cm rise by 2100 relative to levels in 2000 for the 2 °C scenario and levels approximately 10 cm lower for the 1.5 °C scenario. For the 1.5 °C scenario, the rate of sea-level rise in 2100 would be reduced by about 30% compared to the 2 °C scenario.

The Integrated Climate Assessment – Risks, Uncertainties, and Society (ICA-RUS) research project conducted in Japan published its interim report on global risk management strategies and multi-sector impact analyses for a 1.5 °C warmer world as well as for other levels of climate change (mitigation scenarios of 2.0 °C and 2.5 °C temperature increases and a business-as-usual scenario of a 3.7 °C increase). Figure 3.3 shows examples of the analyzed sector risks (ICA-RUS Project Team 2015).

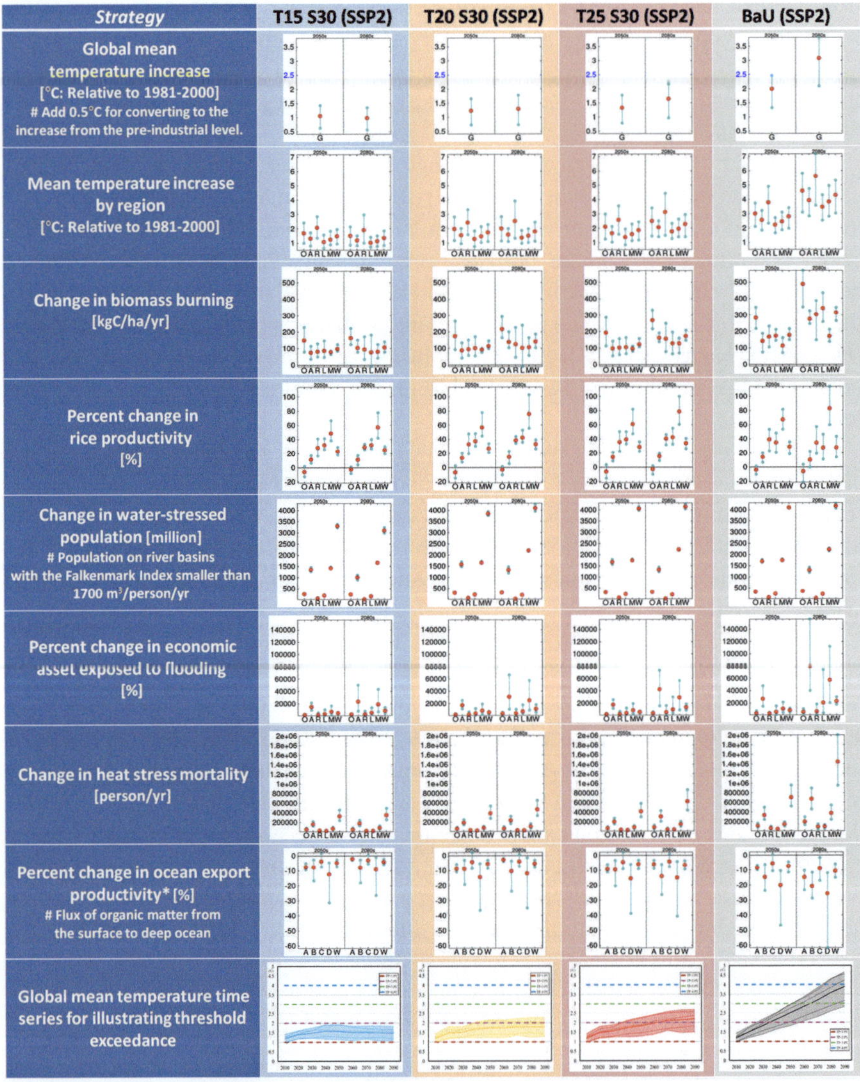

Fig. 3.3 Changes in temperature (*top* two and *bottom* panels) and selected impact variables with climate uncertainties for different strategies, regions, and two time horizons (2050s and 2080s) (*O* OECD, *A* Asia, *R* Eastern Europe and Russia, *L* Latin America, *M* Middle East and Africa, *W* world. *For oceanic data; *A* western Bering Sea, *B* central Equatorial Pacific, *C* eastern Equatorial Pacific, *D* northern Indian Ocean, *W* world ocean)

The team assessed the risk management implications of setting the 1.5 °C, 2.0 °C, and 2.5 °C goals at about 50% probability on each of the sectors of agriculture, ecosystems, water resources, floods, health, and tipping elements and found that the difference in the impacts between any two goals was generally smaller than that between any goal and the business-as-usual (BaU) scenario (i.e., no action on climate change) and also than the range of impacts caused by uncertainty in the behavior of the climate. The interim report tentatively concluded that taking definitive action and devising ways of dealing with climate uncertainty are of greater importance than the specific choice of goal (1.5 °C, 2.0 °C, or 2.5 °C) in terms of the global risks, while also noting that a more comprehensive assessment could alter this finding.

There have been additional studies that do not explicitly focus on the 1.5 °C risks, which would be relevant to the update of the assessment of negative anthropogenic impacts or the revision of the RFCs. As an example, while the impacts of climate change on the global economy have been important for several decades, there are still quite significant uncertainties in the projections. In the time period between IPCC-AR4 and AR5, it even appeared that progress in the field had somewhat stagnated in terms of narrowing the ranges of uncertainty. However, since the publication of AR5, several influential papers on the issue have been published.

Burke et al. (2015) reported nonlinear changes in the overall economic productivity with increasing temperature for all countries, with productivity peaking at an annual average temperature of 13 °C and declining strongly at higher temperatures based on statistical analyses. The authors confirmed that the relationship was valid for both agricultural and non-agricultural activities in both low- and high-income countries. Based on a novel empirical foundation for modeling economic loss in response to climate change, they estimated a reduction in global incomes of 23% under an unmitigated warming world scenario at the end of the century relative to the level expected with no climate change, assuming that the future mimicked past adaptation. In contrast to prior estimates, the study reported that the expected global losses were approximately linear to increasing global mean temperature, with median losses many times larger than those indicated by leading models.

Dietz et al. (2016) assessed the impacts of climate change on the present market value of global financial assets. They found that the expected "climate value at risk" (climate VaR) of current global financial assets was 1.8% following a business-as-usual emissions path and that this amounted to US$2.5 trillion taking a representative estimate of global financial assets. Furthermore, they also estimated that the 99th percentile climate VaR was 16.9%, or US$24.2 trillion, suggesting the possibility of a substantial write-off of the fundamental value of financial assets. However, cutting emissions to limit warming to no more than 2 °C was estimated to reduce the climate VaR by an expected 0.6 percentage points and the 99th percentile climate VaR by 7.7 percentage points. Based on these findings, the authors claimed that limiting warming to no more than 2 °C made financial sense to risk-neutral investors and even more so to the risk averse.

Although it would not bc appropriate to revise the RFCs based solely on the studies arbitrarily discussed above, which predict more severe impacts for specific increases in global temperature, it would be worth noting that the assessment of the RFCs in AR5 is based only on published literature available at the time of writing the report, and we should continue updating the RFCs to refine our evaluation of the detrimental anthropogenic impacts and long-term climate stabilization goals.

3.4 Future Research Needs

This chapter has provided an overview of the assessments of the risks from climate change relevant to the long-term global climate goals. The RFC framework is useful for the discussion of policy on the long-term goals, such as the 2 °C goal. However, several limitations of the assessment framework have also been mentioned. One of these is the non-explicit consideration of the future changes in exposure and vulnerability, which renders the existing RFCs somewhat subjective. Even with the same degree of climate change, if exposure (E) and/or vulnerability (V) vary, the risk levels could also change. IPCC-AR5 (Chapter 19 of the Working Group II report, Oppenheimer et al. (2014)) merely explained the concept of the V- and E-dependent RFCs using a schematic graph and did not fill in specific details because very limited research data (sector impact projections explicitly considering alternative future socioeconomic developments) were available at the time of writing the report. Since the release of IPCC-AR5, the numbers of studies focusing on changes in both climate and socioeconomic conditions have increased (e.g., Jongman et al. 2015, Hasegawa et al. 2014, Hanasaki et al. 2013). Internationally coordinated development of new socioeconomic scenarios (Shared Socioeconomic Pathways (SSPs); Riahi et al. (2017)) is expected to promote such studies for the next IPCC report.

The use of multiple scenarios for attaining long-term temperature goals can also influence the assessment of the risks from climate change attached to the goals. For example, both theoretical studies and those based on analyses using models would show that the 2 °C goal could be attained even with a rather modest near-term GHG mitigation scenario, such as that of the INDCs submitted to COP21, which would then require stronger and more challenging mitigation efforts in the subsequent decades. While the trajectory of climate change might not significantly differ among different scenarios for attaining the same long-term temperature goal, the risks associated with climate change could significantly differ due to the different V and E conditions caused by the choice of mitigation scenario or different atmospheric CO_2 concentrations and resulting fertilization effects. As a specific example, with an increased dependency on bioenergy with carbon capture and storage (BECCS) technology in the latter half of this century, following a weaker near-term mitigation effort, competition for land and/or water between food production and bioenergy production would be expected to increase resulting in an increased risk of food deficits. Here again, the new socioeconomic scenarios mentioned above may be useful for conducting analyses of the risks from climate change with a more consistent consideration of the relevant factors.

Regarding the assessment of risk levels based on surveys of the existing literature and graphical representation of the assessed results, it should be borne in mind that such a representation is based on the subjective judgment of experts. If greater rigor and transparency could be introduced into the judgment process, the results of the assessment would be more widely recognized as a reliable basis for discussion and would, therefore, be more useful.

Regarding differences in the risk levels between the 1.5 °C and 2 °C scenarios, while some studies emphasize the need to increase our understanding of the thresholds and subsequent impacts of large-scale singular events such as ice sheet disintegration (ICA-RUS Project Team 2015), others tend to focus on the changes in extreme weather events and their impacts (Mitchell et al. 2016). More precise analysis of the impacts of changes in extreme events for a small temperature increase, such as 1.5 °C, would increase the need for initial-value ensemble experiments of climate projection and the more effective utilization of ensemble projection scenarios. However, Schleussner et al. (2016) suggested that significant differences in risk levels would be expected in various sectors, not just in those affected by the change in extreme events. Thus, it is apparent that scientific opinion on the reduction of risks achieved by meeting the 1.5 °C goal is still divided. A comprehensive assessment of the status of our scientific understanding requires us to wait for the completion of the IPCC special report on the 1.5 °C world, due in 2018.

References

Burke M et al (2015) Global non-linear effects of temperature on economic production. Nature 525:235–239

Dietz S et al (2016) 'Climate value at risk' of global financial assets. Nat Clim Change. doi:10.1038/nclimate2972

Fujimori S, Su X, Liu J, Hasegawa T, Takahashi K, Masui T, Takimi M (2016) Implications of Paris Agreement in the context of long-term climate mitigation goals. SpringerPlus 5(1):1620

Hanasaki N et al (2013) A global water scarcity assessment under shared socio-economic pathways – Part 2: Water availability and scarcity. Hydrol Earth Syst Sci 17(7):2393–2413

Hasegawa T et al (2014) Climate change impact and adaptation assessment on food consumption utilizing a new scenario framework. Environ Sci Technol 48:438–445

ICA-RUS Project Team (2015) ICA-RUS report 2015: alternatives left to humanity faced with global climate risks (Ver. 1). http://www.nies.go.jp/ica-rus/report/ICA-RUS_REPORT_2015_eng.pdf

IPCC (2014) Summary for policymakers. In: Climate change 2014: impacts, adaptation, and vulnerability. Contribution of Working Group II to the Fifth Assessment Report of the Intergovernmental Panel on Climate Change. Cambridge University Press, Cambridge/New York, pp 1–32

Jongman B et al (2015) Declining vulnerability to river floods and the global benefits of adaptation. PNAS 112(18):E2271–E2280

Levin K, Fransen T (2015) INSIDER: why are INDC studies reaching different temperature estimates? World Resources Institute. http://www.wri.org/blog/2015/11/insider-why-are-indc-studies-reaching-different-temperature-estimates
Mitchell D et al (2016) Realizing the impacts of a 1.5°C warmer world. Nat Clim Chang 6:735–737
Oppenheimer M et al (2014) Emergent risks and key vulnerabilities. In: Field CB et al (eds) Climate change 2014: impacts, adaptation and vulnerability. Part A: Global and sectoral aspects. Contribution of Working Group II to the Fifth Assessment Report of the Intergovernmental Panel on Climate Change. Cambridge University Press, Cambridge, pp 1039–1099
Riahi K et al (2017) The shared socioeconomic pathways and their energy, land use, and greenhouse gas emissions implications: an overview. Global Environ Change 42:153–168
Schleussner CF et al (2016) Differential climate impacts for policy-relevant limits to global warming: the case of 1.5°C and 2°C. Earth Syst Dynam 7:327–351
Schneider SH (2009) Science as a contact sport. National Geographic Society, Washington, DC
Smith JB et al (2009) Assessing dangerous climate change through an update of the Intergovernmental Panel on Climate Change (IPCC) "reasons for concern". PNAS 106(11):4133–4137
UNFCCC (2015) Synthesis report on the aggregate effect of the intended nationally determined contributions. http://unfccc.int/resource/docs/2015/cop21/eng/07.pdf
USAID (2016) Analysis of intended nationally determined contributions (INDCs). http://mitigationpartnership.net/sites/default/files/analysis_of_intended_nationally_determined_contributions_indcs.pdf

Chapter 4
Temporal and Spatial Distribution of Global Mitigation Cost: INDCs and Equity

Jing-Yu Liu, Shinichiro Fujimori, and Toshihiko Masui

Abstract Each country's Intended Nationally Determined Contribution (INDC) pledges an emission target for 2025 or 2030. Here, we evaluated the INDC intergenerational and interregional equity by comparing scenarios with INDC emission target in 2030 and with an immediate emission reduction associated with a global uniform carbon price using AIM/CGE (Asian-Pacific Integrated Model/Computable General Equilibrium). Both scenarios eventually achieve 2 °C target. The results showed that, as compared with an immediate emission reduction scenario, the intergenerational equity status is not favorable for INDC scenario and the future generation suffers more from delayed mitigation. Moreover, this conclusion was robust to the wide range of inequality aversion parameter that determines discount rate. On the other hand, the INDC scenario has better interregional equity in the early part of the century than does the immediate emission reduction scenario in which we assume a global carbon price during the period up to 2030. However, interregional equity worsens later in the century. The additional emission reduction to the INDC in 2030 would improve both inter- and interregional equity as compared to the current INDC. We also suggest that countries should commit to more emission reductions in the follow-up INDC communications and that continuous consideration for low-income countries is needed for global climate change cooperation after 2030.

Keywords Paris Agreement • INDC • Climate change • Generational equity • Mitigation cost • CGE

J.-Y. Liu (✉) • T. Masui
Center for Social and Environmental Systems Research, National Institute for Environmental Studies (NIES), 16–2 Onogawa, Tsukuba, Ibaraki 305–8506, Japan
e-mail: liu.jingyu@nies.go.jp

S. Fujimori
Center for Social and Environmental Systems Research, National Institute for Environmental Studies (NIES), 16–2 Onogawa, Tsukuba, Ibaraki 305–8506, Japan

International Institute for Applied Systems Analysis, Schlossplatz-1, Laxenburg 2361, Austria
e-mail: fujimori.shinichiro@nies.go.jp

S. Fujimori et al. (eds.), *Post-2020 Climate Action*,
DOI 10.1007/978-981-10-3869-3_4

4.1 Introduction

In the IPCC's fifth assessment report (AR5) (IPCC 2015), Working Group III noted several scenarios corresponding to the 450 ppm CO_2 equivalent concentration stabilization. These scenarios mostly have immediate GHG emission reductions, and CO_2 emissions become zero or negative in the latter half of this century. The Paris Agreement reaffirmed that the ultimate climate goal is the 2 °C target, which corresponds to about 450 ppm CO_2 equivalent concentration stabilization in 2100 (UNFCCC 2015). In addition, each country is to submit an INDC (Intended Nationally Determined Contribution) pledge of emission targets for 2025 or 2030. The summation of all countries' submitted emission targets in 2030 is expected to be larger than the least-cost 2 °C scenarios shown in AR5, leading to a projected level of 55 $GtCO_2eq$ in 2030. To achieve the 2 °C target, an additional emission reduction of approximately 15 $GtCO_2eq$ will be required to fill the gap between the post-INDC trajectory and the least-cost 2 °C scenarios after 2030 (UNFCCC 2015).

The spatial and temporal distribution of mitigation costs would be different in a mitigation scenario consistent with the INDCs and achieving the 2 °C target and in the least-cost 2 °C scenarios. This difference implies that the INDCs influence equity both intergenerationally and interregionally.

The discount rate used in projections has a notable influence on the conclusions and policy implications of generational equity issues. Social cost of carbon varies with discount rate and people have conflicting opinions about the urgency for the need for climate policy depending on the discount rate (Anthoff et al. 2009). There have been discussions about how to determine the discount rates. The IPCC's second assessment report (SAR) (IPCC 1995) summarized two major approaches of discounting future values, namely, the prescriptive and descriptive approaches. The former begins with ethical considerations, whereas the latter begins with evidence from decisions that people and governments actually make. AR5 (IPCC 2015) looked deeper into the prescriptive approach and reviewed the use of the Ramsey rule of discounting. AR5 concluded that a relative consensus has emerged in favor of setting the pure rate of time preference at 0 and the inequality aversion parameters somewhere from 1 to 3. The role of the pure rate of time preference and inequality aversion in the discount rate calculation will be explained in Sect. 4.2.3.1.

Several studies have examined numerical INDC assessment. For example, Iyer et al. (2015) compared INDC scenarios with an alternative emission path in which nations cannot undertake mitigation within the INDC period. They found that the use of INDCs made important contributions toward achieving the 2 °C target. Meinshausen et al. (2015) assumed one major emitting economy would take the lead in the international emission allocation scheme as an INDC approach, and the results showed that there would be a large gap between what emission target was required for this leading country and its current INDC pledge. Fawcett et al. (2015) studied Paris pledges from a probabilistic perspective and found that INDCs contribute both to reducing the risks of severe outcomes of global warming and to increasing the probability of limiting warming to 2 °C. However, no studies have yet focused on a quantitative generational equity assessment of the INDC and post-INDC time intervals. AR5 also summarizes the effects on the mitigation cost due to delayed mitigation action (WG3 SPM, Table 4.2). Obviously, emission reductions in a former period contribute to a reduction of the mitigation costs in a later period. Tavoni et al. (2013) and Kober et al. (2014) cover, to some extent, the issue of different mitigation costs across regions under LIMITS projects (http://www.feem-project.net/limits/). Although these studies provide meaningful scientific knowledge with regard to mitigation cost distribution and policy suggestions, their scenario frameworks are stylized for an analysis of 2020 climate action and not for 2030 or for the INDCs.

To address this gap, we attempted to answer three research questions. (1) Does the use of INDCs lead to more mitigation costs for future generations as compared with the least-cost 2 °C scenarios presented in AR5? (2) How do the INDCs affect interregional equity? (3) Does greater reduction of emissions in INDCs help to improve generational equity? Here, we have chosen to focus on the 2 °C goal as a climate mitigation target. However, older generations might change this climate goal if they realize that the cost of mitigation is larger than the benefit. For this reason, we have framed the paper's focus as clearly policy relevant (Paris Agreement).

Section 4.2 presents the model we used and the scenario framework. The analytical method applied to analyze the results derived from the model is also introduced. Section 4.3 presents the model results for some of the main indicators and the results for the intergenerational and interregional equity analysis. In Sect. 4.4, we discuss the interpretations and implications of the results, limitations of the study, and future work in this area.

4.2 Methodology

4.2.1 AIM/CGE

We used AIM/CGE (Asia-Pacific Integrated Model/Computable General Equilibrium), which has been widely used in climate mitigation and impact assessment (Hasegawa et al.2015, 2016; Mittal et al. 2016; Dai et al. 2016; Fujimori et al. 2014a). AIM/CGE is a recursive dynamic general equilibrium model that includes 17 regions and 42 industrial classifications. Energy sectors, including power sectors, are disaggregated in detail. Moreover, to assess bioenergy and land use competition appropriately, agricultural sectors are also highly disaggregated (Fujimori et al. 2014b). This CGE model was developed based on the "standard CGE model" (Lofgren et al. 2002), and details of the model structure and mathematical formulas are described in the AIM/CGE basic manual (Fujimori et al. 2012). The details of the model are shown in Chaps. 12 and 13 of this book.

The implementation of mitigation is represented by either a global emission constraint or country-level emission constraints. A carbon tax is employed to meet the global or country-level emission constraints, and it is therefore either globally uniform or variable across regions, depending on the type of constraint. The carbon tax makes the price of fossil fuel goods higher when emissions are constrained and promotes energy savings and the substitution of fossil fuels by lower emission energies and also acts as an incentive to reduce the nonenergy-related emissions. Gases other than CO_2 are weighted by global warming potential and summed as GHG emissions in CO_2 equivalents. The revenue from the carbon tax is assumed to be received by households.

4.2.2 Scenarios

Table 4.1 outlines the four scenarios simulated in this study: baseline, 450CO2e, INDC_450CO2e, and SINDC_450CO2e. The baseline scenario entails no emission constraints and follows current trends in energy technologies. The 450CO2e scenario includes emission constraints that approximately meet the emission radiative forcing target of 2.6–2.8 W/m^2 in 2100, which can be interpreted as being in the range of the 2 °C target. The scenario employs a uniform global carbon price subject to particular emission targets discussed later. The INDC_450CO2e scenario also meets a similar radiative forcing target. INDC pledges are satisfied before 2030. The resulting gap between INDC_450CO2e and 450CO2e in cumulative CO_2 emissions prior to 2030 is addressed by further reducing emissions from the 450CO2e emission trajectory after 2030 in INDC_450CO2e. The SINDC_450CO2e scenario is similar to the INDC_450CO2e scenario, but the INDC target for each country is 20% higher as compared with the target in INDC_450CO2e. The emission pathway after 2030 is determined in the same manner as in the INDC_450CO2e scenario. Other socioeconomic policies, such as renewable energy targets, are not treated in this study. However, the energy system responds to the carbon price associated with the chosen emission constraints.

The simulation period is 2005–2100. The socioeconomic assumptions behind all of the scenarios follow SSP2 (shared socioeconomic pathways) (O'Neill et al. 2014) socioeconomic assumptions. Socioeconomic assumptions generally include assumptions not only on demographics and economic scale but also on other aspects, such as changes in energy technologies and dietary preference changes. The results might be sensitive to the SSP socioeconomic assumptions. Here SSP2 is used as it is seen to be the continuing of the current social, economic, and technological trend, leaving the world face moderate challenges to mitigation and adaptation (O'Neill et al. 2014). Population and gross domestic product (GDP) growth are assumed to be the same as in Samir and Lutz (2014) and Dellink et al.

Table 4.1 Scenario descriptions

	Emission constraints		Radiative forcing in 2100 (W/m^2)
	Before 2030	After 2030	
Baseline	No emission constraints		7.28
450CO2e	Global emission constraint		2.86
INDC_450CO2e	Each country's emission constraint	Global emission constraint	2.83
SINDC_450CO2e	Each country's emission constraint + additional emission reduction	Global emission constraint	2.79

(2015). Other energy, land use, and agricultural assumptions are based on Fujimori et al. (2014c). Countries are grouped into three regions: developing countries, industrial countries, and transition countries.

4.2.3 *Analytical Methods*

After we obtained the scenario results, we analyzed the mitigation costs with respect to generational equity. Here, consumption loss or its rate is treated as a representative metric of mitigation costs. To make current and future mitigation costs comparable for intergeneration comparisons and to be able to aggregate the costs across time for both intergenerational and interregional analyses, a discount rate is introduced.

We made three types of analysis in Sect. 4.3. First, we divided the years 2020–2100 into two generations to analyze intergenerational equity. The current generation was defined as 2020–2060 and the future one as 2060–2100. Then, we compared the net present value (NPV) of both generations in terms of absolute consumption loss in the various scenarios. Using the discount rate, we converted future mitigation costs into an equivalent present value for 2020 and aggregated the values. The second and third are for interregional equity by evaluating the relationship between the consumption loss rate and GDP per capita across regions in each year as well as their average values for the entire period.

4.2.3.1 Ramsey Rule and the Discount Rate

The discount rates were determined by Eq. 4.1 based on the Ramsey rule (Ramsey 1928):

$$\rho_{t,r} = \delta + \eta g_{t,r} \tag{4.1}$$

Here $\rho(t,r)$ is the discount rate at time t for region r, and δ is the pure rate of time preference for the present. We assumed that the pure rate of time preference was 0 following the practice of previous studies (Cline 1992; Arrow 1999; Lebègue 2005), because there is broad consensus for a zero or near-zero pure rate of time preference for the present (IPCC 2015). The parameter $g(t,r)$ is the consumption growth rate in region r, and η can be viewed as a measure of intertemporal inequality aversion (Dasgupta 2008; Gollier 2008; Garnaut 2008; Arrow et al. 2004). AR5 (IPCC 2015) explained η with the following thought experiment. Imagine an economic policy, the aim of which is to increase consumption by 1 unit for every person in a poor group by reducing consumption for every wealthy person by x units. The maximum value of x that a person would accept to implement the policy is a metric of that person's inequality aversion. Values of 1–3 were often used for inequality aversion, with 2 being the most common value (IPCC 2015; Arrow et al. 2013). This implies that the x from the thought experiment ranges from 2 to 8. Arrow et al. (2013) suggested that η reflects the maximum sacrifice one generation should make to transfer income to another generation.

Figure 4.1 shows how the discount rate varies over time with different values of inequality aversion in baseline. The discount rates are almost the same across scenarios since the GDP loss is small enough to change discount rates significantly. All of the curves show a declining trend after 2020.

4.2.3.2 Equity Principles

Three equity principles are operationalized in our study. First, the use of a discount rate involves equity. The discount rate is determined on the basis of people's inequality aversion, as noted above. Different inequality aversion values are used in the analysis, and their effects on the results are discussed. Second, intergenerational equity was measured by the mitigation cost for current and future generations. In our calculations, we assumed that equity was improved once the relative difference of the NPVs of both generations' consumption losses decreased. Finally, interregional equity was measured by mitigation cost differences and, here, consumption loss rate is used as an indicator of mitigation cost.

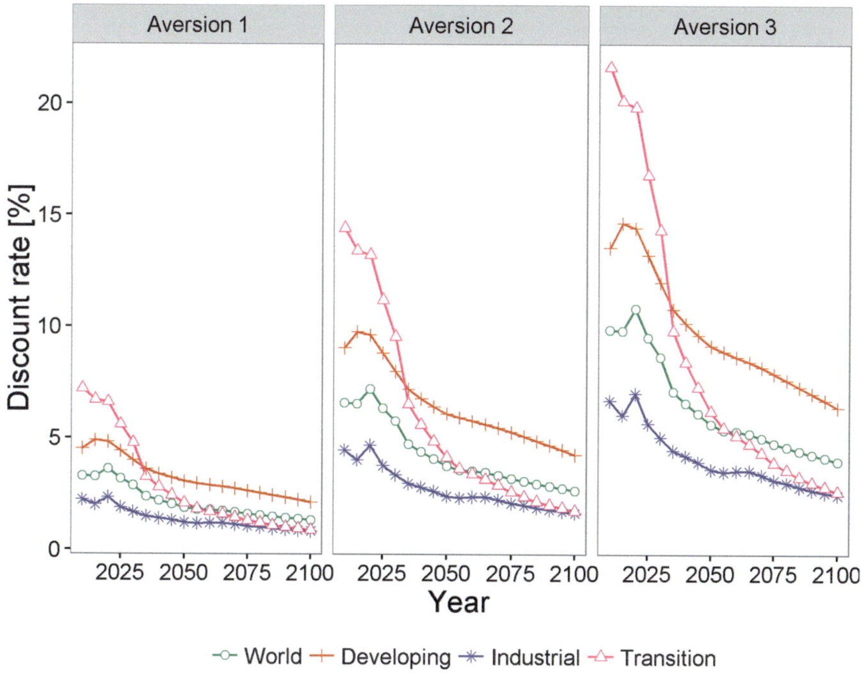

Fig. 4.1 Global and regional discount rates for inequality aversions of 1, 2, and 3 in the baseline scenario

4.3 Results

4.3.1 Main Indicators

4.3.1.1 GHG Emissions

Figure 4.2 shows the emission paths for the baseline, 450CO2e, and INDC_450CO2e scenarios. To meet the 2 °C target, the 450CO2e scenario assumes a gradually declining global emission path. Total global emissions are 48.9 $GtCO_2eq$ in 2020 and 8.76 $GtCO_2eq$ in 2100. In the INDC_450CO2e scenario, each country's emissions between 2020 and 2030 are consistent with its INDC pledge. After 2030, the gap in cumulative CO_2 emissions from before 2030 as compared with emissions in the 450CO2e scenario will be filled by allocating additional emission reductions to the latter part of the period. After 2080, total global emissions reach a lower limit and then remain steady.

Emissions in developing countries, transition countries, and globally are greater in the INDC_450CO2e scenario than those in the 450CO2e scenario from 2020 to 2030, whereas they are about the same in industrial countries. Industrial countries'

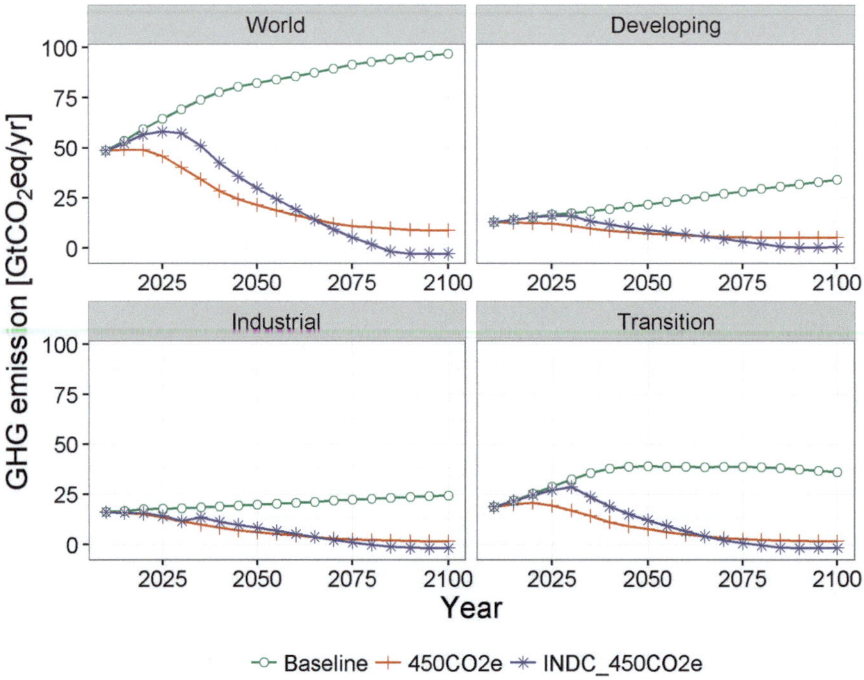

Fig. 4.2 Global and regional GHG emission paths for scenarios baseline, 450CO2e, and INDC_450CO2e

emissions increase slightly after 2030, primarily because the global uniform carbon price after the INDC period is much lower than it is during the INDC period.

4.3.1.2 Carbon Price

In the 450CO2e scenario, a uniform global carbon price is assumed throughout the period. In the INDC_450CO2e scenario, we set country-level carbon prices from 2020 to 2030 and a uniform global carbon price from 2030 to 2100. Carbon price climbed dramatically after 2075 in the INDC_450CO2e scenario (Fig. 4.3a), primarily because greater emission reductions are required in the latter part of the century and the marginal cost of emission reduction is rising. Developing countries and transition countries have lower carbon prices in the INDC_450CO2e scenario than in the 450CO2e scenario from 2020 to 2030, whereas industrial countries have higher prices (Fig. 4.3b).

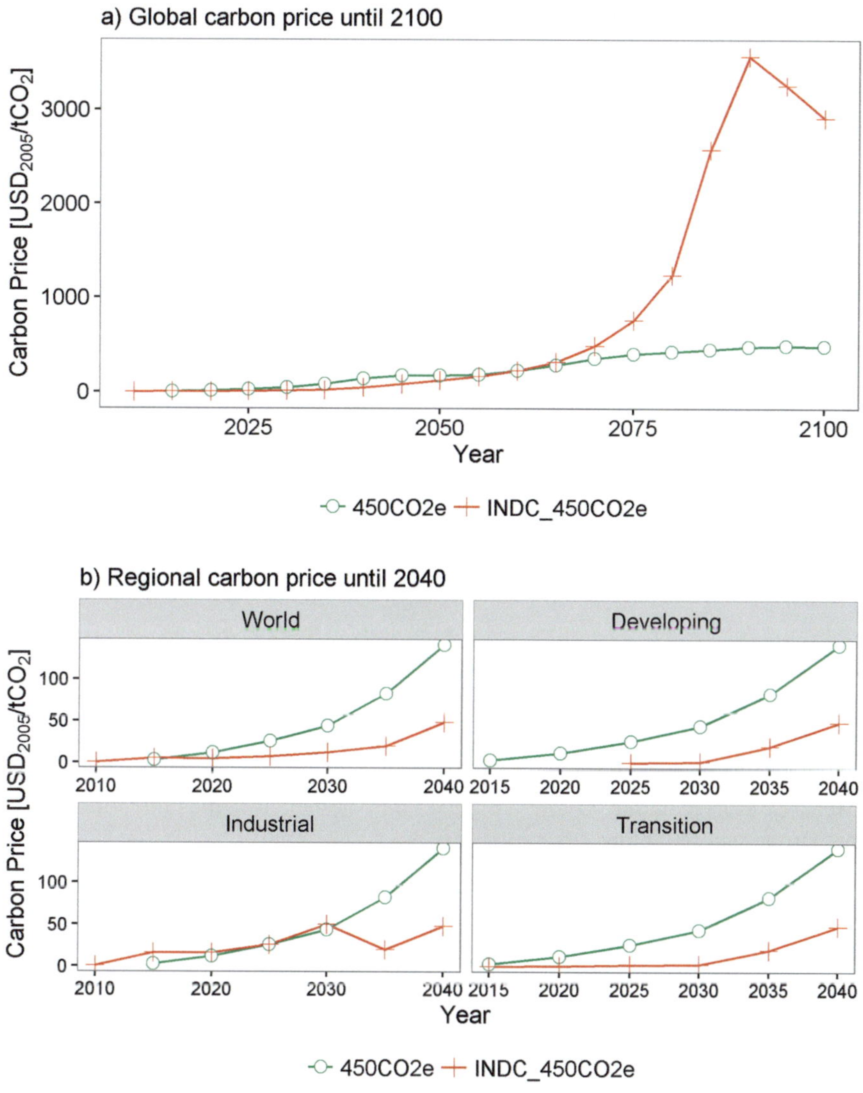

Fig. 4.3 (**a**) Global carbon price through 2100 and (**b**) regional carbon price through 2040 for scenarios 450CO2e and INDC_450CO2e

4.3.1.3 Mitigation Cost

Figure 4.4 presents regional and global consumption loss rates as an indicator of mitigation costs. Global mitigation costs in the INDC_450CO2e scenario are less than those in the 450CO2e scenario before 2060, but they are higher afterward.

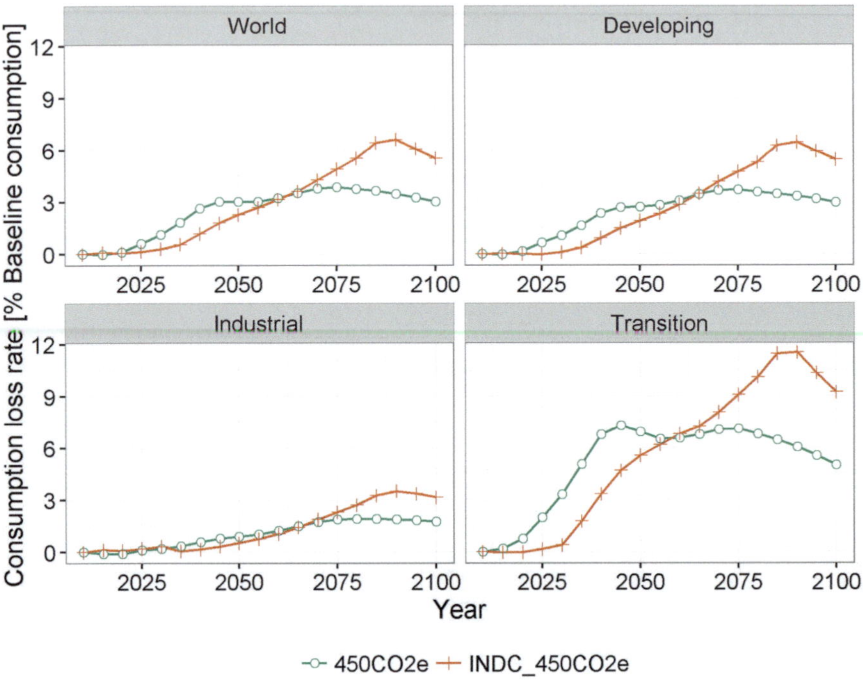

Fig. 4.4 Global and regional mitigation costs for scenarios 450CO2e and INDC_450CO2e

Because the discount rate is not accounted for in these figures, the mitigation cost of the current and the future generations are not comparable.

4.3.2 Intergenerational Equity

We calculated the NPV of consumption losses for the current (2020–2060) and future (2060–2100) generations, as well as the relative differences between the two generations in each scenario. The relative difference was calculated as (NPV of the current generation – NPV of the future generation)/NPV of the future generation. This difference reflects the relative gap in mitigation costs between the two generations. Hence, it is an indicator of intergenerational equity – smaller values indicate greater intergenerational equity. It is important to note that it is only the absolute value of the relative difference that matters, not whether the value is positive or negative.

Figure 4.5 shows the consumption losses for the two generations and the relative differences between the two generations' global consumption losses. The discounted consumption loss is lower in the current generation and higher in the future in the INDC_450CO2e scenario as compared with the 450CO2e scenario for

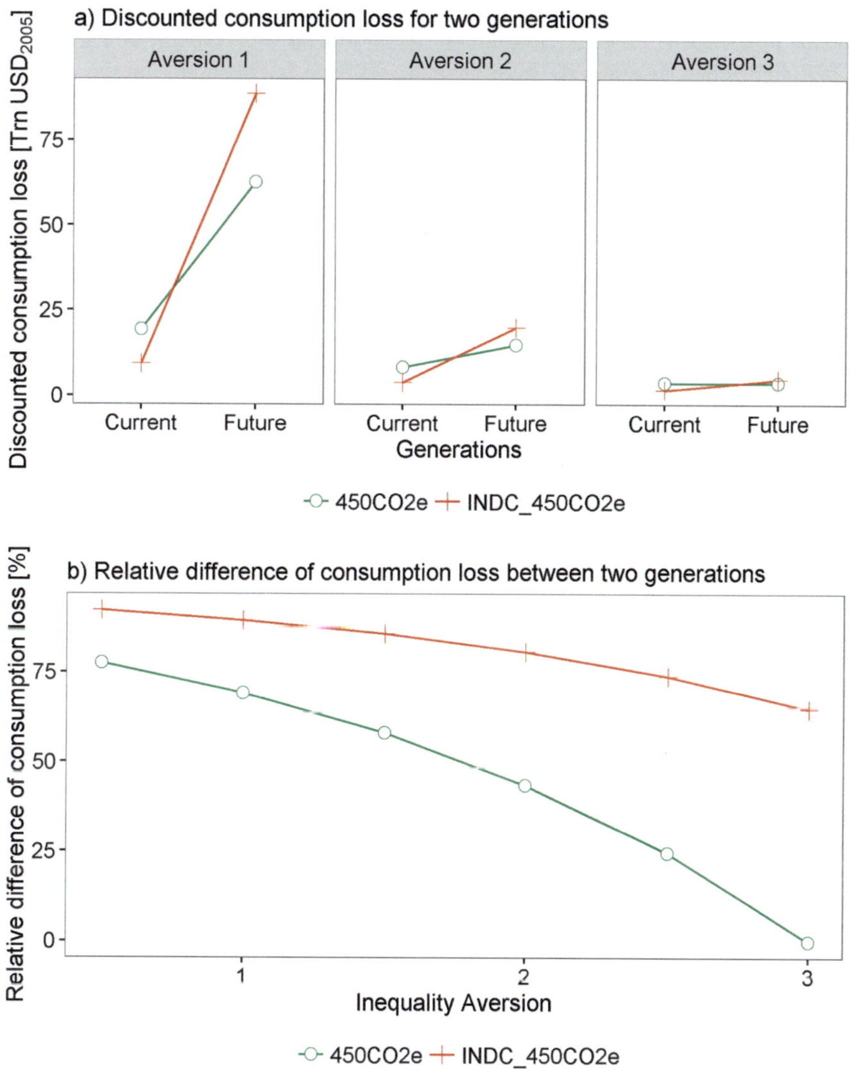

Fig. 4.5 (**a**) Discounted consumption loss for two generations and (**b**) relative difference of consumption losses ([NPV current generation – NPV future generation]/NPV future generation) for scenarios 450CO2e and INDC_450CO2e. 1, 2, and 3 represent inequality aversion

all inequality aversion values. The relative difference decreases as inequality aversion increases in both scenarios, but the relative difference in the 450CO2e scenario is always lower and approaches 0 when the inequality aversion is 3. Except for the case when the inequality aversion is 3 in the 450CO2e scenario, both scenarios favor the current generation over the future generation. In addition, the

INDC_450CO2e scenario always has worse intergenerational equity, regardless of the inequality aversion value. It should be noted that the absolute difference is small when inequality aversion is high, which will not be apparent in the relative differences figure.

4.3.3 Interregional Equity

Figure 4.6 shows scatter plots for the NPV consumption loss rates of the entire time period of 17 regions versus GDP per capita for the 450CO2e and INDC_450CO2e scenarios. We used the NPV consumption loss rate for each region to represent the average mitigation cost throughout the entire time period. The trend lines of the scatter plots show the trend of the relationship between income and mitigation cost for the different scenarios. A negative sloping trend line indicates that mitigation causes low-income countries to have higher consumption loss rates. The slope of the trend line in scenario INDC_450CO2e is more negative than that of scenario 450CO2e when the inequality aversion is small (i.e., 1). When the inequality aversion is 2 or 3, the slope in the INDC_450CO2e scenario is greater than that of the 450CO2e scenario. This occurs because the impact of the latter part of the century become less important to the NPV of the entire time period when the discount rate is relatively high, and the earlier part of the century is more heavily weighted.

Focusing on the confidence intervals (>95%) of the slope parameters, the aversion 3 case is relatively small (Table 4.2) and there is no great overlap between $450CO_2$ and INDC450CO2e. Therefore, under the assumption of equality aversion parameter 3, their difference is sufficiently large. However, it would be difficult to distinguish differences between these two scenarios with small aversion parameters 1.

We also evaluated variations in interregional equity over time from the regression parameters of the inter-region scatter plots in different years (Fig. 4.7). The discount rate is not used here. In the first part of the period, scenario INDC_450CO2e has better interregional equity compared to scenario 450CO2e. After about 2060, however, the situation is reversed. Focusing on the confidence intervals (>95%) of the slope parameters in Fig. 4.7, in the early and late parts of this century, i.e., before 2045 and after 2075, there is no great overlap between the 450CO2e and INDC450CO2e scenarios. Therefore, the slopes of the curves differ significantly. Between 2045 and 2075, however, the confidence intervals are large and we could not distinguish the slopes of the curves in the two scenarios.

We assumed a uniform carbon price across regions for the entire period in the 450CO2e scenario and from 2030 to 2100 in the INDC_450CO2e scenario. A global carbon price would have a greater adverse impact on low-income countries than on high-income countries. Before 2030, INDC pledges ensure that interregional equity improves because low-income regions gain opportunities to

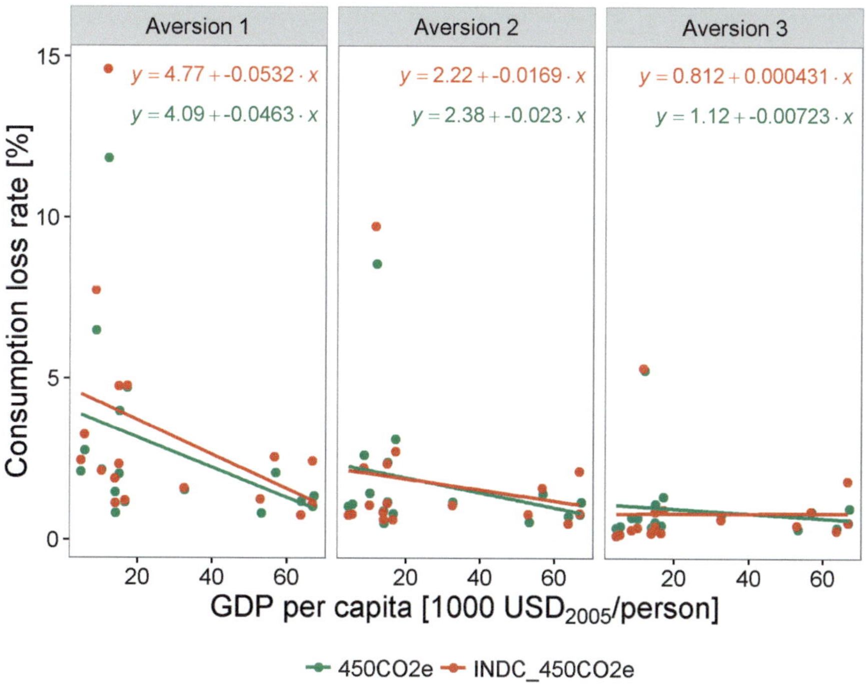

Fig. 4.6 Interregional consumption loss accounted for as NPV for r scenarios 450CO2e and INDC_450CO2e, with inequality aversions of 1, 2, and 3

Table 4.2 Statistical parameters of the estimated slopes for NPV

Inequality aversions	Scenarios			*t* value	*n*	Confidential interval	
Aversion 1	450CO2e	−0.0463	0.0280	−1.6390	17	±	0.0115
	INDC450CO2e	−0.0532	0.0350	−1.5172	17	±	0.0133
Aversion 2	450CO2e	−0.0230	0.0202	−1.1390	17	±	0.0057
	INDC450CO2e	−0.0169	0.0235	−0.7158	17	±	0.0042
Aversion 3	450CO2e	−0.0072	0.0126	−0.5720	17	±	0.0018
	INDC450CO2e	0.0004	0.0136	0.0316	17	±	0.0001

develop their economies by pledging lower emission reductions during this period. During this period, a uniform carbon price in 450CO2e scenario harms the interregional equity. The low-income countries continue to benefit from this economic development from 2030 to 2055 even though the global carbon price increases. Thus, before 2060, interregional equity is better in the INDC_450CO2e scenario, but the carbon price after 2060 becomes much higher in the INDC_450CO2e scenario than in the 450CO2e scenario, and mitigation costs

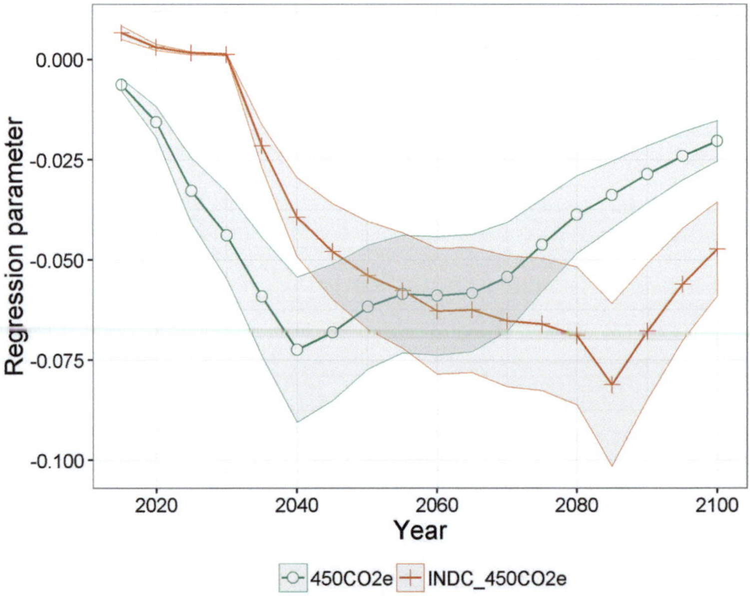

Fig. 4.7 The slope of regressed parameters for each year through 2100 for scenarios 450CO2e and INDC450CO2e. Each undiscounted year's consumption loss rate was regressed for each year

increase for the low-income countries. At this point, the impact of the global carbon price becomes dominant, and the INDC_450CO2e scenario performs worse than the 450CO2e scenario in terms of interregional equity.

4.3.4 Stringent INDC Scenario Analysis

In the SINDC_450CO2e scenario, we assumed each region had more stringent INDC emission targets to explore how it would affect intergenerational and interregional equity. Increasing the emission reduction target in the INDC period reduced the relative difference of consumption losses between current and future generations compared with the INDC_450CO2e scenario (Fig. 4.8).

That the SINDC_450CO2e scenario has a regression line slope that is less negative than the slope of the INDC_450CO2e scenario regression line when the inequality aversion is below 2 (Fig. 4.9) indicates greater interregional equity. This is primarily because the carbon price is lower in SINDC_450CO2e than it is in INDC_450CO2e, particularly in the latter half of the century.

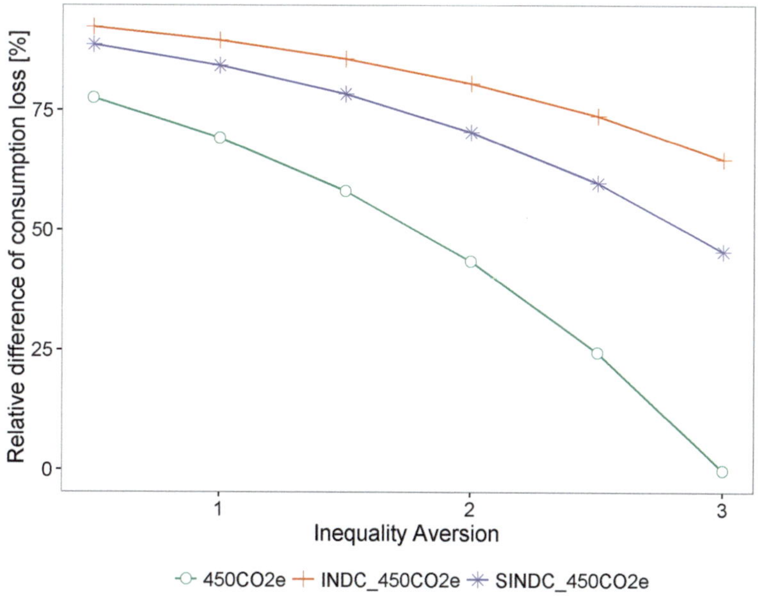

Fig. 4.8 Mitigation cost comparison between current and future generations for inequality aversions of 1, 2, and 3 in scenarios 450CO2e, INDC_450CO2e, and SINDC_450CO2e. The definition of relative difference of consumption loss is the same with that in Fig. 4.5b

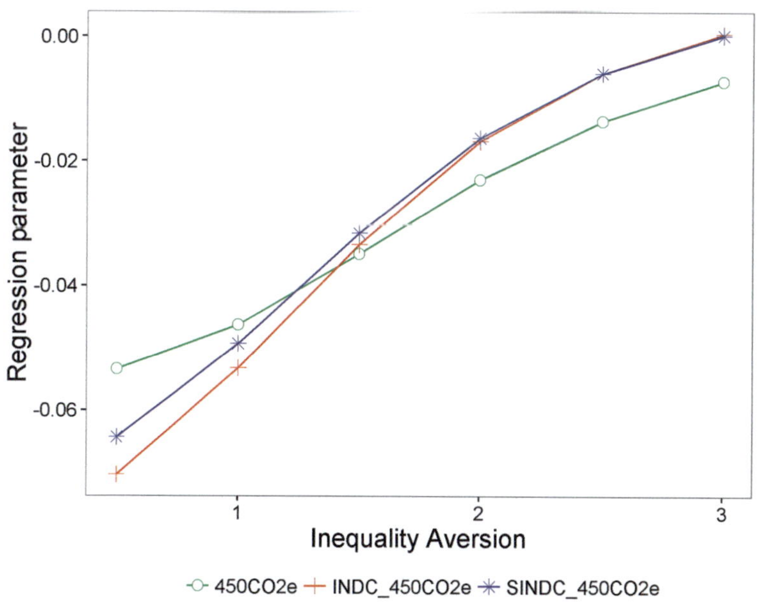

Fig. 4.9 The slope of the regressed parameters for different inequality aversion parameters for scenarios 450CO2e, INDC450CO2e, and SINDC450CO2e. Consumption loss was accounted for as NPV and the different aversionparameters were regressed

4.4 Discussion and Conclusions

This study assessed intergenerational and interregional equity based on the temporal and spatial distribution of global mitigation costs using AIM/CGE.

4.4.1 Interpretations and Policy Implications

There are four points to be discussed. First, we observed that intergenerational equity was worse in scenario INDC_450CO2e than it was in 450CO2e, mainly because INDC_450CO2e has a limited amount of emission reductions through 2030 and postpones the reduction efforts into the future as compared with 450CO2e. The intergenerational equity status for the INDC and post-INDC emission pathway is not favorable in this sense, because the future generation suffers from the larger emission reductions required in the latter part of the century. This conclusion was robust for inequality aversion parameters (i.e., implied discount rates) ranging from 1 to 3. This would suggest that additional emission reduction efforts in the near term are desirable to achieve the 2 °C target from the viewpoint of intergenerational equity.

Second, scenario INDC_450CO2e performs better in terms of interregional equity during the entire period on average as compared with scenario 450CO2e if the inequality aversion is large. Equity was defined such that high-income regions and low-income regions have close consumption loss rates. Under this principle, INDC_450CO2e has better interregional equity in the early part of the century as compared with that of 450CO2e in which a global carbon price is assumed during the period up to 2030. However, this is not the case for the latter part of the century because of the drastic increase of mitigation costs in this period.

Third, the modified INDC scenario, SINDC_450CO2e, had increased emission reductions prior to 2030. Intergenerational equity improved as compared with INDC_450CO2e, although it still was worse than intergenerational equity with 450CO2e. Furthermore, its average interregional equity was also better than that of INDC_450CO2e. The Paris Agreement states that emission targets will be reviewed and revised in 2020 (UNFCCC 2015). This implies that there is still room to enforce stronger emission reduction for 2030. Our results indicate that each country should adjust its target and commit to greater emission reductions in the follow-up INDC communications.

Moreover, although we did not consider a specific burden-sharing scheme after 2030 in this study, we surmise from the interregional equity results that a global uniform carbon price is relatively harmful for developing countries. Considering interregional equity, global carbon pricing should be avoided. However, a global carbon price is the best solution to achieve our climate goals efficiently. Therefore, we may be better off if we find intermediate solutions. Also, it seems that ongoing consideration for low-income countries will be needed as part of global climate change cooperation after 2030.

4.4.2 Limitations and Future Steps

There are four points with respect to limitations of this study and future steps. First, the 21st Conference of Parties (COP21) reached an agreement to pursue efforts to limit the global temperature increase to 1.5 °C above preindustrial levels. How would the status of generational equity change assuming we follow the INDC pledges and the achievement of the 1.5 °C target by the year 2100? We have not yet addressed this issue but regard it as a next step in our study of equity in climate change policy.

Second, in this study we only considered distribution equity in terms of the distribution of global mitigation costs. Historical responsibilities, compensatory justice, and other factors that may also be relevant to the equity discussion (IPCC 1995) were not considered. For example, many studies working on equity issues discuss cumulative emission allocation schemes (Matthews et al. 2014; Raupach et al. 2014; Glen et al. 2015; Zhou and Wang 2016; Tavoni et al. 2015), which are not discussed in this paper. Meanwhile, a focus of recent studies has been the emission transfer from developed countries to developing countries through international trade (Peters and Hertwich 2008; Jakob and Marschinski 2013; Springmann 2014; Kander et al. 2015; Liu et al. 2016). This emission accounting method is certainly relevant to the equity debate, but it was not considered in this study.

Third, here we only considered mitigation costs, negative impacts of climate change, and adaptation costs are obviously important elements of the generational equity issue. Although mitigation cost assessment is a prioritized area in terms of INDC and emission reduction targets, impact and adaptation aspects need to be considered in future studies.

Finally, we assumed that each year's emissions after 2030 were adjusted to fill the gap between the immediate emission reduction scenario and the INDC emission targets during the INDC period. In future studies, we may try to use an emission INDC pathway derived from some type of intertemporal model, for example, the DICE model.

Acknowledgments This study is supported by The Environment Research and Technology Development Fund 2–1402 and S14–5 of the Ministry of the Environment, Japan, and JSPS KAKENHI (Grant Number JP16K18177). The authors are most grateful to the generosity of these funds.

References

Anthoff D, Tol RS, Yohe G (2009) Discounting for climate change. ESRI working paper No. 276

Arrow KJ (1999) Discounting, morality, and gaming. In: Portney PR, Weyant JJ (eds) Discounting and intergenerational equity. Resources for the Future. Washington, DC, pp 13–21. ISBN:0915707896

Arrow K et al (2004) Are we consuming too much? J Econ Perspect 18(3):147–172

Arrow KJ et al (2013) How should benefits and costs be discounted in an intergenerational context? The views of an expert panel (December 19, 2013). Resources for the future discussion paper No. 12–53. Available at SSRN: http://ssrn.com/abstract=2199511 or http://dx.doi.org/10.2139/ssrn.2199511

Cline WR (1992) The economics of global warming. Institute for International Economics, Washington, DC. ISBN:9780881321326

Dai H et al (2016) Key factors affecting long-term penetration of global onshore wind energy integrating top-down and bottom-up approaches. Renew Energy 85:19–30

Dasgupta P (2008) Discounting climate change. J Risk Uncertain 37(2–3):141–169

Dellink R et al (2015) Long-term economic growth projections in the shared socioeconomic pathways. Glob Environ Chang 42:200–214

Fawcett AA et al (2015) Can Paris pledges avert severe climate change? Science 350 (6265):1168–1169

Fujimori S, Masui T, Matsuoka Y (2012) AIM/CGE [basic] manual. Center for Social and Environmental Systems Research, NIES, Tsukuba

Fujimori S et al (2014a) The effectiveness of energy service demand reduction: a scenario analysis of global climate change mitigation. Energy Policy 75:379–391

Fujimori S et al (2014b) Land use representation in a global CGE model for long-term simulation: CET vs. logit functions. Food Sec 6(5):685–699

Fujimori S, Masui T, Matsuoka Y (2014c) Development of a global computable general equilibrium model coupled with detailed energy end-use technology. Appl Energy 128:296–306

Garnaut R (2008) Measuring the immeasurable: the costs and benefits of climate change mitigation. Asian-Pac Econ Lit 22(2):1–13

Glen PP et al (2015) Measuring a fair and ambitious climate agreement using cumulative emissions. Environ Res Lett 10(10):105004

Gollier C (2008) Discounting with fat-tailed economic growth. J Risk Uncertain 37(2–3):171–186

Hasegawa T et al (2015) Consequence of climate mitigation on the risk of hunger. Environ Sci Technol 49(12):7245–7253

Hasegawa T, Fujimori S, Takahashi K, Yokohata T, Masui T (2016) Economic implications of climate change impacts on human health through undernourishment. Climatic Change 136 (2):189–202

IPCC (1995) Climate change 1995–economic and social dimensions of climate change: contribution of working group III to the second assessment report of the intergovernmental panel on climate change. Cambridge University Press, New York

IPCC (2015) Climate change 2014: mitigation of climate change, vol 3. Cambridge University Press, Cambridge

Iyer GC et al (2015) The contribution of Paris to limit global warming to 2 °C. Environ Res Lett 10 (12):125002

Jakob M, Marschinski R (2013) Interpreting trade-related CO_2 emission transfers. Nat Clim Chang 3(1):19–23

Kander A et al (2015) National greenhouse-gas accounting for effective climate policy on international trade. Nat Clim Chang 5(5):431–435

Kober T, Van Der Zwaan BCC, Rösler H (2014) Emission certificate trade and costs under regional burden-sharins regimes for a 2 °C climate change control target. Clim Chang Econ 05(01):1440001

Lebègue R (2005) Révision du taux d''actualisation des investissements publics. Rapport Du Groupe Présidé Par Daniel Lebègue. Commissariat Général Au Plan, Paris. Available at: http://www.plan.gouv.fr/intranet/upload/actualite/Rapport%20Lebegue%20Taux%20actualisation%2024-01-05.pdf

Liu Z et al (2016) Targeted opportunities to address the climate-trade dilemma in China. Nat Clim Chang 6(2):201–206

Lofgren H, Harris RL, Robinson S (2002) A standard computable general equilibrium (CGE) model in GAMS, vol 5. International Food Policy Research Institute, Washington, DC

Matthews HD et al (2014) National contributions to observed global warming. Environ Res Lett 9 (1):014010

Meinshausen M et al (2015) National post-2020 greenhouse gas targets and diversity-aware leadership. Nat Clim Chang 5(12):1098–1106

Mittal S et al (2016) Bridging greenhouse gas emissions and renewable energy deployment target: comparative assessment of China and India. Appl Energy 166:301–313

O'Neill BC et al (2014) A new scenario framework for climate change research: the concept of shared socioeconomic pathways. Clim Chang 122(3):387–400

Peters GP, Hertwich EG (2008) CO_2 embodied in international trade with implications for global climate policy. Environ Sci Technol 42(5):1401–1407

Ramsey FP (1928) A mathematical theory of saving. Econ J 38(152):543–559

Raupach MR et al (2014) Sharing a quota on cumulative carbon emissions. Nat Clim Chang 4 (10):873–879

Samir K, Lutz W (2017) The human core of the shared socioeconomic pathways: population scenarios by age, sex and level of education for all countries to 2100. Glob Environ Chang 42:181–192

Springmann M (2014) Integrating emissions transfers into policy-making. Nat Clim Chang 4 (3):177–181

Tavoni M et al (2013) The distribution of the major economies' effort in the Durban platform scenarios. Clim Chang Econ 04(04):1340009

Tavoni M et al (2015) Post-2020 climate agreements in the major economies assessed in the light of global models. Nat Clim Chang 5(2):119–126

UNFCCC (2015) Adoption of the Paris Agreement FCCC/CP/2015/L.9/Rev.1

Zhou P, Wang M (2016) Carbon dioxide emissions allocation: a review. Ecol Econ 125:47–59

Chapter 5
The Effectiveness of the International Emissions Trading under the Paris Agreement

Shinichiro Fujimori, Izumi Kubota, Hancheng Dai, Kiyoshi Takahashi, Tomoko Hasegawa, Jing-Yu Liu, Yasuaki Hijioka, Toshihiko Masui, and Maho Takimi

Abstract Under the Paris Agreement, parties would set and implement their own emissions targets as nationally determined contributions (NDCs) to tackle climate change risk. The international carbon emissions trading (ET) is expected to reduce global mitigation costs. Here we show the benefit of ET under both NDCs. The results show that the global welfare loss, which was measured based on estimated household consumption change in 2030, decreased by 80% (from 0.5 to 0.1%), as a consequence of achieving the NDCs through ET. These results indicate that ET is a valuable option for the international system, enabling the NDCs to be cost-effectively achieved and more ambitious targets.

Keywords Emissions trading • Paris Agreement • Computable general equilibrium model • Welfare change

---

S. Fujimori (✉) • T. Hasegawa
Center for Social and Environmental Systems Research, National Institute for Environmental Studies (NIES), 16–2 Onogawa, Tsukuba, Ibaraki 305–8506, Japan

International Institute for Applied Systems Analysis, Schlossplatz-1, Laxenburg 2361, Austria
e-mail: fujimori.shinichiro@nies.go.jp

I. Kubota • K. Takahashi • J.-Y. Liu • Y. Hijioka • T. Masui
Center for Social and Environmental Systems Research, National Institute for Environmental Studies (NIES), 16–2 Onogawa, Tsukuba, Ibaraki 305–8506, Japan

H. Dai
College of Environmental Sciences and Engineering, Peking University, Room 426, Environment Building, Beijing 100871, China

M. Takimi
Mizuho Information & Research Institute, Inc., 2–3 Kanda-Nishikicho, Chiyoda-ku, Tokyo 101–8443, Japan

S. Fujimori et al. (eds.), *Post-2020 Climate Action*,
DOI 10.1007/978-981-10-3869-3_5

Abbreviations

AIM	Asia-Pacific Integrated Model
CES	Constant elasticity of substitution
CGE	Computable general equilibrium
COP	Conference of the Parties
ET	Emissions trading
INDCs	Intended nationally determined contributions
NDCs	Nationally determined contributions
OECD	Organisation for Economic Co-operation and Development
PA	Paris Agreement
UNFCCC	United Nations Framework Convention on Climate Change

5.1 Introduction

In 2015, the Conference of the Parties (COP) 21 to the United Nations Framework Convention on Climate Change (UNFCCC) adopted the Paris Agreement (PA) (United Nations Framework Convention on Climate Change, (UNFCCC) 2015). The PA provides a framework for global actions to address climate change in the period after 2020. The objective of the agreement was to maintain the increase in global temperatures well below 2 °C above preindustrial levels, while making efforts to limit the increase to 1.5 °C.

The PA requires parties to prepare nationally determined contributions (NDCs), indicating an individual country's emissions reduction commitments, the measures to be taken to achieve their objectives, and a requirement to report on progress. To raise the level of ambition over time, parties must submit updated NDCs every 5 years. Each party's new NDC must be more ambitious than its previous NDC. Over 180 parties to the UNFCCC communicated their intended nationally determined contributions (INDCs) for 2025/2030 before COP21.

There have already been several assessments related to INDCs published in scientific papers reports and on websites (Meinshausen et al. 2015; Fawcett et al. 2015; Iyer et al. 2015; International Energy Agency, (IEA) 2015; Climate Action Tracker 2015; Lomborg 2015; Benveniste et al. 2015; Kitous and Keramidas 2015; Hof et al. 2015; International Institute for Applied Systems Analysis (IIASA) 2015). Some propose alternative scenarios to achieve the 2 °C goal because the INDC-based emissions are larger than those in the 2 °C scenarios. Some are comparable with the recent Fifth Assessment Report of the Intergovernmental Panel on Climate Change (IPCC AR5) scenario database (Fujimori et al. 2015) and make allocations based on multiple effort-sharing schemes. The consensus across the assessments at this stage is that current INDCs are not in line with the 2 °C goal, which was also stated in the PA (United Nations Framework Convention on Climate Change, (UNFCCC) 2015). To achieve the 2 °C goal, either a further emissions reduction in 2030 or more drastic and rapid reductions are required afterward.

Combating climate change will require the mobilization of substantial resources. Success will depend on the establishment of mechanisms and approaches that incentivize the mobilization of resources for cost-effective and ambitious mitigation action at all levels. Cooperation among parties and private and public sector stakeholders is considered crucial. It is well known that the international carbon emissions trading (ET) system is an economically cost-effective way to reduce global total mitigation cost (Böhringer et al. 2009; Zhang et al. 2017; Carbone et al. 2009; Böhringer and Welsch 2004; Weyant and Hill 1999; Hasegawa et al. 2014).[1] Under the Kyoto Protocol, there are several such systems incorporating market mechanisms, namely, international emissions trading, the clean development mechanism, and joint implementation. They enable parties to reduce emissions cost-effectively and encourage the private sector to contribute to global emissions reduction.

However, there are also some difficulties with implementing the market mechanism. For example, systems for monitoring, reporting, and verification (MRV) of the emissions reduction are needed, but this imposes certain costs. Another issue is that if we establish a carbon market, we need to prepare the market infrastructure, with a positive carbon price being a necessity. For regarding the post-2020 climate actions, there have been some developments regarding the international transaction of carbon credit under bilateral agreements such as the Joint Crediting Mechanism by the government of Japan. Article 6 of the PA provides a foundation to undertake international transfers of mitigation outcomes between parties. However, there have been no studies to clarify the effectiveness of an ET system in the context of INDCs.

Here, we estimate the effectiveness of ET under the current INDCs. The more stringent emissions reduction targets lead to larger costs and therefore ET becomes important. We used the Asia-Pacific Integrated Model/Computable General Equilibrium (AIM/CGE) model to achieve this goal.

5.2 Materials and Methods

5.2.1 Model

This study used the AIM/CGE model. The CGE model used in this study is a recursive, dynamic, general equilibrium model that covers all regions of the world and is widely used in climate mitigation and impact studies (Hasegawa et al. 2015; Mittal et al. 2016; Hasegawa et al. 2016; Fujimori et al. 2014; IIASA 2017). The main inputs of the model are the socioeconomic assumptions of drivers of GHG emissions such as population, gross domestic product (GDP), energy technology, and dietary preference. The production and consumption of all goods and GHG

[1]The financial mechanism in the context of supporting the least developed countries and technological transfer.

emissions are the main outputs as the result of price equilibrium. Here, population and GDP assumptions under Shared Socioeconomic Pathways 2 (SSP2) were used as the basic drivers, and other technological assumptions were based on (the energy technology also follows this assumption) (Fujimori et al. 2017; Hasegawa et al. 2015). The details of the model are shown in Chap. 13 of this book.

If emissions trading is allowed, every region is assumed to import or export emission rights until each region's emission price reaches the international emissions price. This trading can be described by the following equation, which is treated as a part of the formula for the mixed complementarity problem (MCP):

$$\begin{aligned} ET_r^{\text{imp}} &\geq 0 \perp PGHG_r \geq PET \\ ET_r^{\text{exp}} &\geq 0 \perp PET \geq PGHG_r, \end{aligned} \tag{5.1}$$

where ET_r^{imp} is the net emission imports of region r, ET_r^{exp} is the net emission exports of region r, and PET is the international emission price.

5.2.2 Scenario Framework

In our model, the emissions targets as pledged in the INDCs bind the emissions in individual countries, and the carbon price works to achieve the targets. This, in turn, generates climate mitigation costs, which are measured by changes in macroeconomic indicators, such as GDP and consumption, compared to the baseline. We set eight scenarios as shown in Table 5.1. Baseline has no climate policy (carbon pricing policy), whereas the other scenarios do. INDC corresponds to the unconditional emissions targets submitted to the UNFCCC.

The emissions reduction percentages compared to the baseline for each scenario are presented in the Fig. 5.1. We considered with (w/) and without (w/o) emissions trading options for all scenarios. If emissions trading was allowed, every region was assumed to import or export emission rights freely until each region's emissions price reached the international level.

Table 5.1 Scenario list

Scenario name	Emissions target	Emissions trading	Global emissions
Baseline	None	None	
INDC_w/o ET	Based on INDCs	Without	Around 57GtCO_2 eq in 2030 derived from INDCs
INDC_w/ET		With	

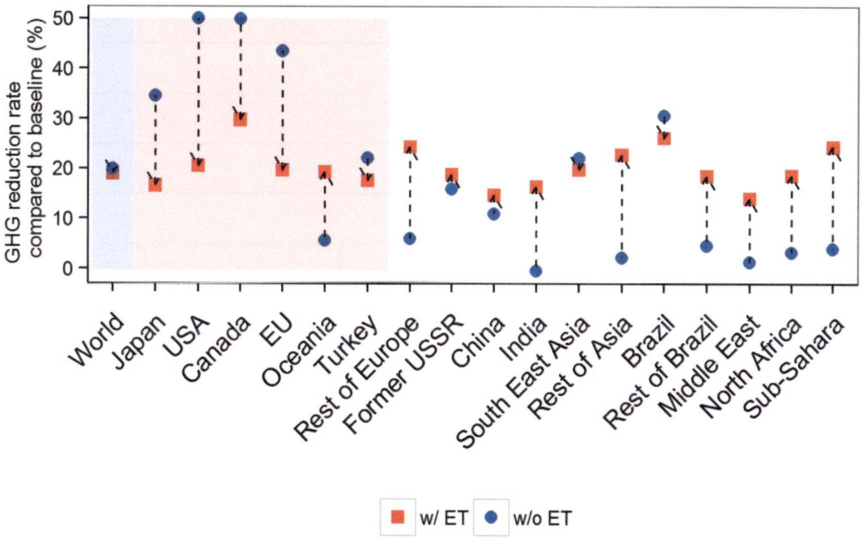

Fig. 5.1 GHG reduction rates by regions for INDC scenarios. *Blue* area is the world total and *red* area is OECD countries

5.3 Results

5.3.1 Welfare Change and Mitigation Cost Under INDCs

ET reduced global welfare loss (accounted for by Hicks' equivalent variation) in 2030 significantly by 80% and 50% (equivalent to around 230 and 210 billion US$), respectively, as shown in Fig. 5.2 (GDP loss is shown in SI Section 4). The INDC w/o ET resulted in a 0.4% welfare in 2030 globally, but in the w/ ET scenario, the loss became 0.1%. The OECD countries tended to have larger losses in the w/o ET scenario, whereas their losses substantially decreased in the w/ ET scenario. For example, Japan, the USA, and EU had 0.7%, 0.8%, and 0.5% welfare loss in 2030, respectively, whereas ET decreased their losses to −0.2%, 0.2%, and −0.1%, respectively (the negative value was due to changes in the international price and trade conditions). A similar trend was observed in the GDP loss rates.

In contrast, the situation varied in developing countries. For example, China and India would face a negative economic impact due to ET, while Africa and South Asia would experience positive impacts. As shown in Fig. 5.3, the carbon prices in developing countries were low (almost negligible) in the w/o ET scenario. If the domestic carbon market were to be opened to the international market, the carbon prices in these regions would be elevated to 10\$/tCO_2. While this is low compared to the values obtained in long-term stringent mitigation studies (Clarke et al. 2014), there would still be some macroeconomic effects. It would decrease capital productivity and could result in these regions losing their international competitiveness, but this would depend on the energy mix and economic structure. The impacts would differ

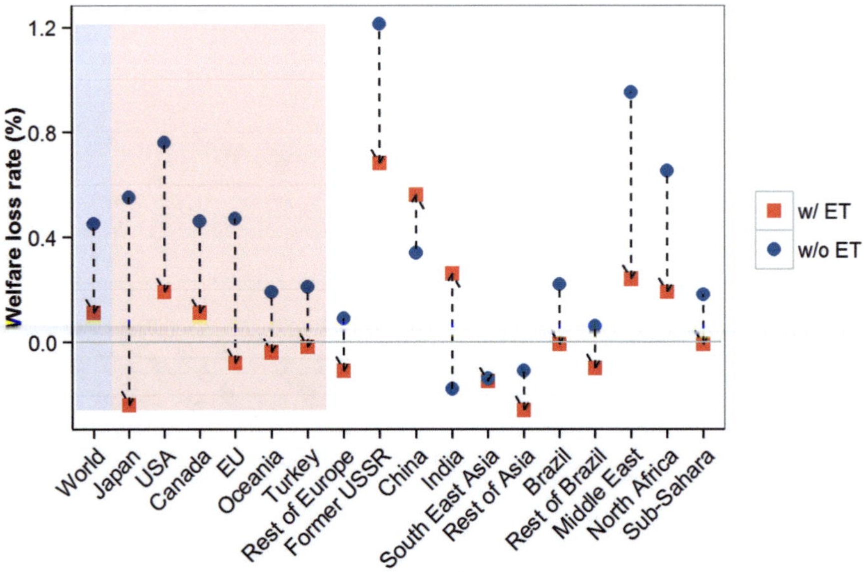

Fig. 5.2 Welfare loss rates in the year 2030 compared to the baseline scenarios for all regions in the INDC_w/ET and INDC_w/o ET scenarios. The *blue* area is the global total and the *red* area is OECD countries

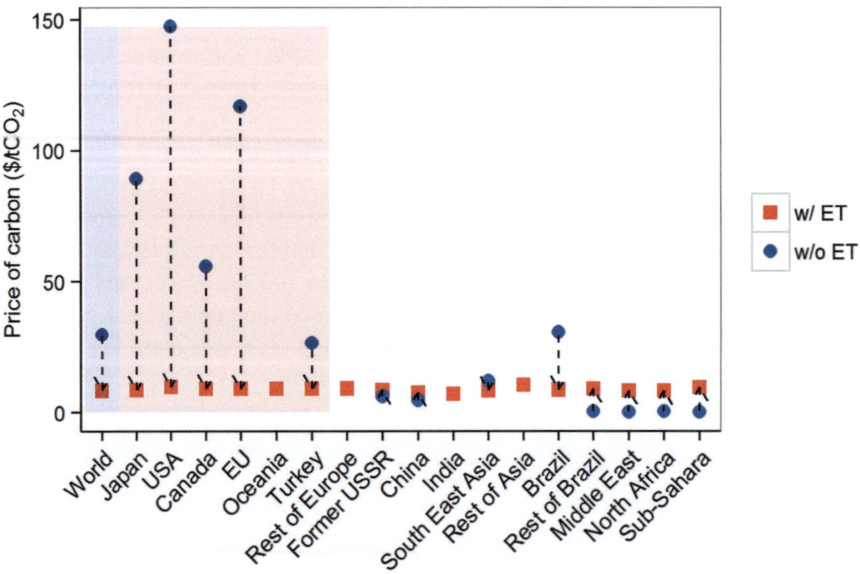

Fig. 5.3 Carbon prices for all regions in the INDC_w/ET and INDC_w/o ET scenarios. The *blue* area is the global total and the *red* area is OECD countries

depending on the magnitude of the carbon price. For example, in countries with a high coal consumption, there would be a relatively large impact on the economy at a certain carbon price. China and India are such examples (see SI in Section 5 showing the sources of primary energy supply and power generation in representative regions).

5.3.2 Financial Flow

In terms of the financial flow associated with ET, money was transferred from OECD to non-OECD countries because OECD countries imported emissions from non-OECD countries. These imports helped to reduce the mitigation cost in OECD countries. The global financial flows were about $40 billion in 2030. As shown previously, the carbon prices in OECD countries were high in the w/o ET scenario, and these countries faced relatively large challenges in meeting their emissions target. Hence, they purchased emissions rights from non-OECD countries. In OECD countries, the monetary flow in the w/ ET scenario (Fig. 5.4) and the welfare loss rates in the w/o ET scenario in the USA and EU were remarkably high (Fig. 5.2). The monetary flow in Japan was not as high, but there were large welfare losses in the w/o ET scenario (Fig. 5.2). This is because the scale of the economy (GDP) in the USA and EU is 3.5 and 3.0 times larger than in Japan, respectively. In non-OECD countries, India, sub-Saharan Africa, and China were the main exporters. Their exported monetary amount ranged from US$5 to 10 billion/year.

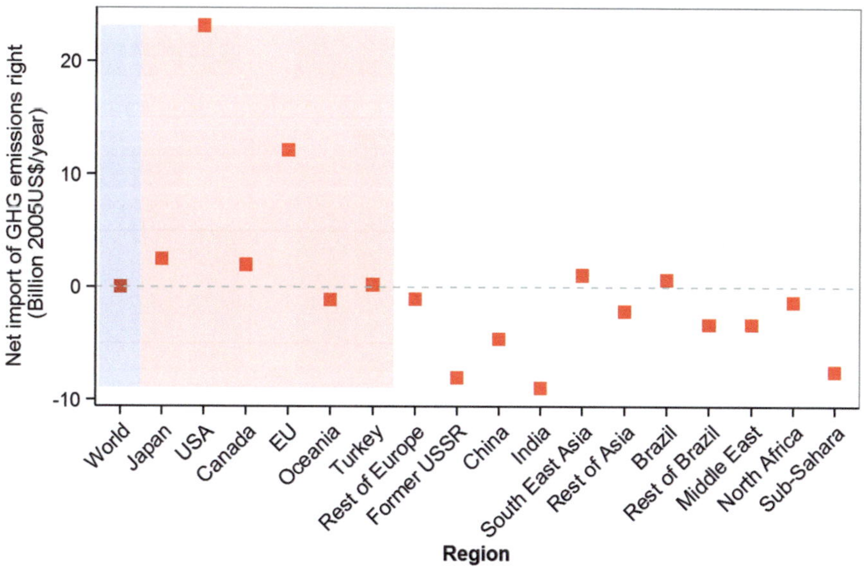

Fig. 5.4 Monetary transfer associated with emissions trading in the INDC_w/ET scenario in 2030. The *blue* area is the global total and the *red* area is OECD countries

5.3.3 Energy Supply and Power System in 2030 for INDCs and Baseline

The high carbon price influences the energy system. Figure 5.5 illustrates power energy sources of the world, OECD, non-OECD regions in the top, and selected OECD countries (Japan, the USA, and EU) and non-OECD countries (China, India,

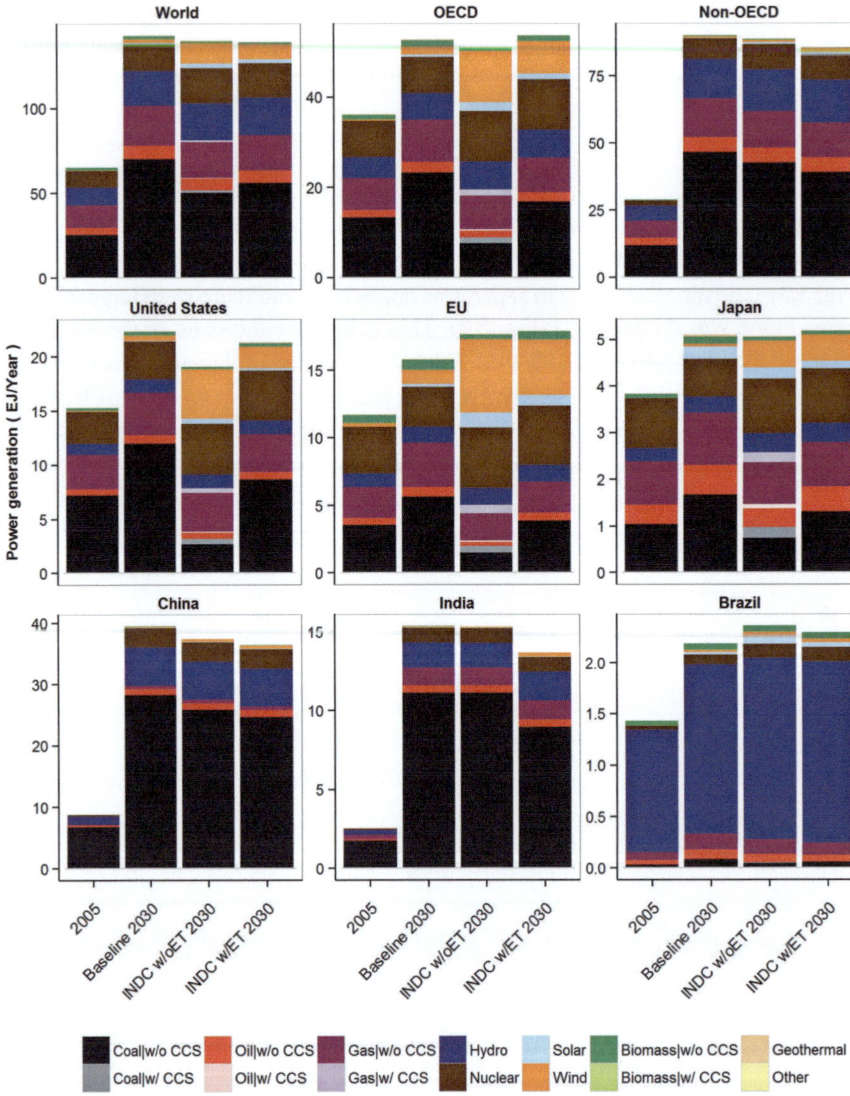

Fig. 5.5 Power generation by energy sources for selected regions in 2005 and 2030

and Brazil) in the bottom as an example of energy structure change. OECD regions basically decrease the electricity in without ET scenario compared to the baseline scenario. On the other hand, non-OECD regions don't have much change in that scenario. OECD regions in w/ ET scenarios increase their electricity relative to w/o ET scenario. This is because the carbon price as well as electricity price in w/ ET scenario becomes lower than those in the w/o ET scenario. Regarding the energy composite, OECD countries in the w/o ET scenario have a larger renewable share than w/ ET or baseline scenarios where carbon prices are high ($>100\$/tCO_2$). Similarly, CCS is also slightly introduced in OECD countries w/ ET scenario in the fossil-fired power generation, while other two scenarios don't, since the carbon prices don't reach the point in which CCS can be implemented.

5.3.4 Discussion and Conclusion

We estimated the benefit of ET under the current INDCs for 2030. The results indicated that ET is a useful option for the international system to efficiently achieve the near-term climate target. The climate mitigation costs under current INDCs in OECD countries would reduce significantly with ET. However, some regions would face negative economic impacts due to the high carbon price. In the more ambitious reduction target scenarios, ET played an essential role. Without ET, OECD countries could face significant macroeconomic losses and somewhat unrealistically high carbon prices in most burden sharing cases.

ET is an attractive measure to achieve the INDC targets efficiently, with the resulting carbon market being around 200 billion US$. However, there are at least two factors to consider. First, who will receive the benefit? The welfare loss rates in OECD countries decreased remarkably by implementing ET, whereas this was not always the case for non-OECD countries. In some developing countries, welfare would decrease despite the revenue obtained from exporting the carbon emissions. These phenomena imply that the market distortion induced by the carbon price is the primary factor in changing the macroeconomic performance rather than the monetary flow associated with ET. While it is economically efficient globally, whether it is acceptable to implement the ET system or not, in the real world, it might be dependent to a large extent on the equity issue.

In addition to the model simulation, there are at least three factors to consider in terms of the international policy framework. The first is how to avoid the double counting of mitigation efforts. The second is whether a cooperative mechanism, such as the Joint Crediting Mechanism or a multilateral mechanism, should be adopted. The third point is whether or not a nonmarket mechanism can also be adopted. The PA includes a combination of the second and third points. International negotiations are required to establish the rules for the market and nonmarket mechanisms, and the first point would then be more important to consider.

We need to recognize that this study had several caveats. First, here we deal with 17 aggregated regions in the modeling framework, but it implicitly assumes that the emissions right can be transferred under the aggregated regions and sectors. Therefore, the mitigation cost in without ET scenarios could be actually higher than the estimates in this study. Although it would be not so critically important to see the global overview, we should be careful to the specific aggregated region's results, and regional aggregation may influence the results. Second, one of the underlying assumptions was that advanced technologies can be accessible anywhere in the world. This assumption enables developing countries to reduce emissions at a low carbon price. This would sometimes be true because wages and many costs in developing countries are cheaper than in developed countries. However, the technology is only applicable with a certain skilled labor and access to such know-how. Therefore, as stated in the PA, the transfer of technologies to developing countries is necessary. Third, to implement the ET system is sometimes costly and we did not consider such transaction costs. As we discussed earlier, MRV would be one of the major costs. The information obtained from this study should be interpreted as the maximum potential ET benefit, if the ET works at its most efficient. Therefore, the results of this study should be considered compared to the cost of ET (opportunity cost), and the final cost and benefit should be evaluated. The fourth point is a more technical issue regarding the modeling. The revenues generated from GHG emission taxes and ET exports are assumed to be received by a representative household. If the emission price is relatively high, as in this study, the amount of money generated by the emission tax becomes enormous. The saving and expenditure behavior of the institutional sectors receiving the tax revenue could then have a large influence on macroeconomic performance and GHG emission prices. If this revenue was used for green investments or investing in other energy technologies, different results would have been obtained. Forth, we just show the climate mitigation cost in this study. The climate change impact risk must be reduced by certain mitigation efforts and should consider that aspects, although such risk might not be obvious in 2030 because of the climate internal variability.

Future studies are expected to follow two directions. One is to incorporate the 1.5 °C goal. This would require not only a 2030 but also a 2050 or century-scale assessment. After the PA, this research topic has attracted much attention, but has not been sufficiently analyzed. The other direction would be to make a more realistic assessment of energy technologies and international cooperation.

References

Benveniste H, Criqui P, Boucher O, Breon F-M, Guivarch C, Prados E et al (2015) The INDC counter, aggregation of national contributions and 2°C trajectories. Rapport du groupe interdisciplinaire sur les contributions nationales Institut Pierre Simon Lapl, http://hal.univ-grenoble-alpes.fr/hal-01245354

Böhringer C, Welsch H (2004) Contraction and convergence of carbon emissions: an intertemporal multi-region CGE analysis. J Policy Model 26:21–39

Böhringer C, Löschel A, Moslener U, Rutherford TF (2009) EU climate policy up to 2020: an economic impact assessment. Energy Econ 31:S295–S305

Carbone JC, Helm C, Rutherford TF (2009) The case for international emission trade in the absence of cooperative climate policy. J Environ Econ Manag 58:266–280
Clarke L, Jiang K, Akimoto K, Babiker M, Blanford G, Fisher-Vanden K et al (2014) Assessing transformation pathways. In: Climate change 2014: mitigation of climate change. Contribution of Working Group III to the fifth assessment report of the Intergovernmental Panel on Climate Change. Cambridge University Press, Cambridge/New York
Climate Action Tracker (2015) Tracking INDCs, http://climateactiontracker.org/
Fawcett AA, Iyer GC, Clarke LE, Edmonds JA, Hultman NE, McJeon HC et al (2015) Can Paris pledges avert severe climate change? Science 350:1168–1169
Fujimori S, Kainuma M, Masui T, Hasegawa T, Dai H (2014) The effectiveness of energy service demand reduction: a scenario analysis of global climate change mitigation. Energy Policy 75:379–391
Fujimori S, Masui T, Matsuoka Y (2015) Gains from emission trading under multiple stabilization targets and technological constraints. Energy Econ 48:306–315
Fujimori S, Hasegawa T, Masui T, Takahashi K, Silva DH, Dai H et al (2017) SSP3: AIM implementation of shared socioeconomic pathways. Global Environ Chang 42:268–283
Hasegawa T, Fujimori S, Shin Y, Takahashi K, Masui T, Tanaka A (2014) Climate change impact and adaptation assessment on food consumption utilizing a new scenario framework. Environ Sci Technol 48:438–445
Hasegawa T, Fujimori S, Shin Y, Tanaka A, Takahashi K, Masui T (2015) Consequence of climate mitigation on the risk of hunger. Environ Sci Technol 49:7245–7253
Hasegawa T, Fujimori S, Takahashi K, Yokohata T, Masui T (2016) Economic implications of climate change impacts on human health through undernourishment. Clim Chang 136(2): 189–202
Hof A, van Soest H, van den Berg M, de Boer HS, den Elzen M, Harmsen M et al (2015) Raising the ambition level of INDCs allows for a smoother energy transition. Assessment of the implications of INDCs for achieving the 2°C climate goal. PBL Beeldredactie: PBL Netherlands Environmental Assessment Agency
IIASA (2017) Shared socioeconomic pathways (SSP) Database Version 0.9.3
International Energy Agency (IEA) (2015) World energy outlook special report 2015: energy and climate change. International Energy Agency, Paris
International Institute for Applied Systems Analysis (IIASA) (2015) IAMC AR5 scenario database
Iyer GC, Edmonds JA, Fawcett AA, Hultman NE, Alsalam J, Asrar GR et al (2015) The contribution of Paris to limit global warming to 2°C. Environ Res Lett 10:125002
Kitous A, Keramidas K (2015) Analysis of scenarios integrating the INDCs. JRC Policy Brief: European Commission
Lomborg B (2015) Impact of current climate proposals. Global Policy 7:109–118
Meinshausen M, Jeffery L, Guetschow J, Robiou du Pont Y, Rogelj J, Schaeffer M et al (2015) National post-2020 greenhouse gas targets and diversity-aware leadership. Nat Clim Chang 5:1098–1106
Mittal S, Dai H, Fujimori S, Masui T (2016) Bridging greenhouse gas emissions and renewable energy deployment target: comparative assessment of China and India. Appl Energy 166:301–313
United Nations Framework Convention on Climate Change (UNFCCC) (2015) Adoption of the Paris Agreement. Proposal by the President (1/CP21)
Weyant JP, Hill JN (1999) Special issue the costs of the kyoto protocol: a multi-model evaluation introduction and overview. Int Assoc Energy Econ. 28790 Chagrin Blvd, Ste 210, Cleveland, OH 44122 USA
Zhang X, Qi T-Y, Ou X-M, Zhang X-L (2017) The role of multi-region integrated emissions trading scheme: a computable general equilibrium analysis. Appl Energy 185:1860–1868

Chapter 6
Achieving Carbon Emissions Peak in China by 2030: The Key Options and Economic Impacts

Hancheng Dai and Toshihiko Masui

Abstract This study investigates the options and sectors that are essential for China to achieve carbon emissions peak by 2030. A dynamic computable general equilibrium (CGE) model is used to generate 14 scenarios from a scenario matrix incorporating three levels of carbon constraints and four options of low-carbon measures. Results suggest that if there is no policy intervention, China's total CO_2 emissions would reach 22.9 Gt in 2030. To cut carbon emissions required by the latest Intended Nationally Determined Contributions (INDC) or the global two-degree target, China should not only rely on optimizing industry structure or restricting its industry output, but more importantly, it should rely on low-carbon technologies in the power and end-use sectors as well as low-carbon style consumption. We also depict how the mitigation costs can be lowered through various low-carbon countermeasures, with which the carbon mitigation cost of achieving China's INDC target in 2030 could be reduced from 155 USD/ton-CO_2 to 35 USD/ton-CO_2. The corresponding GDP loss could fall from 6.3% to merely 0.67%, and welfare will not be affected significantly.

Keywords Carbon emissions peak • Computable general equilibrium model • INDC (Intended Nationally Determined Contributions) • China

6.1 Introduction

Increases in atmospheric greenhouse gases are understood to be largely attributed to human activities such as fossil fuel combustion, land-use changes, and agricultural practices (IPCC 2007). Although the bulk of cumulative greenhouse gas emissions

H. Dai (✉)
College of Environmental Sciences and Engineering, Peking University, Room 426, Environment Building, Beijing 100871, China
e-mail: dai.hancheng@pku.edu.cn

T. Masui
Center for Social and Environmental Systems Research, National Institute for Environmental Studies (NIES), 16–2 Onogawa, Tsukuba, Ibaraki 305–8506, Japan

S. Fujimori et al. (eds.), *Post-2020 Climate Action*,
DOI 10.1007/978-981-10-3869-3_6

have been contributed by industrialized countries, increasing emissions will come from developing countries. Recently, China has become a major economy and – consequently – the top energy consumer and emitter of greenhouse gases in the world. Thus, any meaningful climate stabilization will be impossible without China's participation (Paltsev et al. 2012).

As concerns mount about the environment, energy security, and mounting international pressures on climate mitigation, the Chinese government has made substantial efforts to promote energy-saving and climate mitigation policies. A target of 20% reduction in energy intensity over 2006–2010 was set in the 11th Five-Year Plan (FYP), which was followed by another target of 16% and 17% reduction in energy and carbon intensity, respectively, in the current 12th FYP (2011–2015). China also committed to lower its carbon intensity by 40–45% by 2020, a decrease from 2005 levels, and to increase its share of non-fossil fuels to ~15% (National Development and Reform Commission of the People's Republic of China 2012). In the latest submission of Intended Nationally Determined Contributions to the United Nations Framework Convention on Climate Change (UNFCCC), China pledged to further cut its CO_2 emissions per unit of GDP by 60–65% from 2005 level by 2030 and increase non-fossil fuel sources in primary energy consumption to about 20% by 2030.

China needs a low-carbon economy pathway in the long term. A low-carbon economy is defined as a social/economic entity that demonstrates a high level of energy efficiency, uses low-carbon energy sources and production technologies, and adopts patterns of consumption and behavior that are consistent with low greenhouse gas emissions (Skea and Nishioka 2008). Numerous international studies have described carbon mitigation policy aimed to achieve low-carbon transition. At the international level, previous studies were conducted for the UK (Dagoumas and Barker 2010), Australia (Foran 2011), Japan (Ashina et al. 2012), South Korea (Jones and Yoo 2010), India (Shukla et al. 2008), and Thailand (Selvakkumaran et al. 2015; Thepkhun et al. 2013). Local level studies include the Shiga Prefecture of Japan (Gomi et al. 2010) in China. Most studies suggest that these countries and local entities have the technological potential to reduce CO_2 emissions while satisfying the expected demand for energy services. Various policies were identified to promote low-carbon development, including adjusting the economic structure through the promotion of high-value-added and low-energy-intensive industries, curbing the production of energy-intensive products, implementing comprehensive energy-saving policies, developing renewable electricity, promoting advanced fossil fuel with carbon capture and storage (CCS) technology and nuclear and natural gas, and making use of market-based instruments such as cap-and-trade schemes and carbon taxes, dematerialization of industry, sustainable consumption and low-carbon lifestyles, low-carbon transport, and end-use device efficiency improvements. As for China, numerous studies explored the future energy consumption and emissions (Bretschger and Zhang 2014; Chai and Xu 2014; Dai et al. 2016; Jiang and Hu 2009; Li et al. 2015; Wang and Watson 2010; Zhang et al. 2013, 2014; Zhou et al. 2013) at the national level. Other studies investigated the role of market-based instruments such as carbon cap and emissions trading policy in climate mitigation in Guangdong Province (Cheng et al. 2015; Wang et al. 2015). Chai and Xu (2014) suggest that peaking China's emissions below 12 gigaton of CO_2 ($GtCO_2$) and 8.5 tCO_2 per capita between 2025 and 2030 will be affordable considering the trade-offs between emissions and economic growth. Wang and Zou (2014) argue that the GDP loss of reaching emissions peak ranges from 3.0 to 8.2% under different scenarios.

The research gap of the previous studies is that the impacts of the above countermeasures on the macro-economy are not explicitly quantified. Thus, the objectives of this study are (1) to quantify the effectiveness of different carbon reduction countermeasures in achieving China's INDC target and two-degree target and (2) to assess the economic impacts of achieving the above levels of carbon reduction targets. For these purposes, a dynamic hybrid computable general equilibrium (CGE) model was developed for China in this study.

The remainder of this paper is structured as follows: In Sect. 6.2, the CGE model is described and socioeconomic features of China are introduced. Also, reference and mitigation scenarios depicting potential outcomes by 2030 are described. In Sect. 6.3, simulation results are presented for the economy, energy, and carbon emissions and economic impacts of carbon reduction. Section 6.4 summarizes the major findings and concludes with policy implications. Finally, Sect. 6.5 concludes.

6.2 Methodology

6.2.1 The CGE Model

We developed a recursive dynamic two-region CGE model which is extended from the one-region static (Dai et al. 2011) and dynamic (Dai et al. 2012) versions. This model selects 2005 as the base year and is solved at 1-year steps toward 2030 by GAMS/MPSGE (Rutherford 1999).

As introduced in detail in the appendix, the model includes a production block, a market block with domestic and international transactions, as well as government and household income and expenditure blocks. Forty-one sectors in the model are classified into basic, land-reliant, resource-reliant, and energy transformation sectors (Table 6.1 in the appendix). It is a hybrid CGE model in the sense that various key technologies for energy transformation sectors are formulated, such as non-fossil electricity generation, alternative fuel production (bio-liquid and biogas), and CCS technology. The model divides China into Jiangxi Province and the rest of China, which interact with each other through domestic commodity trade. Labor is assumed to be fully mobile across industries within a region but immobile across regions, whereas vintage capital is assumed to be immobile across either regions or industries. Note that although this is a two-region model, only the results of the whole China will be shown in this study.

6.2.2 Data

The data required by the model include input-output table (IOT) (National Bureau of Statistics of China (NBS) 2011) and energy balance table (EBT) (National Bureau of Statistics of China (NBS) 2008); carbon emission factors; energy prices of coal, oil, and gas; and cost information of renewable energy technology. All the datasets are converted to the base year of 2007. It is well known that IOT and EBT are inconsistent when it comes to energy consumption across sectors, and the energy data from EBT is regarded as more reliable than IOT. A novel characteristic of this CGE model is that the IOT is reconciled based on the energy information from EBT. Therefore, we used the linear least square method, as shown in Eqs. (6.1, 6.2, 6.3 and 6.4), to adjust the sectoral energy consumption share in the IOT that is consistent with EBT. Furthermore, based on the share, the monetary energy input of each sector in the IOT ($EN^{IOT}_{r,en,i}$) is calculated (Eq. 6.2). This step is crucial since base-year sectoral energy consumption data affect future projection and sectoral carbon abatement cost significantly:

$$\varepsilon = \sum\nolimits_{r,en,i} \left(Shr^{IOT}_{r,en,i} - Shr^{EBT}_{r,en,i} \right)^2 \tag{6.1}$$

$$Shr^{IOT}_{r,en,i} = \frac{EN^{IOT}_{r,en,i}}{TCON^{IOT}_{r,en}} \tag{6.2}$$

$$Shr^{EBT}_{r,en,i} = \frac{EN^{EBT}_{r,en,i}}{TCON^{EBT}_{r,en}} \tag{6.3}$$

$$\sum\nolimits_i ENr,en,i^{IOT} = \sum\nolimits_i ENr,en,i^{EBT} \times P_{r,en} \tag{6.4}$$

where ε is error to be minimized, *en* is energy commodities (coal, gas, oil, electricity), *i* is sector classification in Table 6.1, *r* is region, $Shr^{IOT}_{r,en,i}$ is share of energy consumption across sectors in IOT (%), $Shr^{EBT}_{r,en,i}$ is share of energy consumption across sectors in EBT (%), $ENr\,,en\,,i^{IOT}$ is energy consumption of *en* in sector *i* in IOT (USD), $ENr\,,en\,,i^{EBT}$ is energy consumption of *en* in sector *i* in IOT (USD), $TCON^{IOT}_{r,en}$ is total energy consumption of *en* in IOT (USD), $TCON^{EBT}_{r,en}$ is total energy consumption of *en* in IOT (USD), and $P_{r\,,en}$ is price of energy *en* (USD/PJ).

6.2.3 Scenario Matrix Design

All scenarios share the common assumptions in terms of population, investment, autonomous energy efficiency improvement (AEEI), total factor productivity (TFP), and so on (Table 6.1). GDP growth rates will gradually slow down over the following decades. China's future population follows the United Nations' 2008 medium population projection (UN 2008). An AEEI is assumed separately for electricity and liquid, solid, and gas fuels. We also assume that, in the future, the economy will decouple from energy-intensive products, such as cement, iron, and steel, and chemical products and paper. In addition, the extraction cost of coal, crude oil, and natural gas will gradually increase due to resource depletion.

The scenario design considers two dimensions: the stringency level of carbon constraint and the availability of low-carbon options. By taking into account the above dimensions, there are 14 scenarios in total as shown in Table 6.2.

For the former dimension, three levels of carbon constraints are set for China. Level 1 (CAP1) is in line with China's both Copenhagen commitment and INDC, which suggests that the carbon intensity in terms of GDP will fall by around 45% and 65% as compared to 2005 level in 2020 and 2030, respectively. Under the GDP assumption in this study, the emissions in 2030 will reach 13.4 $GtCO_2$. Moreover, the emissions would peak in 2025 at around 14.1 $GtCO_2$. Level 2 (CAP2) reflects greater efforts than the current INDC, which requires China to achieve an emissions peak in 2020 and reduce its emissions by 1% per year after 2020. By contrast, level 3 (CAP3) is the most stringent constraint since it is consistent with the two-degree target which requires the global GHG emissions in 2050 halving to the 1990s level. The global emission allowance is allocated to China based on the assumption that each country's per capita CO_2 emissions converge in 2050. As a result, the emission allowance in 2030 for China is around 6.5 billion tons or close to the 2005 level.

In addition to the carbon constraints, different low-carbon options are also considered in the mitigation scenarios, for instance, low carbon household consumption (LC), non-fossil energy development (NFE), and deployment of CCS technology (CCS), which are added one after another as shown in Table 6.2. In the case of high-carbon household consumption patterns, it is assumed that people tend to follow the materialized lifestyles, spending less on food and housing and more on clothing, household facilities, and services, transport, and communications, and they have better health care and education opportunities. By contrast, in the low-carbon patterns, a dematerialized lifestyle is preferred, in which less is spent on clothing, household facilities, and transport and more is spent on service commodities. More detailed information about settings in future Chinese household expenditure patterns can be found in Table 6.2 in the appendix and Dai et al. (2012). In the NFE-related scenarios, the non-fossil fuel supply potential and investment cost, including nuclear, hydro, solar, wind, and biomass power, follows the assumption in the report released by Jiang and Hu (2009). In the CCS-related scenarios, CCS technology is allowed to enter the market with the reduction cost of 150 USD/ton-CO_2, and the annual penetration rate is limited to 5%.

6.3 Results

6.3.1 An Overview of the Future Socioeconomic Trends

Before showing the main results, we describe a broad picture of socioeconomic circumstances from 2005 to 2030 in China, as shown in Fig. 6.1. It reveals that China's economy will keep growing and its energy consumption and emissions will rise continuously in the noCAP HC scenario where no climate policy intervention is in place. China's per capita GDP will grow by approximately 7.0 times over the next 25 years, rising to around 12,000 USD in 2030. As a result, China's consumption of primary energy, final energy, and electricity in 2030 will increase to 9.96 billion ton of coal equivalent (btce) and 6.94 btce and 15.57 petawatt hour (PWh), equivalent to an increase of 4.22, 4.44, and 6.24 times from the 2005 level, respectively. Consequently, the emissions will reach 22.93 $GtCO_2$, and carbon intensity in terms of GDP will improve merely by 31.7% in 2020 and 46.9% in 2030 as compared to 2005, which is far less than pledged in the Copenhagen commitment and the latest INDC.

In contrast, in the noCAP LC, CAP1–3 scenarios, the climate policy intervention is implemented. It shows that energy, emissions, and economy will be affected to a various extent. First of all, low-carbon consumption alone could lead to a reduction of 13.4% in primary energy consumption and 16.1% in CO_2 emissions in 2030 and contribute to 5.3 and 7.3 additional percentages of carbon intensity improvement in 2020 and 2030, respectively. This finding is consistent with another study of us (Dai 2012; Dai et al. 2012), but is ignored or not quantified in many mitigation studies that focus on supply-side mitigation. Furthermore, under the three levels of carbon constraints, China's emissions will peak in 2025 in CAP1 and 2020 in CAP2 and CAP3 scenarios. Carbon intensity targets in 2020 and 2030 could be achieved in CAP1 scenarios and exceeded in CAP2 and CAP3 scenarios. Moreover, the energy consumption will fall significantly compared with the non-CAP scenarios due to the increased cost of fossil energy. Impacts on the economy are dependent on the low-carbon options and will be elaborated in the following sections.

6.3.2 Carbon Emission Reduction: The Key Options and Sectors

Conceptually, in a CGE model, carbon reduction could be attributed to non-price- and price-induced factors. The non-price factors include technological advances such as energy efficiency improvement and deployment of non-fossil energy power generation technology and CCS technology. On the other hand, the price-induced factors are caused by the carbon price signal following a complex causal chain. When a carbon cap is imposed on the economy, carbon emission allowance will become a scarce resource and have a value – the carbon price. The carbon price will

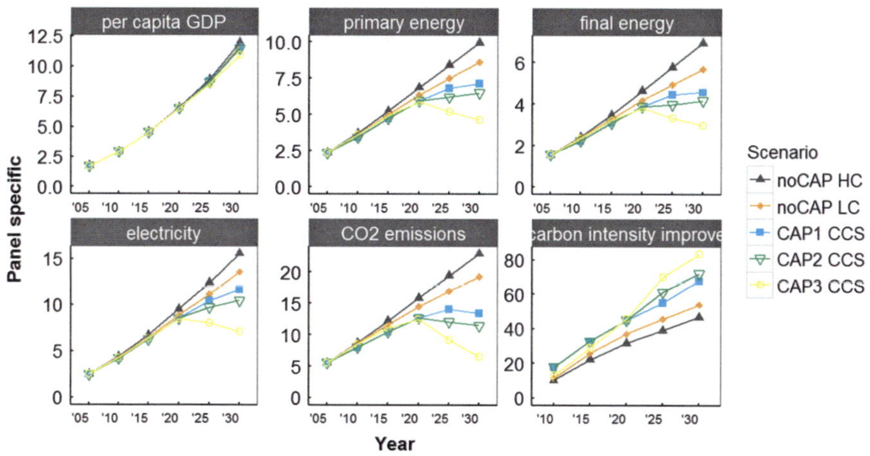

Fig. 6.1 Trajectory of main indicators of economy, energy, and emissions of China

increase both production prices of all sectors to a different extent (depending on each sector's carbon intensity in terms of production). Consequently, the consumption price will be distorted, leading to widespread effects in both demand and supply sides of an economy as a response to the price signal. On the demand side, consumers will shift from relatively higher-carbon-intensive products and energy to low-carbon ones; on the supply side, the producers will adjust their activity to lower the supply of energy- and carbon-intensive products and put more investment in low-carbon industries; on the macroeconomic level, the economy structure could be decarbonized through decoupling from carbon-intensive industries. Based on the decomposition analysis as explained in Appendix 2, the contribution of various low-carbon countermeasures is quantified as shown in Fig. 6.2. It demonstrates that decarbonizing the power sector is the most important approach under any mitigation target to be taken by China. For example, the effects of reducing electricity demand and developing renewable and nuclear power contribute to 56% of the total reduction in CAP1–3 scenarios. Besides, as the mitigation target strengthens, the order of the other key policies will change. For instance, under the CAP1 scenario, adjusting industry structure (contributing to 0.80 $GtCO_2$ or 10.3% of the total reduction of 7.63 $GtCO_2$, hereafter % only) and improving end-use energy efficiency (0.77 $GtCO_2$, 10.0%) are the most important measures next to decarbonizing power, followed by CCS technology (0.59 $GtCO_2$ or 7.7%), fuel switch (0.48 $GtCO_2$ or 6.3%), and lowering the activity level of the energy-intensive industries. By contrast, under the more stringent CAP2 and CAP3 scenarios, improving end-use energy efficiency (1.37 $GtCO_2$, or 14.1% in CAP2, and 2.98 $GtCO_2$, or 20.3% in CAP3) will by far be the most important measures, followed by CCS technology (0.87 $GtCO_2$, or 9.0% in CAP2, and 0.99 $GtCO_2$, or 6.8% in CAP3), while adjusting industry structure, restricting energy-intensive industry output, and fuel switch become the secondary effective measures.

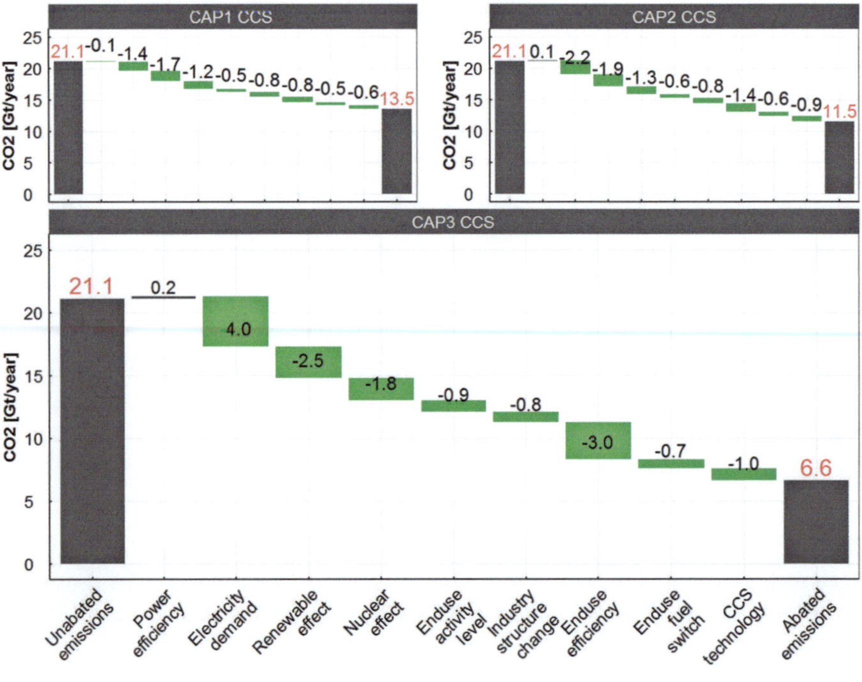

Fig. 6.2 Contribution of low-carbon countermeasures to carbon reduction in 2030

From the perspective of sectoral emissions, several sectors are the key contributors to carbon reduction (Fig. 6.3). Interestingly, it is found that the order of sectoral contribution is consistent among CAP1–3 scenarios. Apparently, the dominating contributor is the power sector which accounts for around 55% of the total reduction. The second largest contributor is household sector, which is responsible for 14.3% (4.93 $GtCO_2$), 12.9% (6.10 $GtCO_2$), and 11.1% (9.03 $GtCO_2$) of the total reduction in CAP1, CAP2, and CAP3 scenarios, respectively. The third to seventh contributors are metal production, machinery, transport, nonmetal production, and coal mining sectors; all together they contribute to around 23% of the total reduction. The remaining 11% of the total reduction is contributed by the other sectors.

6.3.3 *Energy Mix Change*

Climate policy will have significant impacts on the energy system, especially coal consumption for power generation. In 2030 (Fig. 6.4 left), compared with the noCAP HC scenario, low-carbon consumption leads to primary energy saving of 1333 mtce (or 13%), and under CAP1, CAP2, and CAP3 scenarios, primary energy saving would be 2947 (30%), 3658 (37%), and 5610 (56%) mtce, respectively.

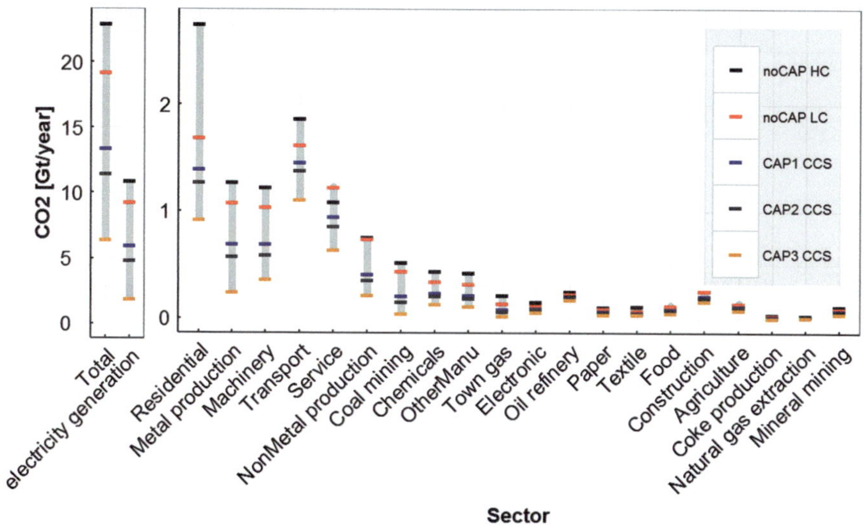

Fig. 6.3 Sectoral carbon emissions under different scenarios in 2030

Most strikingly, coal consumption in 2030 will reduce from 6812 mtce in the noCAP HC scenario to 1254 mtce in the CAP3RE scenario, which is lower than the current level. As a result of non-fossil energy penetration and carbon emission constraints, the primary energy structure will be diversified. However, without carbon constraints the share of non-fossil energy in the primary energy mix will stay mostly less than 10% from 2005 to 2030 (Fig. 6.4 right), which is half of the governmental target in 2030. Only if renewable energy deployment targets are coupled with carbon emission targets, the share of non-fossil energy could reach around 20% in 2030 in CAP1 and CAP2 scenarios and 30% in CAP3 scenario.

6.3.4 Economic Impacts of Climate Mitigation

Further analysis conducted from a macroeconomic perspective suggests that low-carbon countermeasures are not only effective for reducing carbon emissions but also are helpful for reducing the economic costs of carbon reduction, which are measured by carbon mitigation cost (or carbon price), sectoral output change, GDP change, and welfare change in the CGE model. As shown in Fig. 6.5 (upper), the carbon mitigation cost of achieving each level of carbon reduction is strongly dependent on the low-carbon options. Without the low-carbon options, carbon price in 2030 will reach 155, 223, and 661 USD/ton-CO_2 under CAP1, CAP2, and CAP3 scenarios, respectively. The carbon price will increase the energy costs and production prices of all sectors significantly; consequently, the demand for the carbon-intensive products will decrease, leading to a reduction in their output level.

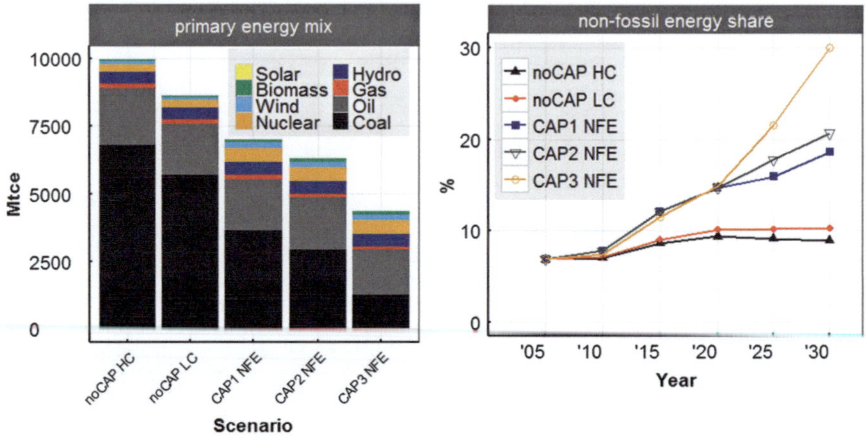

Fig. 6.4 Primary energy mix in 2030 (*left*) and the share of non-fossil fuel in primary energy mix (*right*)

As shown in Fig. 6.6, the output loss is closely related to the energy intensity of each sector. In the CAP3 LC scenario in which the carbon price is nearly 500 USD/tCO_2, the energy-intensive industry will suffer a significant loss in their output. For instance, compared to the reference scenario, the output loss of mineral mining is over 65%, losses of metal production and chemical sectors are over 40%, and the losses of paper and pulp, electronic, nonmetal production, machinery, and other manufacturing and transport sectors are more than 20%. On the other hand, the impacts on service sector will be relatively milder with its output decreasing by 7.8%. The negative impacts on industry sectors will lower the return of capital and labor wage, resulting in GDP loss and people's welfare loss. The GDP loss in 2030 would be 6.3%, 9.1%, and 21.5%, and welfare loss would be 3.0%, 6.1%, and 18.4% (Fig. 6.5 lower) in the CAP1–3 scenarios without any low-carbon options, respectively. Apparently, as a developing country with improving the people's living standard as the priority in its policy agenda, China could not afford such high cost, especially the mitigation cost aimed at achieving the two-degree target.

However, after the low-carbon countermeasures are introduced, carbon price, output loss, GDP loss, and welfare loss will decrease significantly to moderate levels. In the CCS case, the carbon price in 2030 could fall to 35, 58, and 208 USD/ton-CO_2 under CAP1–3 CCS scenarios, respectively. As a result, the sectoral output loss will decrease as well (right panel of Fig. 6.6). Most noticeably, the loss of mineral mining drops to around 40%; the loss of chemical sector drops to 22%; the loss of paper and pulp, electronic, and transport sectors drops to around 10%; and the loss of service sector will be only 1.6%. Accordingly, the GDP loss in the CAP1–3 CCS scenarios would be 0.67%, 1.44%, and 6.33%, respectively, which is much lower than the aforementioned. Moreover, in the CAP1 and CAP2 CCS scenarios, the welfare will not reduce but increase slightly since we assume the

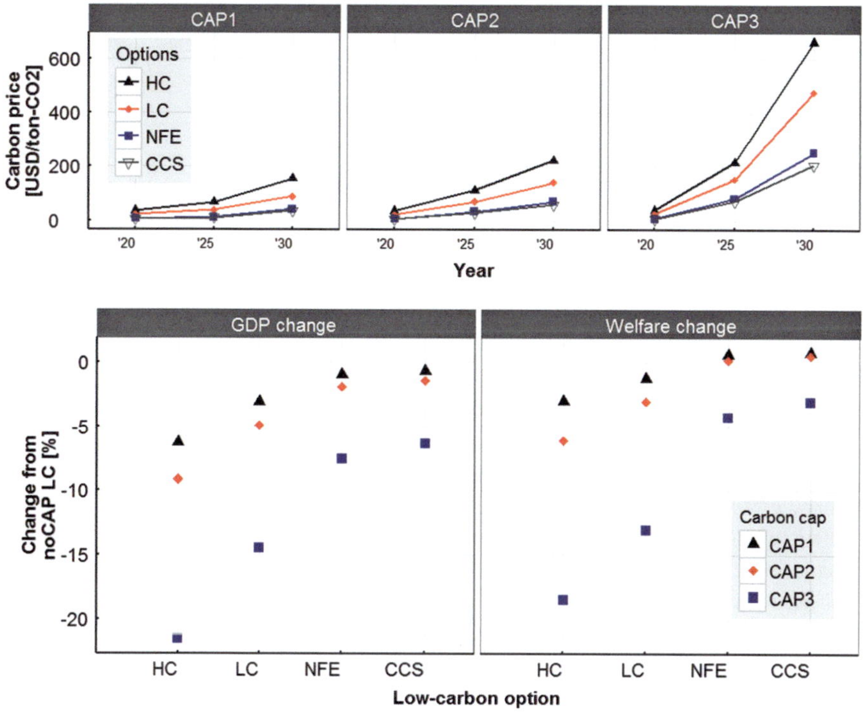

Fig. 6.5 Carbon price (*upper*) and impacts on the economy in 2030 (*lower*) under different carbon caps and low-carbon options

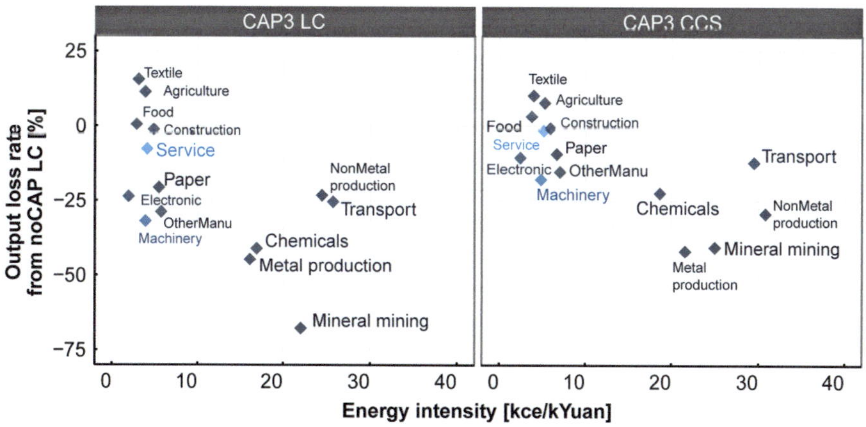

Fig. 6.6 The relationship of loss in output value and energy intensity of different sectors

carbon tax revenue is recycled to the household. Even in the CAP3 CCS scenario, welfare loss reduces to 3.06% from 18%.

6.3.5 Co-benefits of Climate Mitigation

Cutting carbon emissions is believed to reduce air pollutant emissions and improving energy security. Greenhouse gases and air pollutants are mainly emitted from the combustion of fossil fuels. As a result, measures taken for controlling fossil energy consumption could reduce CO_2, SO_2, NOx, and other air pollutants simultaneously. In this study, seven air pollutants are considered: carbon monoxide (CO), ammonium hydroxide (NH_3), non-methane volatile organic compounds (NMVOC), methane (CH_4), nitrous oxide (N_2O), nitrogen oxide (NOx), and sulfur dioxide (SO_2) which are emitted from energy combustion and industry processes. Energy combustion emissions are calculated based on emission factors from IPCC (2006), whereas process emissions proportional to activity levels (such as cement production) for the whole China in the base year are obtained from EDGAR 4.1 emission inventory data. As shown in Fig. 6.7, total air pollutant emissions will increase in 2030 in the reference scenario (2030 noCAP LC) due to energy use growth and output increases. The figures also indicate the main contributors of emission reduction in the mitigation scenario (2030 CAP3 CCS). For example, SO_2 emissions in 2030 fall below 2005 levels, which can be attributed to the power sector (accounting for 79% of the total reduction). NOx reduction is mainly contributed by power and transportation sectors. CO emission reduction is mostly contributed by household sectors (39%) and metal production (19%), followed by the transportation sector (8%). CH_4 emission falls due to decreases in coal mining (88%). Moreover, the main causes of NMVOC reduction are oil refinery (38%), household (27%), chemical (13%), and service (12%) sectors. Carbon reduction has little influence on emissions of N_2O and NH_3 because their emission sources are in the agriculture sector whose outputs are not affected much by carbon mitigation.

Energy security improvement is another co-benefit of climate mitigation (Fig. 6.8). China had been self-sufficient with respect to oil supply until it became a net oil importer in 1993. Given stagnant domestic production (Computer-Aided of Engineering (CAE) 2011) and continuous increases in demand for crude oil, it is believed that oil import dependency will undoubtedly increase in the future. This study shows that in the reference scenario the import dependency of China will increase to 76% in 2030 (Fig. 6.7). Nevertheless, in the mitigation scenario, oil importation in 2030 would reduce by 61%, causing the corresponding oil importation dependency to fall to 50%.

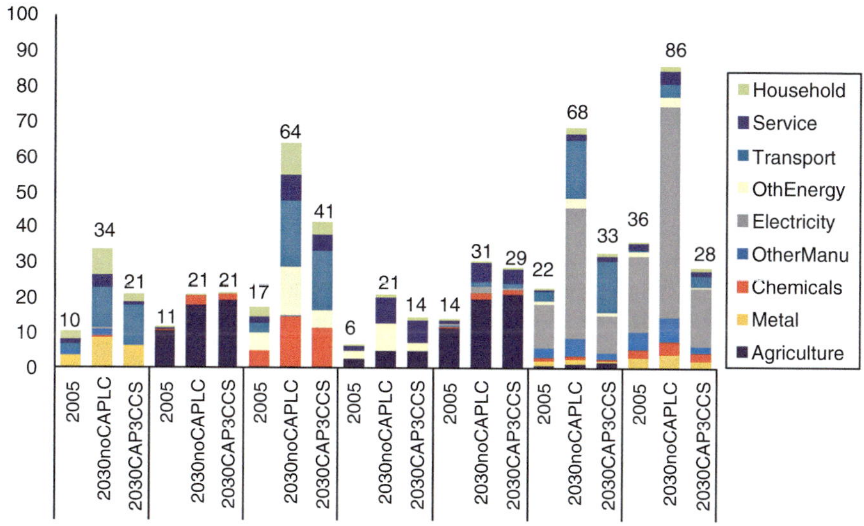

Fig. 6.7 Co-benefits of climate mitigation on air pollution in 2030 (Unit: CH_4 and CO, 10 million tons; NH_3, NMVOC, NOx, and SO_2, million tons; N_2O, 100 kilotons)

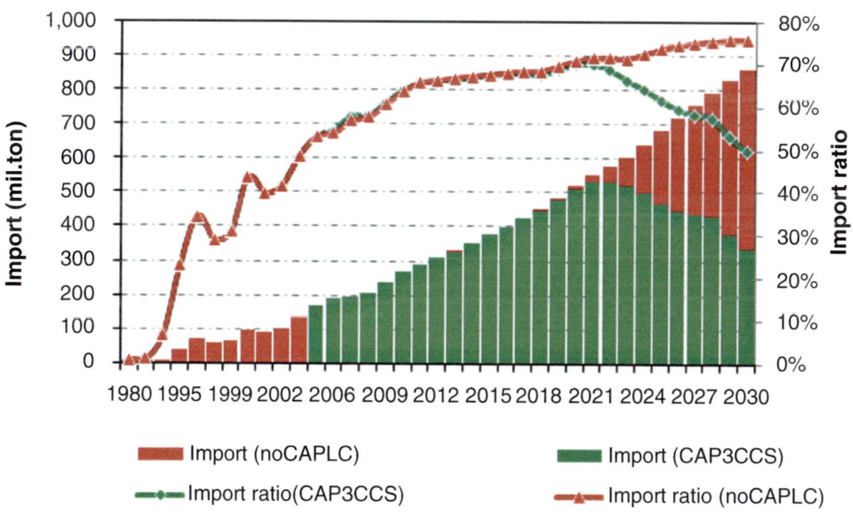

Fig. 6.8 Oil import and import dependency under reference and mitigation scenarios (Data source: National Bureau of Statistics of China (NBS) 1996 and model simulation)

6.4 Discussion

6.4.1 Key Findings and Policy Implications

This study offers important insights for China to achieve its INDC and more ambitious mitigation targets by 2030. Under the INDC target, measures in the power sector, including reducing electricity demand and developing renewable and nuclear power, are the most important approaches, which could contribute to 56% of the total required reduction. Furthermore, adjusting industry structure (contributing to 10.3% of the total reduction) and improving end-use energy efficiency (10.0%) are the second most important measures, followed by CCS technology (7.7%), fuel switch (6.3%), and lowering the activity level of the energy-intensive industries. However, if more ambitious targets are taken, technological approaches such as improving energy efficiency and CCS technology will be more important than adjusting industry structure.

We also investigated the economic cost of carbon reduction and how the cost could be lowered by low-carbon options. Without additional low-carbon countermeasures, the carbon price (155 USD/ton-CO_2 under INDC target and over 660 USD/ton-CO_2 under two-degree target in 2030) and GDP losses (6.3% under INDC and 21.5% under two-degree targets in 2030) would be considerably high. However, if all low-carbon countermeasures are introduced, the carbon price could be reduced to 35 USD and 207 USD, and the GDP loss would be reduced to 0.67% and 6.3% under INDC and two-degree targets, respectively.

Furthermore, this study confirms that a low-carbon economy brings about considerable co-benefits in terms of air pollutant reduction and energy security enhancement. Many air pollutants will be reduced by a large percentage as a result of steep carbon reductions, which is beneficial to human health. In addition, the oil importation dependency will fall, and the primary energy mix will be diversified toward less use of coal and greater use of natural gas, biomass and nuclear fuels, hydro power, and wind and solar energy, thus improving the energy security significantly.

To mitigate the economic costs, low-carbon countermeasures and practices evaluated in this study are strongly recommended, such as developing non-fossil energy, promoting more efficient end-use technology, encouraging individuals to adopt low-carbon consumption lifestyles, restricting the output of energy-intensive industrial products, continuing to optimize the industry structure, introducing emissions trade schemes among provinces, and employing CCS technology. Many of these countermeasures complement each other. For example, if household expenditures shift from material products and transport to service-oriented goods, then the demand for energy- and carbon-intensive commodities will decrease, and their production will decrease as well. Consequently, the share of the secondary industry will fall, and the demand for electricity and fossil energy will be lower.

In order to facilitate the penetration of low-carbon technologies, the following policies are recommended, including subsidizing renewable energy instead of fossil energy, making long-term plans for developing non-fossil energy, and removing the barriers for private investment in developing non-fossil energy. In addition, other policies are essential such as establishing a carbon market for inter- and intra-province carbon emissions trade and facilitating the transfer of clean and non-fossil energy technologies as well as financial resources from rich coastal provinces to the resource-rich underdeveloped western provinces.

6.4.2 *Limitations and Future Work*

Many aspects remain to be improved in future work. For instance, more low-carbon technologies should be represented in a top-down-type economic model. This model mainly considers power generation technology. In fact, the transport sector is a key area in which important low-carbon technologies such as hybrid and electric vehicles can be used. In addition, because China is a country with great socioeconomic diversity among provinces, to promote low-carbon economic development at the provincial level, a multi-region model that divides China into each of the 31 provinces is more appropriate.

6.5 Conclusion

This study uses a CGE model to investigate the options and sectors that are essential for China to achieve carbon emissions peak by 2030. Our simulation reveals that as China becomes a moderately developed country in 2030, its per capita emissions will be 15.7 tCO_2, and the total CO_2 emissions would soar up to 22.9 $GtCO_2$ if there is no intervention of mitigation policy. To cut the emissions required by China's latest INDC or even more ambitious mitigation targets such as two-degree target, China should not only rely on optimizing industry structure or restricting its industry output, but more importantly it should rely on low-carbon technologies in the power and end-use sectors as well as low-carbon style consumption. Under the INDC target, developing non-fossil energy contributes to 37% of the total required reduction, and reducing electricity contributes to 18%. Furthermore, adjusting industry structure (contributing to 10.3% of the total reduction) and improving end-use energy efficiency (10.0%) are the second most important measures, followed by CCS technology (7.7%), fuel switch (6.3%), and lowering the activity level of the energy-intensive industries.

We also depict how the mitigation costs can be lowered through various low-carbon countermeasures such as household consumption pattern changes, non-fossil energy development, and employment of CCS technology. With such low-carbon options, the carbon mitigation cost of achieving China's INDC target in 2030 could be reduced from 155 USD/ton-CO_2 to 35 USD/ton-CO_2. The corresponding GDP loss could fall from its high value of 6.3% to merely 0.67%, and welfare of household sector will not be affected significantly. Even under the more ambitious two-degree target, the carbon price and GDP loss in 2030 could be reduced to 207 (from 661) USD/ton-CO_2 and 6.3% (from 21.5%), respectively. Moreover, cutting carbon emissions has enormous co-benefits in terms of improving air quality and energy security.

In order to provide a more comprehensive assessment of mitigation policy, further research should be conducted in low-carbon technologies in other key sectors such as transport sector. Moreover, regional level studies are also important.

Acknowledgments This study was supported by the Environmental Research and Technology Development Fund (S-12-2 and 2-1402) of the Ministry of the Environment, Government of Japan. The authors are grateful for the comments from the anonymous reviewers of this paper.

Appendix 1: The CGE Model

The appendix provides a technical description of the CGE model based on *Dai* (2012).

Production

Each producer maximizes profit subject to the production technology. Activity output of each sector follows a nested constant elasticity of substitution (CES) production function. Each sector has two types of production function: one uses the existing capital stock, and another uses new investment (Dai et al. 2011). The difference between these two subsectors is the efficiency and mobility of capital among the sectors. Inputs are categorized into material commodities, energy commodities, land, labor, capital, and resource.

The producer maximizes its profit by choosing its output level and input use, depending on their relative prices subject to its technology. The producer's problem can be expressed as

$$\max \pi_{r,j} = p_{r,j} \cdot Z_{r,j} - \left(\sum_{i=1}^{N} p_{r,i} \cdot X_{r,i,j} + \sum_{i=1}^{N} \omega_{r,f} \cdot V_{r,f,j} \right) \quad (6.5)$$

subject to

$$Z_{r,j} = \upsilon_{r,j} \left[X_{r,1,j}, X_{r,2,j}, \ldots, X_{r,N,j}, ; V_{r,1,j}, \ldots, V_{r,F,j} \right] \quad (6.6)$$

Table 6.1 Sector/commodity definition in the model and the corresponding household consumption category

Nr.	Basic sectors	Consumption category	Nr.		Consumption category
1	Textiles and clothing	Clothing	23	Railway transport	
2	Chemicals	Furnishings	24	Road transport	
3	Nonmetallic mineral products	Housing	25	Urban public passenger transport services	Transport and communications
4	Other nonmetallic products	Housing			
5	Iron and steel smelting and pressing	Furnishings	26	Water transport	
			27	Air transport	
6	Nonferrous metals	N/A	28	Other transport	
7	Grain mill products	Food	29	Storage and warehousing	Transport and communications
8	Vegetable oil refining	Food			
9	Slaughtering and meat processing	Food		**Energy transformation**	
10	Beverages	Food	30	Gas production and supply	Housing
11	Other food manufacturing	Food	31	Petroleum and nuclear fuel processing	Transport and communications
12	Paper and paper products	Education and recreation	32	Coking	N/A
			33	Electricity	Furnishings
13	Metal products	Furnishings		**Land-reliant sectors**	
14	Transport equipment	Transport and communications	34	Agriculture	Food
15	Machinery	Furnishings	35	Forestry	Furnishings
16	Electronic equipment	Furnishings	36	Animal husbandry (livestock)	Food
17	Other manufacturing	Furnishings		**Resource-reliant sectors**	
18	Water production and supply	Housing			
19	Construction	Housing	37	Mineral mining	Housing
20	Scientific research and education	Education and recreation	38	Other agriculture	Food
			39	Coal mining	Housing
21	Health, social security, and welfare	Health care and medical services	40	Extraction of petroleum	N/A
			41	Natural gas	Housing
22	Services	Miscellaneous goods and services			

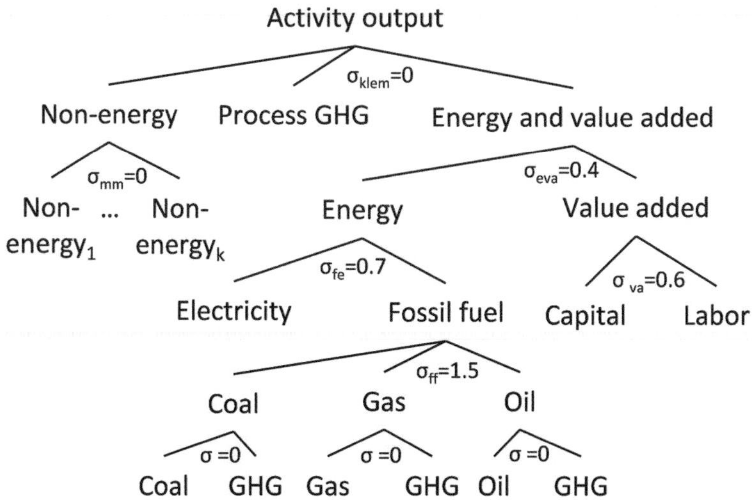

Fig. 6.9 Production tree of basic sectors

where

$\pi_{r,j}$: Profit of j-th producers in region r
$Z_{r,j}$: Output of j-th sector in region r
$X_{r,i,j}$: Intermediate inputs of *i-th* goods in j-th sector in region r
$V_{r,f,j}$: *f-th* primary factor inputs in j-th sector in region r
$p_{r,j}$: Price of the j-th composite commodity
$\omega_{r,f}$: *f-th* factor price in region r
$\upsilon_{r,j}$: *Share parameter in the CES production function*

Industrial sectors are classified into basic, energy transformation, and power generation sectors.

Basic Sectors

For the basic production functions, activity output is determined by a fixed coefficient aggregation of non-energy and energy intermediate commodities and primary factors (Fig. 6.9). The composite of non-energy inputs is in Leontief form. Energy and the value-added bundle are nested by value-added and energy inputs. The value-added bundle is a CES function of primary factors. The composite of energy inputs is a CES aggregation of electricity and fossil fuels. Fossil fuels are further disaggregated into five types:

There are four levels in the above production tree. At each level a virtual firm is assumed, each of which aims to maximize the corresponding profit subject to the production technology.

At the top level, output is a Leontief function of the quantities of value-added and aggregate energy input and aggregate intermediate input, associated with process GHG emissions:

$$\max \pi_{r,j}^{z} = p_{r,j}^{z} \cdot Z_{r,j} - \left(p_{r,j}^{vae} \cdot QVAE_{r,j} + p_{r,j}^{inta} \cdot QINTA_{r,j} + p_{r}^{ghg} \cdot QGHG_{r,j,act,ghg}\right) \quad (6.7)$$

subject to

$$Q_{r,j} = \min\left(\frac{QVAE_{r,j}}{qvae_{r,j}}, \frac{QINTA_{r,j}}{qinta_{r,j}}, \frac{QGHG_{r,j,act,ghg}}{qghg_{r,j,act,ghg}}\right) \quad (6.8)$$

where

$\pi_{r,j}^{z}$: Profit of the j-th firm producing gross domestic output $z_{r,j}$ at the top level
$Z_{r,j}$: Gross domestic output of the j-th firm
$Q_{r,j}$: Output in sector j of region r
$QVAE_{r,j}$: Value-added and energy composite input
$QINTA_{r,j}$: Composite intermediate input
$QGHG_{r,j,act,ghg}$: Process emissions of GHGs per unit of output
$p_{r,j}^{z}$: Price of j-th gross domestic output
$p_{r,j}^{vae}$: Price of composite goods of factor and energy
$p_{r,j}^{inta}$: Price of composite intermediate goods
p_{r}^{ghg}: GHG emission price
$qvae_{r,j}$: Technical coefficient expressing the composite amounts of value-added and energy inputs required per unit of $Q_{r,j}$
$qinta_{r,j}$: Technical coefficient expressing the composite amounts of non-energy intermediate inputs required per unit of $Q_{r,j}$
$qghg_{r,j,act,ghg}$: Technical coefficient expressing the process GHG emissions per unit of $Q_{r,j}$

At the second level of the production tree, there are two virtual firms with profit-maximization problems. First, composite value-added and energy input is CES aggregation of value-added input and total energy input:

$$\max \pi_{r,j}^{vae} = p_{r,j}^{vae} \cdot QVAE_{r,j} - \left(p_{r,j}^{va} \cdot QVA_{r,j} + p_{r,j}^{fe} \cdot QFE_{r,j}\right) \quad (6.9)$$

subject to

$$QVAE_{r,j} = \alpha_{r,j}^{vae} \cdot \left(\delta_{r,j}^{vae} \cdot QVA_{r,j}^{-\rho_{r,j}^{vae}} + \left(1 - \delta_{r,j}^{vae}\right) \cdot QFE_{r,j}^{-\rho_{r,j}^{vae}} \right)^{-\frac{1}{\rho_{r,j}^{vae}}} \quad (6.10)$$

Second, aggregate non-energy intermediate input is defined as Leontief function of disaggregated intermediate input:

$$\max \pi_{r,j}^{inta} = p_{r,j}^{inta} \cdot QINTA_{r,j} - \left(\sum_i p_{r,j}^{q} \cdot QINT_{r,i,j} \right) \quad (6.11)$$

subject to

$$QINTA_{r,j} = \min\left(\frac{QINT_{r,i,j}}{qint_{r,i,j}} \right) \quad (6.12)$$

where

$\pi_{r,j}^{vae}$:	Profit of j-th firm producing composite input of value added and energy
$\pi_{r,j}^{inta}$:	Profit of j-th firm producing composite intermediate input
$QVA_{r,j}$:	Aggregate value-added input
$QFE_{r,j}$:	Aggregate energy input (electricity and fossil energy)
$QINT_{r,i,j}$:	i-th non-energy inputs in j-th firm
$p_{r,j}^{va}$:	Price of composite value-added input
$p_{r,j}^{fe}$:	Price of the composite energy input (including electricity and fossil fuel)
$p_{r,j}^{q}$:	Price of the i-th composite goods
$qint_{r,i,j}$:	The amounts of each input required per unit of composite intermediate input
$\alpha_{r,j}^{vae}$:	Shift (or efficiency) parameter in the CES function
$\delta_{r,j}^{vae}$:	CES share parameter, $0 \leq \delta_{r,j}^{vae} \leq 1,\ \sum_i \delta_{r,j}^{vae} = 1$
$\rho_{r,j}^{vae}$:	The CES substitution parameter, in which the elasticity of substitution between value added and energy, σ, equals $\frac{1}{1+\rho}$
$\sigma_{r,j}^{vae}$:	Elasticity of substitution between value-added bundle and energy

At the third level of the production tree, there are two virtual firms with profit-maximization problems as well. First, composite value-added input is CES aggregation of capital and labor input:

$$\max \pi_{r,j}^{va} = p_{r,j}^{va} \cdot QVA_{r,j} - \left(pl_r \cdot QLAB_{r,j} + pk_{r,j} \cdot QCAP_{r,j} \right) \quad (6.13)$$

subject to

$$QVA_{r,j} = \alpha_{r,j}^{va} \cdot \left(\delta_{r,j}^{cap} \cdot QCAP_{r,j}^{-\rho_{r,j}^{va}} + \delta_{r,j}^{lab} \cdot QLAB_{r,j}^{-\rho_{r,j}^{va}}\right)^{-\frac{1}{\rho_{r,j}^{va}}} \tag{6.14}$$

Moreover, composite energy input is CES aggregation of electricity input and fossil fuel input:

$$\max \pi_{r,j}^{fe} = p_{r,j}^{fe} \cdot QFE_{r,j} - \left(p_{r,\text{"ele"}}^{q} \cdot QELE_{r,j} + p_{r,j}^{fos} \cdot QFOS_{r,j}\right) \tag{6.15}$$

subject to

$$QFE_{r,j} = \alpha_{r,j}^{fe} \cdot \left(\delta_{r,j}^{ele} \cdot QELE_{r,j}^{-\rho_{r,j}^{fe}} + \left(1 - \delta_{r,j}^{ele}\right) \cdot QFOS_{r,j}^{-\rho_{r,j}^{fe}}\right)^{-\frac{1}{\rho_{r,j}^{fe}}} \tag{6.16}$$

where

$\pi_{r,j}^{va}$:	Profit of j-th firm producing composite input of value added
$\pi_{r,j}^{fe}$:	Profit of j-th firm producing composite input of energy
$QCAP_{r,j}$:	Capital input required per unit of value-added input
$QLAB_{r,j}$:	Labor input required per unit of value-added input
$QELE_{r,j}$:	Electricity input required per unit of composite energy input
$QFOS_{r,j}$:	Composite fossil fuel input required per unit of composite energy input
pl_r:	Labor price in region r
$pk_{r,j}$:	Capital price in j-th sector of region r
$p_{r,\text{"ele"}}^{q}$:	Price of the composite goods of electricity
$p_{r,j}^{fos}$:	Price of composite fossil fuel input in j-th sector
$\alpha_{r,j}^{fe}$:	Shift (or efficiency) parameter in the CES function
$\delta_{r,j}^{fe}, \delta_{r,j}^{cap}, \delta_{r,j}^{lab}$:	CES share parameters, $0 \le \delta_{r,j}^{*} \le 1, \ \sum_{i} \delta_{r,j}^{*} = 1$
$\rho_{r,j}^{fe}$:	CES substitution parameter, in which the elasticity of substitution between electricity and composite fossil fuel, σ, equals $\frac{1}{1+\rho}$
$\sigma_{r,j}^{va}$:	Elasticity of substitution between capital and labor
$\sigma_{r,j}^{vae}$:	Elasticity of substitution between electricity and fossil fuel

At the fourth level of the production function, composite fossil fuel is CES aggregation of coal, crude oil, natural gas, coke, petrol oil, and manufactured gas:

$$\max \pi_{r,j}^{fos} = p_{r,j}^{fos} \cdot QFOS_{r,j} - \left(\sum_{fos} p_{r,fos}^{q} \cdot QFF_{r,fos,j}\right) \tag{6.17}$$

subject to

$$QFOS_{r,j} = \alpha^{ff}_{r,j} \cdot \left(\sum_{fos} \delta^{ff}_{r,j} \cdot QFF_{r,fos,j}^{-\rho^{ff}_{r,j}} \right)^{-\frac{1}{\rho^{ff}_{r,j}}} \tag{6.18}$$

where

$\pi^{fos}_{r,j}$: Profit of j-th firm producing composite input of fossil fuel

$QFF_{r,fos,j}$: CES shift (or efficiency) parameter

$p^{q}_{r,fos}$: Price of fossil fuel input

$\alpha^{ff}_{r,j}$: Shift (or efficiency) parameter in the CES function

$\delta^{ff}_{r,j}$: CES share parameter, $0 \leq \delta^{ff}_{r,j} \leq 1, \ \sum_{i} \delta^{ff}_{r,j} = 1$

$\rho^{ff}_{r,j}$: CES substitution parameter, in which the elasticity of substitution among fossil fuels, σ, equals $\frac{1}{1+\rho}$

$\sigma^{ff}_{r,j}$: Elasticity of substitution among fossil fuels

Energy Transformation Sector (Except Power Generation)

Energy transformation sectors include gas production and supply, petroleum and nuclear fuel processing, and coking. The energy bundle is linked at the top level in order to maintain the first law of thermal efficiency of the conversion of primary energy to the secondary energy (Fig. 6.10). Functions at other levels are the same as the basic sectors.

Thus the problem is expressed in

$$\begin{aligned} \max \pi^{z}_{r,j} = \ & p^{z}_{r,j} \cdot Z_{r,j} \\ & - \left(p^{vae}_{r,j} \cdot QVAE_{r,j} + p^{fe}_{r,j} \cdot QFE_{r,j} + p^{inta}_{r,j} \cdot QINTA_{r,j} + p^{ghg}_{r} \cdot QGHG_{r,j,act,ghg} \right) \end{aligned} \tag{6.19}$$

subject to

$$Q_{r,j} = \min \left(\frac{QVAE_{r,j}}{qvae_{r,j}}, \frac{QFE_{r,j}}{qfe_{r,j}}, \frac{QINTA_{r,j}}{qinta_{r,j}}, \frac{QGHG_{r,j,act,ghg}}{qghg_{r,j,act,ghg}} \right) \tag{6.20}$$

where

$QFE_{r,j}$: Aggregate energy input (electricity and fossil energy)

$qfe_{r,j}$: Technical coefficient expressing the aggregate energy inputs required per unit of $Q_{r,j}$

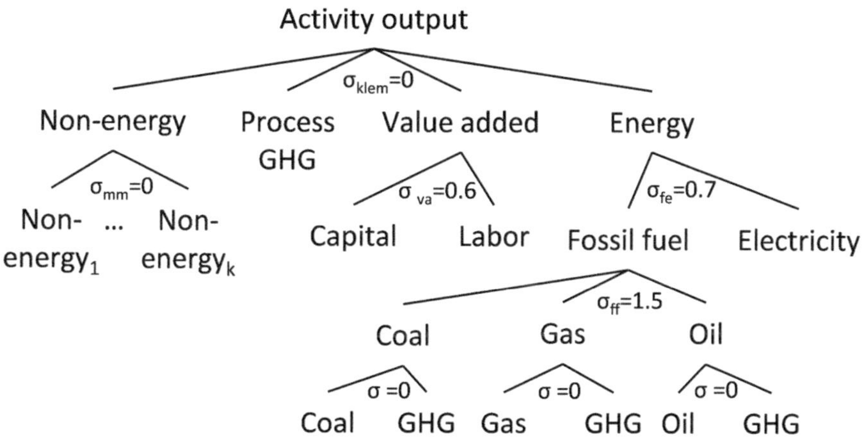

Fig. 6.10 Production tree of energy transformation sectors

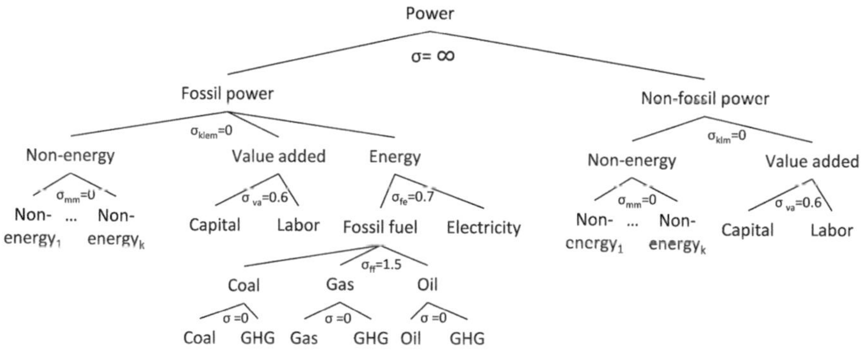

Fig. 6.11 Production tree of electricity generation sectors

Power Generation Sector

Electricity is generated by eight technologies, e.g., coal, gas, oil, nuclear, hydro, wind, solar PV, and biomass power. Disaggregation of the electricity sector into eight technologies in the base year follows the methodology developed by Sue Wing (2006, 2008). Production function of each technology is the same as that of energy transformation sectors. Each technology is perfectly substitutable with another. Electricity output is almost in a linear relationship with energy inputs (Fig. 6.11).

Household Consumption

Household and government are the final consumers. The representative household endows primary factors to the firms and receives income from the rental of primary factors (labor and capital), rents from fixed factors (land and natural resources), and lump-sum transfer from the government (e.g., carbon tax revenue of the government). The income is then used for either investment or final consumption. The objective of household consumption is to maximize utility by choosing levels of goods consumption following Cobb-Douglas preferences, subject to commodity prices and budget constraint. The agent's problem is expressed as

$$\max u_{r,h}\left[X^p_{r,1},\ldots,x^p_{r,i}\right] = A^p_r \cdot \prod_{i=1}^{N} \left(X^p_{r,i}\right)^{\alpha^p_{r,i}} \tag{6.21}$$

subject to

$$\begin{aligned} EH_r &= \sum_i p^q_{r,j} \cdot X^p_{r,i} \\ &= \sum_{f=1}^{F} \omega_{r,f} \cdot V_{r,f} + \sum_j pld_r \cdot QLAND_{r,j} + \sum_{res,j} p^{res}_{r,j} \cdot QRES_{r,j} + T^{cab}_r \\ &\quad - T^d_r - S^p_r \end{aligned} \tag{6.22}$$

$$T^{cab}_r = pghg_{r,\text{“}CO2\text{”}} \cdot \mathrm{TEMS}_{r,\text{“}CO2\text{”}} \tag{6.23}$$

$$T^d_r = \tau^d_r \cdot \sum_f \omega_{r,f} \cdot V_{r,f} \tag{6.24}$$

$$S^p_r = sr^p_r \cdot \sum_f \omega_{r,f} \cdot V_{r,f} \tag{6.25}$$

where

$u_{r,h}$: Utility function of households
EH_r: Household expenditure
$X^p_{r,i}$: Household consumption of i-th commodity
$V_{r,f}$: *f-th* primary factor endowment by household
S^p_r: Household savings
$\mathrm{TEMS}_{r,\text{“}CO2\text{”}}$: CO_2 emissions in region r
$pghg_{r,\text{“}CO2\text{”}}$: Carbon price
T^d_r: Direct tax
τ^d_r: Direct tax rate
sr^p_r: Average propensity to save by the household
$\omega_{r,f}$: Price of the f-th primary factor
A^p_r: Scaling parameter in Cobb-Douglas function
$\alpha^p_{r,i}$: Share parameter in Cobb-Douglas function, $0 \le \alpha^p_{r,i} \le 1,\ \sum_i \alpha^p_{r,i} = 1$

Government

The government is assumed to collect taxes, including a direct tax on household income, ad valorem production tax (indirect tax) on gross domestic output, ad valorem import tariff on imports, and carbon tax. Based on a Cobb-Douglas demand function (Hertel and Tsigas 2004), the government spends its revenue on public services which are provided to the whole society and on the goods and services which are provided to the households free of charge or at low prices (NBS 2006). The model assumes that the revenue from the carbon tax is recycled to the representative agent as a lump-sum transfer:

$$\max u_{r,g}\left[x_{r,1}^{g}, \ldots, x_{r,i}^{g}\right] = A_r^g \cdot \prod_{i=1}^{N} \left(x_{r,i}^{g}\right)^{\alpha_{r,i}^{g}} \tag{6.26}$$

subject to

$$\sum_i p_{r,i} \cdot x_{r,i}^{g} = T_r^d + \sum_j T_{r,j}^{z} + \sum_j T_{r,j}^{m} - S^g \tag{6.27}$$

$$T_{r,j}^{z} = \tau_{r,j}^{z} \cdot p_{r,j} \cdot Z_{r,j} \tag{6.28}$$

$$T_{r,i}^{m} = \tau_{r,i}^{m} \cdot pm_{r,i} \cdot M_{r,i} \tag{6.29}$$

$$S_r^g = sr_r^g \cdot \left(T_r^d + \sum_j T_{r,j}^{z} + \sum_j T_{r,j}^{m}\right) \tag{6.30}$$

where

$u_{r,g}$: Utility function of government
$x_{r,i}^{g}$: Government consumption of i-th commodity
S_r^g: Government savings
$T_{r,j}^{z}$: Production tax on the j-th commodity
$T_{r,j}^{m}$: Import tariff on the j-th commodity
$\tau_{r,j}^{z}$: Production tax rate on the j-th commodity
$\tau_{r,i}^{m}$: Import tariff rate on the i-th commodity
sr_r^g: Average propensity to save by the government
$Z_{r,j}$: Gross domestic output of the j-th commodity
$M_{r,i}$: Import of the i-th commodity
$pm_{r,i}$: Price of the i-th imported commodity
A_r^g: Scaling parameter in Cobb-Douglas function
$\alpha_{r,i}^{g}$: Share parameter in Cobb-Douglas function, $0 \leq \alpha_{r,i}^{g} \leq 1, \ \sum_i \alpha_{r,i}^{g} = 1$

Investment and Savings

Investment is an important part of the final demand. In the CGE model, a virtual agent is assumed for investment which receives all the savings from the household, the government, and the external sector to purchase goods for domestic investment. The virtual investment agent is assumed to maximize the utility based on a Cobb-Douglas demand function subject to its (virtual) income constraint. Mathematically, the investment problems can be described as follows:

$$\max u_{r,v}\left[x^{v}_{r,1},\ldots,x^{v}_{r,i}\right] = A^{v}_{r}\cdot\prod_{i=1}^{N}\left(x^{v}_{r,i}\right)^{\alpha^{v}_{r,i}} \tag{6.31}$$

subject to

$$\sum_{i} p_{r,i}\cdot x^{v}_{r,i} = S^{p}_{r} + S^{g}_{r} + \varepsilon\cdot S^{f}_{r} \tag{6.32}$$

where

$u_{r,v}$: Utility of virtual investment agent
S^{f}_{r}: Current account deficits in foreign currency terms (or alternatively foreign savings)
ε: Foreign exchange rate
$x^{v}_{r,1}$: Demand for the *i*-th investment goods
A^{v}_{r}: Scaling parameter in Cobb-Douglas function,
$\alpha^{v}_{r,i}$: Share parameter in Cobb-Douglas function, $0 \le \alpha^{v}_{r,i} \le 1,\ \sum_{i}\alpha^{v}_{r,i} = 1$

International Transaction

The model is an open economy model that includes the interaction of commodity trade with the rest of the world. Like most other countries' CGE models, this model assumes the small open economy, meaning that an economy is small enough for its policies not to alter world prices or incomes. The implicit implication of small-country assumption is that export and import prices are exogenously given for the economy. In this study, future international prices are fixed to be the same level for non-energy commodities but increase by 3% yearly for energy commodities compared to the 2005 level.

Two types of price variables are distinguished. One is prices in terms of the domestic currency p^{e}_{i} and p^{m}_{i}; the other is prices in terms of the foreign currency p^{We}_{i} and p^{Wm}_{i}. They are linked with each other as follows:

$$p_i^e = \varepsilon \cdot p_i^{We} \tag{6.33}$$

$$p_i^m = \varepsilon \cdot p_i^{Wm} \tag{6.34}$$

Furthermore, it is assumed that the economy faces balance of payments constraints, which is described with export and import prices in foreign currency terms:

$$\sum_i p_i^{We} \cdot E_{r,i} + S_r^f = \sum_i p_i^{Wm} \cdot M_i \tag{6.35}$$

where

$E_{r,i}$: Export of i-th commodity in region r
$M_{r,i}$: Import of i-th commodity in region r
p_i^{We}: Export price in terms of foreign currency,
p_i^e: Export price in terms of domestic currency
p_i^{Wm}: Import price in terms of foreign currency,
p_i^m: Import price in terms of domestic currency

Substitution Between Imports and Domestic Goods

The Armington assumption is adopted, i.e., the domestic and imported goods are imperfectly substitutable for each other, which implies that households and firms don't directly consume or use imported goods but instead a so-called Armington composite goods, which is made up of imported and locally produced goods as well as goods produced in other provinces by a two-level nested CES function (Fig. 6.12).

Import activity is described by the bottom nesting of Fig. 6.12. In the CGE model, the Armington composite goods at this level are created by virtual firms which maximize their profits by choosing a proper combination of imported and locally produced goods. The solution of their profit-maximization problem leads to their input demands for imported and domestic goods, which depend on the corresponding relative prices of domestic and imported goods. Mathematically, this problem can be expressed as

$$\text{Max } \pi_{r,i}^{md} = p_{r,i}^{md} \cdot Q_{r,i}^{md} - \left[\left(1+\tau_{r,i}^m\right) \cdot p_i^m \cdot M_{r,i} + p_{r,i}^d \cdot D_{r,i}^d\right] \tag{6.36}$$

subject to

$$Q_{r,i}^{md} = \alpha_{r,i}^{md} \cdot \left(\delta_{r,i}^m \cdot M_{r,i}^{-\rho_{r,i}^{md}} + \delta_{r,j}^d \cdot D_{r,i}^{-\rho_{r,i}^{md}}\right)^{-\frac{1}{\rho_{r,i}^{md}}} \tag{6.37}$$

Fig. 6.12 Nesting of imported goods, locally produced goods, and goods produced in other provinces

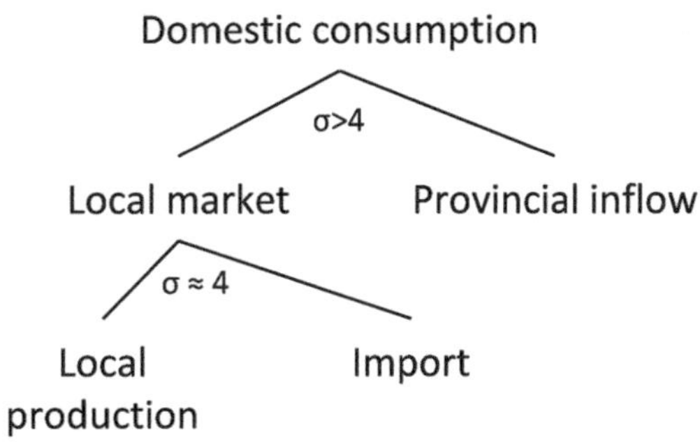

where

$\pi^{md}_{r,i}$: Profit of the firm producing the i-th Armington composite goods of import and locally produced goods

$Q^{md}_{r,i}$: The i-th Armington composite goods of import and locally produced goods

$D^{d}_{r,i}$: The i-th locally produced goods

$p^{md}_{r,i}$: Armington price of the i-th imported and locally produced goods

$p^{d}_{r,i}$: Price of the i-th locally produced goods

$\tau^{m}_{r,i}$: Import tariff rate on the i-th commodity

$\alpha^{md}_{r,i}$: Shift (or efficiency) parameter in the Armington composite goods production function

$\delta^{m}_{r,i}, \delta^{d}_{r,j}$: Input share parameters in the Armington composite goods production function ($0 \leq \delta^{m}_{r,i} \leq 1, 0 \leq \delta^{d}_{r,j} \leq 1, \delta^{m}_{r,i} + \delta^{d}_{r,i} = 1$)

$\rho^{md}_{r,i}$: The CES substitution parameter, in which the elasticity of substitution between imported and domestic goods, σ, equals $\frac{1}{1+\rho}$

Then the composite imported and locally produced goods will be further aggregated with the goods produced in other provinces to form the final Armington composite goods that are consumed by households and the government and as intermediate inputs by firms, which will be introduced in section F.

Transformation Between Exports and Domestic Goods

On the supply side, the produced commodities are distributed to the international market, local market, and market in other provinces by a two-level nested constant elasticity of transformation function. Similar to the treatment of import, a virtual firm is assumed for each commodity which transforms the gross domestic output into exports and domestic goods as follows:

$$\text{Max } \pi^{dx}_{r,i} = \left(p^{e}_{i} \cdot E_{r,i} + p^{dd}_{r,i} \cdot D^{s}_{r,i}\right) - \left(1 + \tau^{z}_{r,i}\right) \cdot p^{z}_{r,i} \cdot Q^{dx}_{r,i} \tag{6.38}$$

subject to

$$Q^{dx}_{r,i} = \alpha^{dx}_{r,i} \cdot \left(\delta^{e}_{r,i} \cdot E^{\rho^{dx}_{r,i}}_{r,i} + \delta^{d}_{r,j} \cdot D^{s}_{r,i} \rho^{dx}_{r,i} \right)^{\frac{1}{\rho^{dx}_{r,i}}} \quad (6.39)$$

where

$\pi^{dx}_{r,i}$: Profit of the firm engaged in the i-th transformation
$Q^{dx}_{r,i}$: Gross domestic output of the i-th goods
$D^{s}_{r,i}$: i-th goods supplied to domestic market
$p^{z}_{r,i}$: Price of the i-th gross domestic output
$p^{dd}_{r,i}$: Price of domestically supplied goods
$\tau^{z}_{r,i}$: Production tax rate on the i-th commodity
$\alpha^{dx}_{r,i}$: Shift (or efficiency) parameter in the transformation function
$\delta^{e}_{r,i}$, $\delta^{d}_{r,j}$: Share parameters in the transformation function ($0 \leq \delta^{e}_{r,i} \leq 1$, $0 \leq \delta^{d}_{r,j} \leq 1$, $\delta^{e}_{r,i} + \delta^{d}_{r,i} = 1$)
$\rho^{dx}_{r,i}$: Transformation elasticity parameter, in which the elasticity of substitution between imported and domestic goods, σ, equals $\frac{1}{\rho - 1}$

It should be noted that the goods supplied to the domestic market at this level, $D^{s}_{r,i}$, will be further distributed to local market and market in other provinces through interprovincial trade, which will be described in the next section.

Interprovincial Trade

An important feature of this model is that it is a country model in which interprovincial trade is treated. Similar to the case of international trade, Armington assumption is adopted to distinguish between locally produced commodity and commodity produced by firms in other provinces, and CES and CET functions are employed to describe commodity inflow from and outflow to all provinces, respectively.

Substitution Commodity Between Local Market and Inflow from Other Provinces

This section describes the top-level nesting of Fig. 6.12 which treats interprovincial inflow of commodity. By this stage the commodity in the local market is an aggregation of locally produced and imported goods, which needs to be further aggregated with goods produced in other provinces to form the final Armington composite goods to be consumed by the final consumers and firms. The treatment is similar to import:

$$\text{Max } \pi^{dd}_{r,i} = p^{a}_{r,i} \cdot Q^{dd}_{r,i} - \left[p^{md}_{r,i} \cdot Q^{md}_{r,i} + \sum_{rr} p^{inf}_{rr,i} \cdot D^{inf}_{rr,r,i} \right] \tag{6.40}$$

subject to

$$Q^{dd}_{r,i} = \alpha^{dd}_{r,i} \cdot \left(\delta^{md}_{r,i} \cdot Q^{md}_{r,i}{}^{-\rho^{dd}_{r,i}} + \sum_{rr} \delta^{inf}_{rr,r,i} \cdot D^{inf}_{rr,r,i}{}^{-\rho^{dd}_{r,i}} \right)^{-\frac{1}{\rho^{dd}_{r,i}}} \tag{6.41}$$

where

$\pi^{dd}_{r,i}$: Profit of the firm producing the i-th Armington composite goods of local market and inflow from other provinces

$Q^{dd}_{r,i}$: Armington composite goods

$D^{inf}_{rr,r,i}$: The i-th goods inflowing from region rr to region r

$p^{a}_{r,i}$: Armington price taken by the final consumers and firms

$p^{inf}_{rr,i}$: Price of the i-th goods inflowing from province rr to region r

$\alpha^{dd}_{r,i}$: Shift (or efficiency) parameter in the Armington composite goods production function

$\delta^{md}_{r,i}$, $\delta^{inf}_{rr,r,j}$: Input share parameters in the Armington composite goods production function ($0 \leq \delta^{md}_{r,i} \leq 1, 0 \leq \delta^{inf}_{rr,r,j} \leq 1, \delta^{md}_{r,i} + \sum_{rr} \delta^{inf}_{rr,r,i} = 1$)

$\rho^{dd}_{r,i}$: The CES substitution parameter, in which the elasticity of substitution between imported and domestic goods, σ, equals $\frac{1}{1+\rho}$

Transformation Between Goods Sold in Local Market and Outflowing to Other Provinces

Goods supplied to the domestic market, $D^{s}_{r,i}$, will be further distributed to local market and market in other provinces through, similar to the treatment of export, a CET function as follows:

$$\text{Max } \pi^{pp}_{r,i} = \left(p^{d}_{i} \cdot D^{local}_{r,i} + \sum_{rr} p^{out}_{rr,i} \cdot D^{out}_{r,rr,i} \right) - p^{dd}_{r,i} \cdot D^{s}_{r,i} \tag{6.42}$$

subject to

$$Q^{pp}_{r,i} = \alpha^{pp}_{r,i} \cdot \left(\delta^{local}_{r,i} \cdot D^{local}_{r,i}{}^{\rho^{pp}_{r,i}} + \sum_{rr} \delta^{out}_{r,rr,i} \cdot D^{out}_{r,rr,i}{}^{\rho^{pp}_{r,i}} \right)^{\frac{1}{\rho^{pp}_{r,i}}} \tag{6.43}$$

where

$\pi^{pp}_{r,i}$: Profit of the firm engaged in the i-th transformation

$Q^{pp}_{r,i}$: Out of the i-th goods supplied to local and other provinces' markets

$D^{local}_{r,i}$: i-th goods supplied to local market

$D^{out}_{r,rr,i}$: i-th goods outflowing from region r to other province rr

$p^{out}_{rr,i}$: Price of the i-th goods outflowing to other province rr

$\alpha^{pp}_{r,i}$: Shift (or efficiency) parameter in the transformation function

$\delta^{local}_{r,i}$, $\delta^{out}_{r,rr,i}$: Share parameters in the transformation function ($0 \leq \delta^{local}_{r,i} \leq 1$, $0 \leq \delta^{out}_{r,rr,j} \leq 1$, $\delta^{local}_{r,i} + \sum_{rr} \delta^{out}_{r,rr,i} = 1$)

$\rho^{pp}_{r,i}$: Transformation elasticity parameter, in which the elasticity of substitution between imported and domestic goods, σ, equals $\frac{1}{\rho - 1}$

Market Clearance Conditions

The above sections describe the behavior of economic agents such as the households, firms, government, and investment agents and the interactions with other provinces and the rest of the world. The final step is to impose the market-clearing conditions to all commodities and factor markets as follows:

$$Q_{r,i} = x^{p}_{r,i} + x^{g}_{r,i} + x^{v}_{r,i} + \sum_{j} x_{r,i,j} \tag{6.44}$$

$$\sum_{j} v_{r,f,j} = V_{r,f} \tag{6.45}$$

Macro Closure

In a CGE model, the issue of macro closure is the choice of exogenous variables among all variables in the model, mainly including investment and saving macro closure and current account balance macro closure. In this model, investment is exogenously assumed. In addition, foreign exchange rate is fixed and thus balance of payment is an endogenous variable.

Appendix 2: The Decomposition Method

The appendix describes how carbon reduction can be attributed to each countermeasure in section 4.3.1 by decomposition techniques. As shown in Eq. 6.46, the total carbon emission (EM) is the summation of emissions from power generation (EM_{ele}) and end-use sectors (EM_{end}) minus emissions captured by CCS technology

(EM_{ccs}). EM_{end} and EM_{ele} can be further broken down into subcomponents. In Eq. 6.47, EM_{end} is a product of the underlying factors of activity levels (GDP), industry structure (STR = VA_i/GDP), end-use energy efficiency (EFF = E_i/VA_i), and fuel-switch factor (FS = EM_i/E_i). Similarly, in Eq. 6.48, emission from power generation (EM_{ele}) is a product of the underlying factors of total electricity demand (ELE_{tot}), non-fossil-fired electricity (NF = ELE_{fos}/ELE_{tot}), and power generation efficiency (EFF = EM_{ele}/ELE_{fos}):

$$EM = EM_{end} + EM_{ele} - EM_{ccs} \tag{6.46}$$

$$\begin{aligned} EM_{end} &= \Sigma_i (GDP * VA_i/GDP * E_i/VA_i * EM_i/E_i) \\ &= ACT * STR * EFF * FS \end{aligned} \tag{6.47}$$

$$EM_{ele} = ELE_{tot} * ELE_{fos}/ELE_{tot} * EM_{ele}/ELE_{fos} = EL * NF * EFF \tag{6.48}$$

where

EM:	Total emission
EM_{end}:	Emission from end-use sectors
EM_{ele}:	Emission from power generation
EM_{ccs}:	Emission captured by CCS technology
VA_i:	Value added of sector i
E_i:	Energy consumption of sector i
EM_i:	Emission from sector i
ELE_{tot}:	Total electricity demand
ELE_{fos}:	Fossil-fired electricity production
ACT:	Activity level effect
STR:	Industry structure effect
EFF:	Energy efficiency effect
FS:	Fuel-switching effect
EL:	Total electricity demand effect
NF:	Non-fossil power effect

A change in emissions from the reference to countermeasure scenarios, $\triangle EM$, can be expressed as a change in emission from end-use sectors, power generation, CCS absorption, and emissions trade (Eq. 6.49). Using the Laspeyres index method (Sun and Ang 2000), a change in emissions from end-use sectors and power generation can be expressed as the joint contribution of the underlying effects indicated by $_f$ in Eqs. 6.50 and 6.51, in which each effect can be derived from multiplication, as shown for total electricity demand in Eq. 6.52. The first part of Eq. 6.52, $\triangle ELE_{tot}$*NF*EFF, can be interpreted as the partial effect of the total electricity demand component on the change of CO_2 emissions between the reference and mitigation scenarios. The following parts capture interactions between the remaining variables and form the so-called residual term:

Table 6.2 Household consumption expenditure parameters under high-carbon and low-carbon scenarios of China in 2030

	2005		2030			Direction	
Per capita expenditure (2005 USD)	970	312	10,280	3307	7700	HC	LC
Food (%)	36.7	45.1	26.5	31.7	27.4	26.0	32.0
Clothing (%)	10.1	5.8	10.1	5.9	9.4	10.0	6.0
Housing (%)	10.2	14.3	9.6	23.0	11.7	10.0	12.0
Furnishings (%)	5.6	4.4	8.2	5.3	7.8	8.0	5.0
Health care and medical services (%)	7.6	6.4	6.4	6.5	6.4	6.4	6.5
Transport and communications (%)	12.6	9.8	19.1	13.3	18.2	19.0	13.0
Education and recreation (%)	13.8	12.0	14.9	11.5	14.4	15.0	12.0
Miscellaneous goods and services (%)	3.5	2.3	5.1	2.7	4.8	5.6	13.5

$$\triangle EM = \triangle EM_{end} + \triangle EM_{ele} - \triangle EM_{ccs} \tag{6.49}$$

$$\triangle EM_{end} = ACT_f + STR_f + EFF_f + FS_f \tag{6.50}$$

$$\triangle EM_{ele} = EL_f + NF_f + EFF_f \tag{6.51}$$

$$EL_f = \triangle ELE_{tot} * NF * EFF + 1/2 * ELE_{tot} * (\triangle NF * EFF + NF * \triangle EFF) + 1/3 * (\triangle NF * \triangle EFF) \tag{6.52}$$

References

Ashina S, Fujino J, Masui T, Ehara T, Hibino G (2012) A roadmap towards a low-carbon society in japan using backcasting methodology: feasible pathways for achieving an 80. Energy Policy 41:584–598

Bretschger L, Zhang L (2014) Carbon policy in a high-growth economy: the case of china. Available at SSRN 2466370

Chai Q-M, Xu H-Q (2014) Modeling an emissions peak in china around 2030: synergies or trade-offs between economy, energy and climate security. Adv Clim Chang Res 5:169–180. doi: http://dx.doi.org/10.1016/j.accre.2015.06.001

Cheng B, Dai H, Wang P, Zhao D, Masui T (2015) Impacts of carbon trading scheme on air pollutant emissions in Guangdong province of china. Energy Sustain Dev 27:174–185

Computer-aided of Engineering (CAE) (2011) China's mid- and long-term energy development strategy: electricity, oil, gas, nuclear and environment, vol 1. Science Press, Beijing. (in Chinese)

Dagoumas S, Barker TS (2010) Pathways to a low-carbon economy for the UK with the macro-econometric e3MG model. Energy Policy 38:3067–3077

Dai H (2012) Integrated assessment of China's provincial low carbon economy development towards 2030: Jiangxi province as an example. Tokyo, Tokyo Institute of Technology. PhD

Dai H, Masui T, Matsuoka Y, Fujimori S (2011) Assessment of china's climate commitment and non-fossil energy plan towards 2020 using hybrid aIM/CGE model. Energy Policy 39:2875–2887. doi:http://dx.doi.org/10.1016/j.enpol.2011.02.062

Dai H, Masui T, Matsuoka Y, Fujimori S (2012) The impacts of china's household consumption expenditure patterns on energy demand and carbon emissions towards 2050. Energy Policy 50:736–750

Dai H, Mischke P, Xie X, Xie Y, Masui T (2016) Closing the gap? Top-down versus bottom-up projections of china's regional energy use and CO_2 emissions. Appl Energy 162:1355–1373

Foran B (2011) Low carbon transition options for Australia. Ecol Model 223:72–80

Gomi K, Shimada K, Matsuoka Y (2010) A low-carbon scenario creation method for a local-scale economy and its application in Kyoto city. Energy Policy 38:4783–4796

Hertel TW, Tsigas ME (2004) Structure of GTAP. In: Hertel TW (ed) Global trade analysis: modeling and applications. Cambridge University Press, New York, p 48

IPCC (2006) IPCC guidelines for national greenhouse gas inventories. IGES, Japan

IPCC (2007) Climate change 2007: the physical science basis. The Intergovernmental Panel on Climate Change

Jiang K, Hu X (2009) China's low carbon development scenario. In: 2050 China energy ST of, report CE, editors. 2050 china energy and CO_2 emissions report. Science Press, Beijing. (in Chinese)

Jones R, Yoo B (2010) Korea's green growth strategy: mitigating climate change and developing new growth engines. OECD Publishing, Paris

Li W, Li H, Sun S (2015) China's low-carbon scenario analysis of CO_2 mitigation measures towards 2050 using a hybrid aIM/CGE model. Energies 8:3529–3555

Masui T, Matsumoto K, Hijioka Y, Kinoshita T, Nozawa T, Ishiwatari S, Kato E, Shukla P, Yamagata Y, Kainuma M (2011) An emission pathway for stabilization at 6 Wm−2 radiative forcing. Clim Chang 109:59–76

National Bureau of Statistics of China (NBS) (1996–2006) China statistical year book: 1996–2006. China Statistics Press, Beijing. (in Chinese)

National Bureau of Statistics of China (NBS) (2006) Input-output tables of China 2002. China Statistics Press, Beijing. (in Chinese)

National Bureau of Statistics of China (NBS) (2008) China energy statistical year book 2008. China Statistics Press, Beijing. (in Chinese)

National Bureau of Statistics of China (NBS) (2011) Input-output tables of china 2007. China Statistics Press, Beijing. (in Chinese)

National Development and Reform Commission of the People's Republic of China (2012) Letter including autonomous domestic mitigation actions 2010

Paltsev S, Morris J, Cai Y, Karplus V, Jacoby H (2012) The role of china in mitigating climate change. Energy Econ 34:S444–S450

Rutherford TF (1999) Applied general equilibrium modeling with mPSGE as a gAMS subsystem: an overview of the modeling framework and syntax. Comput Econ 14:1–46

Selvakkumaran S, Limmeechokchai B, Masui T, Hanaoka T, Matsuoka Y (2015) A quantitative analysis of low carbon society (lCS) measures in Thai industrial sector. Renew Sust Energy Rev 43:178–195

Shukla P, Dhar S, Mahapatra D (2008) Low-carbon society scenarios for India. Clim Pol 8:156–176

Skea J, Nishioka S (2008) Policies and practices for a low-carbon society. Clim Pol 8:5–16

Sue Wing I (2006) The synthesis of bottom-up and top-down approaches to climate policy modeling: Electric power technologies and the cost of limiting US CO_2 emissions. Energy Policy 34(18):3847–3869

Sue Wing I (2008) The synthesis of bottom-up and top-down approaches to climate policy modeling: electric power technology detail in a social accounting framework. Energy Econ 30(2):547–573

Sun J, Ang B (2000) Some properties of an exact energy decomposition model. Energy 25 (12):1177–1188

Thepkhun P, Limmeechokchai B, Fujimori S, Masui T, Shrestha RM (2013) Thailand's low-carbon scenario 2050: the aIM/CGE analyses of cO 2 mitigation measures. Energy Policy 62:561–572

UN (2008) World population prospects: the 2008 revision population database, United Nations

Wang T, Watson J (2010) Scenario analysis of china's emissions pathways in the 21st century for low carbon transition. Energy Policy 38:3537–3546

Wang Y, Zou L-L (2014) The economic impact of emission peaking control policies and china's sustainable development. Adv Clim Chang Res 5:162–168. doi:http://dx.doi.org/10.1016/j.accre.2015.05.001

Wang P, Dai H-C, Ren S-Y, Zhao D-Q, Masui T (2015) Achieving Copenhagen target through carbon emission trading: economic impacts assessment in Guangdong province of China. Energy 79:212–227

Zhang D, Rausch S, Karplus VJ, Zhang X (2013) Quantifying regional economic impacts of CO_2 intensity targets in China. Energy Econ 40:687–701

Zhang S, Bauer N, Luderer G, Kriegler E (2014) Role of technologies in energy-related CO_2 mitigation in china within a climate-protection world: a scenarios analysis using rEMIND. Appl Energy 115:445–455

Zhou N, Fridley D, Khanna NZ, Ke J, McNeil M, Levine M (2013) China's energy and emissions outlook to 2050: perspectives from bottom-up energy end-use model. Energy Policy 53:51–62

Chapter 7
India INDC Assessment: Emission Gap Between Pledged Target and 2 °C Target

P.R. Shukla, Shivika Mittal, Jing-Yu Liu, Shinichiro Fujimori, Hancheng Dai, and Runsen Zhang

Abstract India has pledged to reduce the carbon intensity of GDP by 33–35% in the year 2030 compared to 2005 level in its Intended Nationally-Determined Contributions (INDC). The goal of limiting the global average temperature rise well below 2 °C was reaffirmed in the Paris Agreement adopted under UNFCCC. This chapter assesses three scenarios for India spanning till 2030. First, the reference scenario assumes continuation of the ongoing policies along the conventional path. Second, INDC scenario assumes the successful implementation of INDC decarbonization target. Two-degree (2 °C) scenario assumes an emission constraint aligned with the global of 2 °C stabilization target. The modelling assessment is carried out using a top-down computable general equilibrium AIM/CGE (Asia-Pacific Integrated Model/Computable General Equilibrium) model. The results show that even after full implementation of the INDCs, the emissions would still be 25 billion tonnes of CO_2 equivalent higher than 2 °C stabilization target over the period 2016–2030. Enhanced actions like penetration of renewable technologies, end-use demand management and improvement in energy efficiency could help to close this emission gap.

P.R. Shukla (✉)
Indian Institute of Management, Ahmedabad, India
e-mail: shukla@iimahd.ernet.in

S. Mittal
Chalmers University of Technology, Gothenburg, Sweden

J.-Y. Liu • R. Zhang
Center for Social and Environmental Systems Research, National Institute for Environmental Studies (NIES), 16–2 Onogawa, Tsukuba, Ibaraki 305–8506, Japan

S. Fujimori
Center for Social and Environmental Systems Research, National Institute for Environmental Studies (NIES), 16–2 Onogawa, Tsukuba, Ibaraki 305–8506, Japan

International Institute for Applied Systems Analysis, Schlossplatz-1, Laxenburg 2361, Austria
e-mail: fujimori.shinichiro@nies.go.jp

H. Dai
College of Environmental Sciences and Engineering, Peking University, Room 426, Environment Building, Beijing 100871, China

S. Fujimori et al. (eds.), *Post-2020 Climate Action*,
DOI 10.1007/978-981-10-3869-3_7

Keywords India INDC • Two-degree target • AIM/CGE

7.1 Introduction

Prior to the 2015 Paris Climate Conference (COP21), nations communicated their Intended Nationally Determined Contributions (INDCs) to the UNFCCC which included the emission reduction targets and/or road map of actions following respective national circumstances. The Paris Agreement sets out a goal of limiting the global temperature rise well below two degree Celsius (2 °C) above the pre-industrial level (UNFCCC 2015a). India, the world's third largest emitter, submitted its INDCs to the United Nations Framework Convention on Climate Change (UNFCCC) and signed the Paris Agreement on climate change. India has pledged to reduce the carbon intensity of GDP by 33–35% in the year 2030 compared to the 2005 level. In this chapter, we assess the impact of India's INDCs on energy system. In addition to this, mitigation scenario is developed to understand how well the India's INDCs emission pathway is aligned with pathway pegged to the global 2 °C stabilization target. The scenario assessment is carried out using a top-down computable general equilibrium model (AIM/CGE) (Fujimori et al. 2014).

India's INDCs comprise of varied mitigation and adaption strategies. The mitigation strategy is focused on adding non-fossil fuel-based power generation capacity and enhanced energy efficiency improvements in building and industry sectors (Table 7.1). India has set ambitious mitigation goals to be achieved by 2030, e.g. (i) raising renewable power capacity to 175 gigawatts (GW), (ii) creating an additional carbon sink of 2.5–3 billion tonnes of carbon dioxide through afforestation and (iii) increasing non-fossil fuel-based power generation capacity to 40%.

7.2 Assessment Methodology: Model and Scenarios

7.2.1 Model

7.2.1.1 AIM/CGE

A top-down computable general equilibrium AIM/CGE model is used for scenario assessment. The model structure and formulation of AIM/CGE model is presented in the Chap. 12. The model has been extensively used for estimating the economic impact of climate policies at national level (Tran et al. 2016; Mittal et al. 2016; Dai et al. 2016; Namazu et al. 2013). The assumptions related to socio-economic drivers like population, GDP and emissions constraints are exogenously specified to the model. The modelling assessment spans till 2030 time horizon.

Table 7.1 Mitigation actions in India's INDCs

Sr. no.	Category	Targets
1	Carbon intensity	33–35% reduction by 2030 from 2005 level
2	Non-fossil fuel-based electric power capacity	Increase in total electric power installed capacity from non-fossil fuel to 40% by 2030
3	Wind power capacity	Increase in installed capacity to 60 GW by 2022
4	Solar power capacity	Increase in installed capacity to 100 GW by 2022
5	Biomass power capacity	Increase in installed capacity to 10 GW by 2022
6	Small hydropower capacity	Increase in installed capacity to 5 GW by 2022
7	Nuclear power capacity	Increase in installed capacity to 63 GW by 2032
8	Afforestation	Increase in forest/tree cover to 5 million hectares (mha) and 100 million tonnes CO_2 equivalent carbon sequestration annually
9	Energy efficiency	10% reduction in the current energy consumption by 2019
10	Cess on coal	$2 per tonne of CO_2 emission cess on coal
11	Building sector	Replacement of 200 million incandescent bulbs and reduction of the energy consumption in fans by 50% by penetration of superefficient fan
12	Transport sector	(i) 20% blending of biofuels, both for bio-diesel and bioethanol, by 2017
		(ii) 6–7 million hybrid and electric vehicles per year by 2020
		(iii) Switch from Bharat Stage IV (BS IV) fuels to Bharat Stage V (BS V)/Bharat Stage VI (BS VI) across the country
13	Industry sector	4–5% reduction in specific energy consumption compared to 2012 level using market-based energy efficiency trading mechanism covering eight energy-intensive industrial sectors

Source: UNFCCC (2015b)

7.2.2 *Scenario Description*

7.2.2.1 Reference Scenario

The reference scenario assumes continuation of the current policy dynamics along weak climate policy regime akin to typical business-as-usual (BaU) scenario. The GDP grows at the 8% compounded annual growth rate (CAGR) during the period 2015–2030 similar to the economic growth projection assumed in India's INDC (UNFCCC 2015b). The population projection follows the United Nations median population scenario (UNDP 2015) with population growing at 1.2% CAGR and reaching 1.5 billion by 2030.

7.2.2.2 INDC Scenario

INDC scenario follows the national climate change policies. It assumes the implementation of the upper limit of India's Copenhagen pledge, i.e. 25% reduction in emission intensity by 2020 (GOI 2010), and the upper limit of the INDC pledge, i.e. 35% reduction in emission intensity by 2030 compared to 2005 level. It assumes the implementation of the government initiatives and mitigation strategies listed in India's INDCs. Besides emission intensity reduction target, the scenario also assumes India's implementation strategy to follow additional INDC targets such as increasing the share of non-fossil fuel, i.e. renewables and nuclear-based power generation capacity as well as energy efficiency improvement rate listed in the India's INDC.

7.2.2.3 Two-Degree (2 °C) Scenario

Two-degree scenario considers a stricter climate regime. It assumes more aggressive efforts to reduce greenhouse gas (GHG) emissions in line with the global temperature stabilization target of limiting the temperature below 2 °C. A constraint on GHG emissions is imposed representing the emission allowances allocated to India based on the equal per capita burden sharing scheme. The underlying socio-economic assumptions are similar to reference scenario. The country-specific emission pathways aligned with 2 °C stabilization target are computed by using the convergence approach used by Bows and Anderson (2008) taking 2050 as the convergence year (Eq. 7.1). It implies that all the nations would converge to equal per capita within the same period (2005–2050). The global emission trajectory consistent with the medium probability of limiting the temperature increase below 2 °C is taken from Rogelj et al. (2015). The AIM/CGE model endogenously generates the shadow prices from the constraint placed upon the total GHG emissions in climate stabilization scenario:

$$S_y = \frac{S_0\ (y_c - y) + P_c\ (y - y_0)}{y_c - y_0} \tag{7.1}$$

where y_0 and y_c represent the starting year and convergence year, respectively, S_y is the share of the country's emission in the total emissions in the year y, and S_0 is the share of the country's emission in the total emission in the start year y_0.

7.3 Results and Analysis

7.3.1 *Primary Energy Demand and Energy Mix*

The overall energy demand is expected to increase fivefold from 19 EJ in 2005 to 106 EJ by 2030 in the reference scenario. Compared to the reference scenario, the overall energy demand is 13% lower in 2020 and 43% lower in 2030 in the 2 °C scenario. The energy reduction is driven by the changes in fossil fuel prices in climate constrained scenario resulting in fuel switching and demand reduction. In the INDC scenario, the energy demand is marginally lower (3%) compared to the reference scenario in 2030.

Coal continues to remain the largest energy source in the primary energy mix in all scenarios. Most coal is used for power generation. The share of coal increases from 42% in 2005 to 68% in 2030 in the reference scenario. Targeted mitigation efforts in the INDC and 2 °C scenario lead to diversification of energy sources and contribute to rising share of renewables, gas and nuclear. Coal demand declines substantially in the 2 °C scenario, falling to a third compared to the reference scenario in 2030. The share of gas reaches to nearly a fifth in the 2 °C scenario. In the INDC scenario, the energy transformation is marginal compared to the reference scenario through to 2030. The share of non-biomass renewables increases from 1% in reference scenario to 3% and 7% in the INDC and 2 °C scenario, respectively, contributing mainly to power generation (Fig. 7.1).

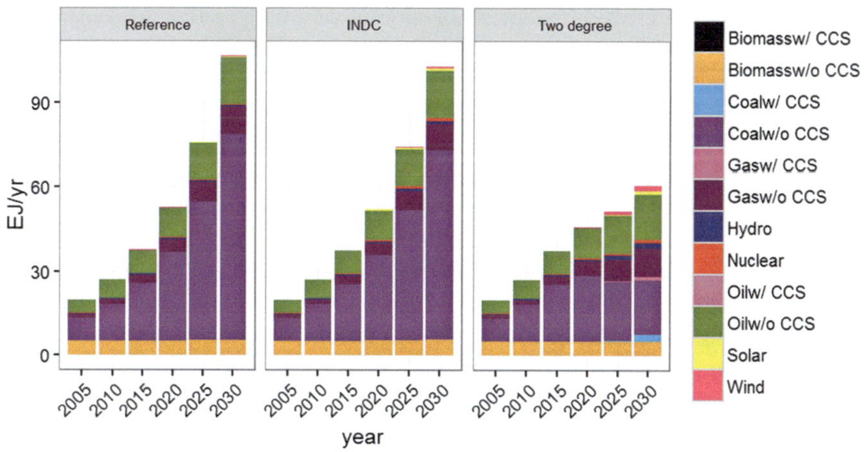

Fig. 7.1 Energy mix in different scenarios

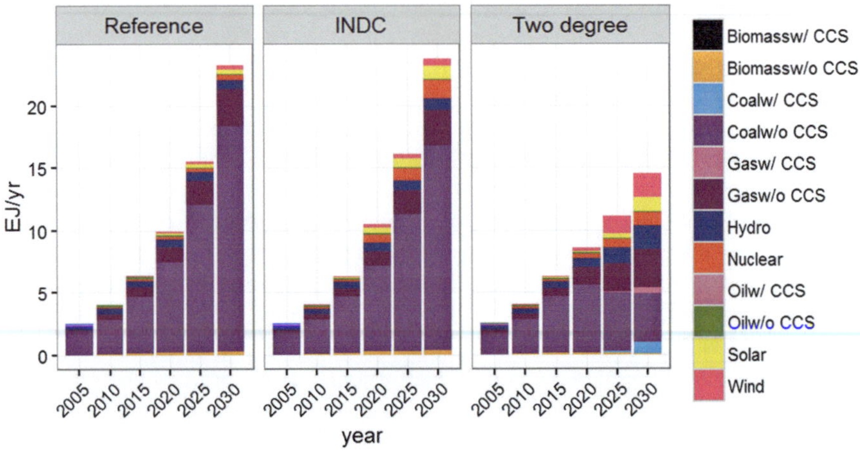

Fig. 7.2 Electricity mix in different scenarios

7.3.2 Power Generation Mix

Driven by rising population, income and ambitious government programs for universal electricity access by 2022, the electricity demand in India increases tenfold from 2.5 EJ in 2005 to 24 EJ in 2030 under reference scenario. Higher penetration of intermittent renewables causes marginal rise in energy consumption by the power sector in the INDC scenario in 2025 due to network losses. In 2 °C scenario, the electricity demand falls by 42% in 2030 compared to the demand in the reference scenario since the internalization of very high carbon price increases electricity prices substantially and reduces electricity demand (Fig. 7.2).

Coal continues to enjoy high share (74%) in the power generation mix in the reference scenario with capacity increasing from 124 GW in 2010 to 800 GW in 2030 (Table 7.2). The gas-based power capacity increases sevenfold from 17 GW in 2010 to 117 GW in 2030. In INDC scenario, the share of renewables in the power generation increases considerably due to the policy push aimed at achieving the decarbonization target pledged in the INDC. The renewable power generation capacity, in 2030 INDC in scenario, increases from 40 GW in 2010 to 287 GW. In the optimal cost 2 °C scenario, the share of electricity generation from coal declines to 34.5%, whereas the share of non-fossil-based power generation increases to 39% in 2030 compared to the reference scenario.

7.3.3 GHG Emissions

GHG emissions are expected to increase from 1824 Mt CO_2 equivalent in 2005 to 9083 Mt CO_2 equivalent in 2030 in the reference scenario (see Table 7.3 for details). GHG emissions in INDC scenario closely follow the reference scenario.

Table 7.2 Electricity generation capacity (GW) in different scenarios

			2030		
		2010	Reference	INDC	2 °C
1	Solar	0.03	22	134	76
2	Wind	11	23	78	169
3	Nuclear	6	60	59	53
4	Coal	124	800	738	216
5	Oil	7	4	4	4
6	Biomass	2	14	17	1
7	Hydro	26	64	58	136
8	Gas	17	117	113	146

Table 7.3 GHG composition across scenarios (Mt CO_2 equivalent/year)

Scenario	Mt CO_2 equivalent	2005	2010	2020	2030
Reference	F gases	5	5	7	10
	CO_2	1257	1825	3830	8004
	CH_4	490	571	736	929
	N_2O	72	83	108	140
INDC	F gases	5	5	7	10
	CO_2	1257	1808	3673	7479
	CH_4	490	571	735	854
	N_2O	72	83	108	117
2 °C	F gases	5	5	7	10
	CO_2	1257	1801	3093	3072
	CH_4	490	571	621	576
	N_2O	72	83	52	47

In the 2 °C scenario, India's GHG emission would require to peak by 2020 to align with the global climate stabilization target. As Fig. 7.3 shows, India's GHG emissions in 2 °C scenario will have to immediately stabilize and thereafter remain steady to a level which is far below the INDC target of 35% emission intensity reduction. The strict GHG emission cap under 2 °C scenario is achieved by significant reduction in final energy demand and efficient use of energy and rapid shift towards low-carbon fuels and technologies like renewables and nuclear.

Under reference scenario, the power sector in 2030 accounts for 68% of energy-related CO_2 emissions, and 20% comes from the industry sector (Fig.7.4). In 2030, 45% emission reduction in INDC scenario is contributed by the power sector via enhanced use of non-fossil-based power generation. The electricity sector is highly decarbonized in the 2 °C scenario. Nearly 65% reduction is contributed by the electricity sector through fuel switching and 19% by the industry sector through efficiency improvement and demand reduction. The transportation sector contributes to 3% direct emission reduction in 2030 due to demand reduction and fuel switching.

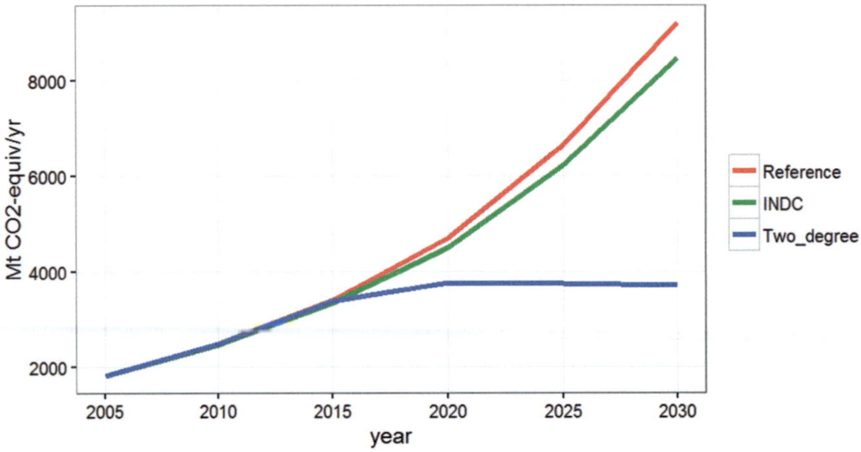

Fig. 7.3 Emission trajectories in different scenarios

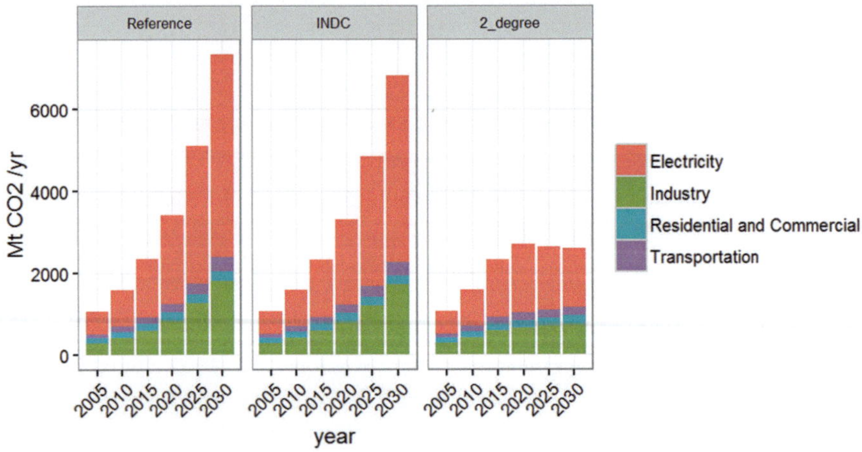

Fig. 7.4 Energy-related emission across scenarios

7.3.4 Cost of Mitigation

The assessment of INDC emission pathway assumes a policy push for varied measures announced in the INDC document such as targets for deployment of renewable energy, demand-side push for energy efficiency, etc. The assessment of 2 °C emission pathway assumes optimal market responses to achieve the long-term 2 °C temperature stabilization target. This section presents the implication of exogenous targets on the economic parameters such as GDP loss relative to GDP

Table 7.4 Implication of 2 °C scenario

2 °C scenario				
Parameter	Unit	2020	2025	2030
GDP loss	%	0.7%	2.4%	4.9%
Carbon price	US$2005/t$CO_2$	15.5	67.5	145.8
Electricity price	US$2005/GJ	16	24	31
Difference w.r.t BaU	US$2005/GJ	4	12.6	20

in the reference scenario and price of tonnes of mitigated carbon and the price of electricity. The GDP losses in 2 °C scenario reach 0.7% in 2020 and rise to nearly 5% in 2030. The AIM/CGE model generates carbon prices endogenously given the emission constraints. The carbon price starts at $15 per t$CO_2$ in 2020 to $146 per t$CO_2$ in 2030 under 2 °C scenario (see Table 7.4). The emissions prices are internalized in the electricity final prices. The results show that electricity prices increase substantially in the 2 °C scenario (see Table 7.4). This leads to reduction in energy consumption as well as switch to cleaner energy sources.

7.3.5 *Emission Intensity*

GHG intensity measured in terms of emission per unit GDP is a widely used indicator for assessing the coupling of GHG emissions and economic growth across countries as well as across scenarios within the same country. The emission reduction targets in India's INDCs are specified in terms of emission intensity reduction of 33–35% in the year 2030 compared to 2005. The results of our INDC scenario targeting 25% reduction (Copenhagen pledge) in the year 2020 and 35% reduction (INDC pledge) in 2030 show that as of now, i.e. in 2015, India's emission intensity reduction trajectory is in line with these pledges. In case of INDC scenario, the 35% emissions reduction (Fig. 7.5) is contributed mainly by energy intensity reduction of 27% between 2005 and 2030. The remaining 8% reduction is contributed by change in fuel and technology mix. Notably, the difference in emission intensity in 2030 between reference scenario and INDC scenario is relatively small.

Under 2 °C scenario, the GHG emission intensity needs urgent and steep reduction leading to 70% decline compared to 2005 by the year 2030. Evidently, in the 2 °C scenario, the strong decoupling between growths of GHG emissions and GDP is contributed more by the rapid change towards low-carbon fuel and technology mix (Fig. 7.5) and to a relatively lesser extent by the energy intensity of GDP (Fig. 7.6). Importantly, the decline in GHG intensity of India in 2030 compared to 2005 will have to be 70% to remain aligned to optimal global actions consistent with 2 °C scenario.

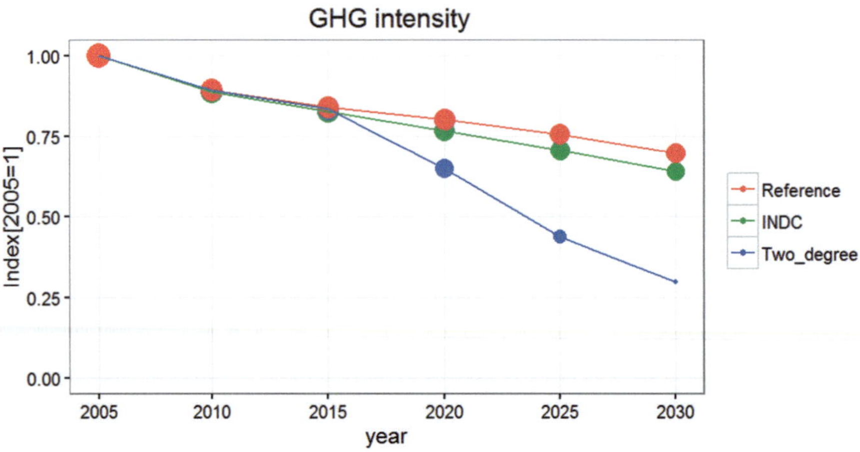

Fig. 7.5 GHG intensity in different scenarios

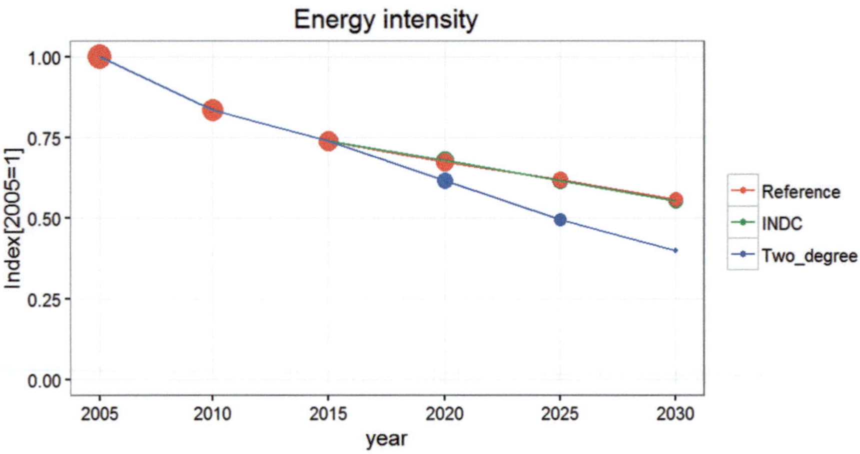

Fig. 7.6 Energy intensity in different scenarios

7.4 Discussion and Conclusion

The results of the study show that reduction in the GHG emissions in INDC would help to mitigate 5 billion tonnes of CO_2 equivalent aggregate emissions compared to reference scenario over the period 2016–2030. The rate of emission reduction in the INDC scenario is slow compared to what could be India's response under an optimal global mitigation regime targeting to restrict global warming to below 2 °C compared to pre-industrial levels. Even after achieving the government targets envisioned in India's INDCs, the emission gap of 25 billion tonnes of CO_2

equivalent over the period 2016–2030 would still remain between the emission reduction pledged in INDC and the 2 °C stabilization target. Filling this gap would need directing enhanced financial flows to support global and national innovations and transfer of low-carbon technologies that deliver mitigations in India at global levels that are compatible with global carbon prices. India's INDCs have placed on table a road map of actions which can support myriad policy initiatives such as feed-in tariffs, renewable purchase obligations and links with global carbon trading markets which can attract private investments towards low-carbon activities. Our modelling assessment shows that the difference between reference and INDC emission pathways is relatively small. The incremental finance and functioning technology transfer regime for low-carbon technologies are therefore vital for any upward revision of INDC decarbonization target which can close the gap with mitigation efforts required in India that are in line with optimal global response for achieving 2 °C target.

The scenario assessment shows that, in the INDC as well as the 2 °C scenario, the investments in renewable energy, end use demand management and energy efficiency are vital constituents of low-carbon action portfolio in the short run. In 2 °C scenario, the energy consumption declines as rise in penetration of energy-efficient technologies is driven by the increase in fossil fuel prices which has to absorb the rising carbon prices. The shares of renewable and nuclear energy therefore rise in the energy mix. The policy reforms like removal of fossil fuel subsidies would further help in bringing down the emissions in the short term and reduce the emission gap between the INDC and two degree targets.

The analysis in the present paper spanned over the INDC period till 2030. The initial results of long-term scenario analysis from our related study spanning the period till 2050 show that longer time horizon is crucial to align short-term and long-term mitigation actions for limiting the temperature rise to below 2 °C level. The analysis in that study assumes that countries, including India, are allocated emissions rights for each year from 2016 to 2050 following per capita emissions rights principle. The emission budget for a country is then the fraction of cumulative global emission budget to meet 2 °C target over the period 2016–2050. In case of 2 °C scenario, as Fig. 7.7 shows, if India's emission pathway till the year 2030 is locked in to INDC pathway, then the annual emission reduction rate post-2030 will have to be unrealistically steep to remain within the cumulative emissions budget.

Evidently therefore, the determination of optimal emission pathway would require long-term study that assesses the technical feasibility and economic implications of achieving high-emission reduction rates over longer time horizon rather than limiting assessment to the INDC period terminating in 2030. The wisdom therefore lies in raising ambition at the earliest, e.g. during the first stocktake in 2018 as per the Paris Agreement (UNFCCC 2015a) rather than wait till 2030. Urgency to raise ambition and related actions is even greater in case the 1.5 °C target is aimed at. The analysis in this paper demonstrates the urgent need to assess the emission gap between the INDCs and 2 °C as well as 1.5 °C emission pathway and use the analysis to identify key areas on which to focus ambitious mitigation strategies so as to close the otherwise impending emission gap.

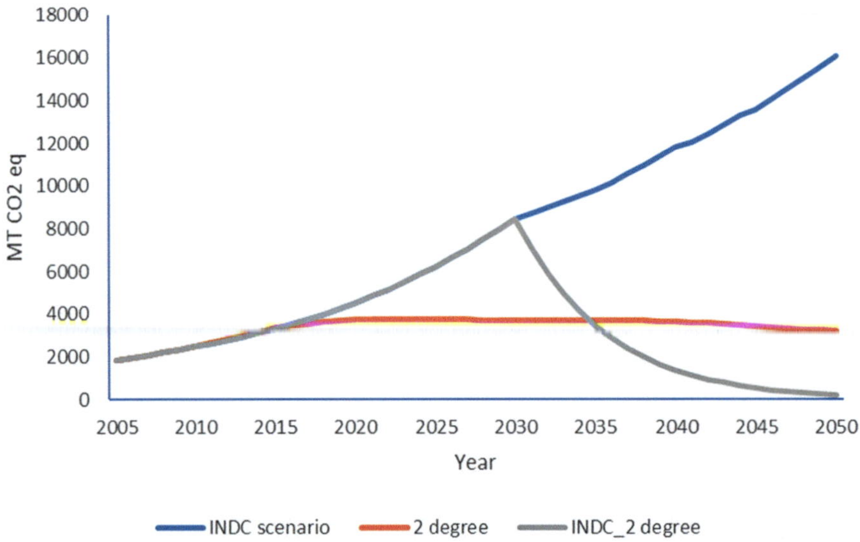

Fig. 7.7 Long-term GHG emissions in different scenarios

References

Bows A, Anderson K (2008) Contraction and convergence: an assessment of the CCOptions model. Climate Change 91:275–290. doi:10.1007/s10584–008-9468-z

Dai H, Xie X, Xie Y, Liu J, Masui T (2016) Green growth: the economic impacts of large-scale renewable energy development in China. Appl Energy 162:435–449. doi:10.1016/j.apenergy.2015.10.049

Fujimori S, Masui T, Matsuoka Y (2014) Development of a global computable general equilibrium model coupled with detailed energy end-use technology. Appl Energy 128:296–306. doi:10.1016/j.apenergy.2014.04.074

GOI (2010) Letter from Ministry of Environment and Forests United Nations Framework Convention on Climate Change

Mittal S, Dai H, Fujimori S, Masui T (2016) Bridging greenhouse gas emissions and renewable energy deployment target: Comparative assessment of China and India. Appl Energy 166:301–313. doi:10.1016/j.apenergy.2015.12.124

Namazu M, Fujimori S, Shukla PR, Matsuoka Y (2013) Two low-carbon development pathways in India. Glob Environ Res, 17: 119–128

Rogelj J, Luderer G, Pietzcker RC, Kriegler E, Schaeffer M, Krey V, Riahi K (2015) Energy system transformations for limiting end-of-century warming to below 1.5 °C. Nat Clim Chang 5(6):519–527. doi:10.1038/nclimate2572

Tran T, Fujimori S, Masui T (2016) Realizing the intended nationally determined contribution: the role of renewable energies in Vietnam. Energies 9(8):587. doi:10.3390/en9080587

UNFCCC (2015a) Adoption of the Paris Agreement. United Nations Office at Geneva, Geneva

UNFCCC (2015b) India's Intended Nationally determined contribution: working towards climate justice United Nations Framework Convention Climate Change

United Nations, Department of Economic and Social Affairs, Population Division (2015) World population prospects: the 2015 revision, methodology of the United Nations population estimates and projections, working paper No. ESA/P/WP.242

Chapter 8
An Assessment of Indonesia's Intended Nationally Determined Contributions

Shinichiro Fujimori, Ucok W.R. Siagian, Tomoko Hasegawa, Bintang B. Yuwono, Rizaldi Boer, Gito Immanuel, and Toshihiko Masui

Abstract This chapter provides an overview of Indonesia's current economy, energy sector, land use, and climate policies. We assessed Indonesia's Intended Nationally Determined Contributions (INDC) using an Asia-Pacific Integrated Model/Computable General Equilibrium model coupled with an agriculture, forestry, and other land use (AFOLU) model. The model shows that the emission reduction target of INDC can be achieved at low economic cost (less than 1% of GDP) and that the mitigation actions required would not harm economic development. While emissions from land use and land use change are high nowadays, the energy sector is expected to grow rapidly and become more important in the future. Therefore, climate-related policymakers should focus equally on land use and the energy sector in the future.

Keywords Indonesia • INDC • Energy • Land use • Mitigation • CO_2 emissions

S. Fujimori (✉) • T. Hasegawa
Center for Social and Environmental Systems Research, National Institute for Environmental Studies (NIES), 16–2 Onogawa, Tsukuba, Ibaraki 305–8506, Japan

International Institute for Applied Systems Analysis, Schlossplatz-1, Laxenburg 2361, Austria
e-mail: fujimori.shinichiro@nies.go.jp

U.W.R. Siagian • B.B. Yuwono
Center for Research on Energy Policy, Institut Teknologi Bandung, Jln Ganesha 10, Bandung, Indonesia

R. Boer • G. Immanuel
Center for Climate Risk and Opportunity Management in Southeast Asia Pacific (CCROM-SEAP), Bogor Agricultural University, Bogor 16143, Indonesia

T. Masui
Center for Social and Environmental Systems Research, National Institute for Environmental Studies (NIES), 16–2 Onogawa, Tsukuba, Ibaraki 305–8506, Japan

S. Fujimori et al. (eds.), *Post-2020 Climate Action*,
DOI 10.1007/978-981-10-3869-3_8

8.1 Introduction

Indonesia has pledged to pursue a low-carbon development strategy in support of the global effort to limit mid-century global average temperature increase to 2 °C. In its Intended Nationally Determined Contributions (INDC) (Indonesia 2016), Indonesia pledged unconditionally to reduce its emissions by 29% below the baseline or business-as-usual (BAU) levels by 2030. This reduction could be increased to 42% with international support (conditional pledge). As the second largest greenhouse gas (GHG) emitter in the country, the energy sector is expected to contribute significantly to emission reductions. The energy sector's targets for 2030 are 17.5% below baseline emission levels for the unconditional pledge and up to 32.7% for the conditional pledge, amounting to reductions of 253 megatonnes (Mt) of carbon dioxide equivalent (CO_2e) and 472 MtCO_2e, respectively.

Indonesia has various energy sector mitigation options to achieve its target, namely, efficiency measures, deployment of renewable energy, nuclear power, and clean coal technology, including carbon capture and storage (CCS). These options in combination are referred to as the "low-carbon energy system." Energy development with these mitigation options is generally more expensive than under baseline conditions (without mitigation actions). In a country with limited wealth such as Indonesia, the economic impact of any mitigation action is an important factor for its selection and eventual implementation.

Indonesia has a unique GHG emission structure. The total GHG emissions were 1.8 gigatonnes (Gt) CO_2e in 2005, approximately 65% of which derived from the agriculture, forestry, and other land use (AFOLU) sectors. The INDC (Indonesia 2016) shows that Indonesia has taken significant steps to cut emissions from land use sectors by reducing deforestation and forest degradation, restoring ecosystem functions, and supporting sustainable forest management, including social forestry, through active participation of the private sector, small and medium enterprises, civil society organizations, local communities, and the most vulnerable groups. Reaching the emission target requires quantitative evaluation and highly efficient mitigation countermeasures, both of which must be prioritized. Meanwhile, energy-related emissions are predicted to increase rapidly due to economic development and industrialization. This begs the fundamental question of how the energy and AFOLU sectors will contribute to the emission reduction target.

In this chapter, we assess Indonesia's INDC using an Asia-Pacific Integrated Model/Computable General Equilibrium (AIM/CGE) model coupled with an AFOLU model. We quantify and compare the energy system and land use situations under baseline and INDC achievement scenarios. Finally, the policy implications of the results are discussed.

8.2 National Circumstances

8.2.1 General National Circumstances

The Republic of Indonesia is the largest archipelago in the world, bridging Asia and Oceania between the Pacific and the Indian Oceans. It consists of approximately 17,000 islands with a population of 250 million people. The majority (almost 80%) of Indonesians live in the western part of the country on the islands of Java and Sumatra (see Fig. 8.1).

About 50 million of Indonesia's 200 million hectares (ha) are devoted to agricultural activities. There are nearly 20 million ha of arable land, of which about 40% is wetland (e.g., rice fields), 40% is dry land, and 15% is used for shifting cultivation.

Unemployment in Indonesia is still relatively high, and poverty remains a challenge. Nevertheless, employment in Indonesia has improved over the past 8 years, decreasing from around 10% in 2004 to around 6% in 2013. In 2014, about 27.7 million people (11% of the population) in Indonesia were considered to live in poverty. According to the country's Medium-Term Development Plan (Bappenas 2014), the government plans to implement various development and welfare programs to reduce the poverty rate to 6.5–8.0% of the population by 2019.

More than 136 million people (57% of the population) live on the island of Java and around half inhabit urban areas; provinces with >50% of their inhabitants living in urban areas include DKI Jakarta (100%), Riau (83%), Banten, Yogyakarta, and West Java (>60%).

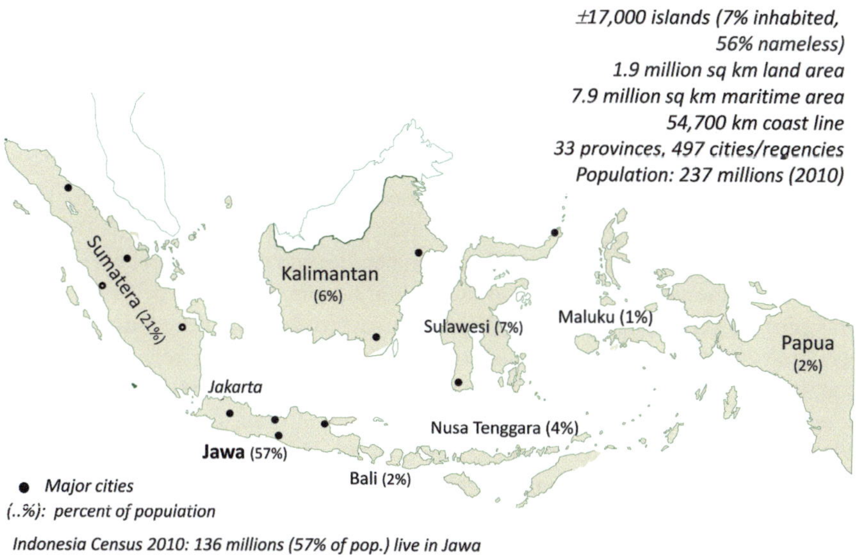

Fig. 8.1 Map of Indonesia with basic statistics

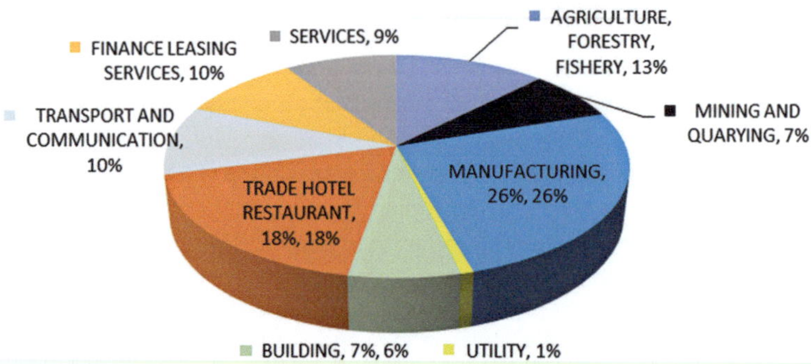

Fig. 8.2 Distribution of Indonesian value added by sectors in 2012

Indonesia's economy has grown rapidly in the last decade. Its gross domestic product (GDP) increased from 2300 trillion Indonesian rupiah (IDR; USD 248 billion) in 2004 to IDR 10,543 trillion (USD 888 billion) in 2014. This corresponds to a per capita GDP increase from IDR 10.5 million (USD 1132) in 2004 to IDR 41.8 million (USD 3632) in 2014. During this period, Indonesian economic growth averaged 5.7% per annum (p.a.). Over the last 60 years, Indonesia's economy has undergone a structural transformation from a largely agriculture-based economy to a largely industrial and service economy. Figure 8.2 shows the aggregate economic structure (in %GDP) in 2012; industry and services accounted for 87% of the economy. The major contributors to the industrial sectors were manufacturing, mining and extraction, and construction, while trading, hotels, restaurants, finance, real estate, transport, and telecommunication were the major contributors to the commerce and service sectors.

Indonesian life expectancy at birth has improved significantly in the past four decades from only 47.9 years in 1970 to 69.7 years in 2011. Statistik Indonesia (Negara 2015) estimated that life expectancy at birth increased to 70.1 years during 2010–2015. Following sustained efforts, the adult literacy rate increased significantly from 79% in 1970 to 95% in 2011.

Prior to 1999, Indonesia had been successful in alleviating poverty. In 1970, 60% of the population (70 million people) was living in absolute poverty. By 1990, that number had dropped to 15% (27 million people) and continued to improve until 1997, when the figure decreased to 20 million people. However, in 1999, for the first time in years, Indonesia experienced a severe 18-month drop in social and economic conditions, resulting in over 100 million people living below the poverty line. Following the economic and political reforms starting in 2000, the number of people living in poverty had dropped to 27.7 million (11% of the population) by 2014. According to the Medium-Term Development Plan (Bappenas 2014), the government of Indonesia plans to implement various development and welfare programs to reduce the poverty rate to 6.5–8.0% by 2019. The employment rate has also been improving over the past 8 years, decreasing from about 10% in 2004 to about 6% in 2013.

Table 8.1 Indonesia's energy resources

Energy resources	Reserves	Resources
Oil, billions of barrels	7.4	
Natural gas, TSCF	150	
Coal, billions of tons	29	119
CBM, TSCF		453
Shale gas, TSCF		574
	Potential power	
Hydro	75,000 MW	
Geothermal	29,000 MW	
Micro-hydro	750 MW	
Biomass	14,000 MWe	
Solar	4.80 kWh/m^2/day	
Wind	3–6 m/s	

Values are averages for the whole country (Nasional 2015)
TSCF trillion standard cubic feet

8.2.2 Energy Resources and Trends

Indonesia has a wealth of energy resources (Table 8.1); oil, gas, and coal have been exploited intensively to supply domestic and foreign markets. Prior to 2004, Indonesia was a net oil exporter and a member of OPEC. However, increasing demand and diminishing reserves have made Indonesia a net oil importer since 2004. Indonesia also has significant renewable resources (Table 8.1). The use of these resources remains limited because oil and electricity have been heavily subsidized, which has prevented renewable resource prices from being competitive.

The energy sector is an important driver of economic activity and also generates government revenues through the sale of natural resources to domestic and foreign markets, royalties, and taxes. A schematic of energy flow in Indonesia is shown in Fig. 8.3.

Energy consumption has been growing in line with economic and population growth. Between 2000 and 2013, the total final energy demand grew on average 3.6% p.a., from 104 megatonnes of oil equivalent (Mtoe) to 164 Mtoe. The development of Indonesia's energy supply and demand during 2010–2013 (Pusdatin-ESDM 2013) is summarized in Figs. 8.3, 8.4, 8.5, 8.6 and 8.7. Industrial, residential, and transport sectors dominated the final energy consumption (Fig. 8.3). High consumption growth occurred in transport (6.9% p.a.) and commercial (4.6% p.a.) sectors, which had higher growth than industrial (1.8% p.a.) and residential (0.9% p.a.) sectors. Energy demand is still dominated by oil, which accounts for around 37.0% of the total consumption, followed by biomass (26.1%), gas (15.6%), coal (11.4%), and electricity (9.9%) (Fig. 8.4). High growth in demand occurred for coal (10.8% p.a.), electricity (6.8% p.a.), and gas (4.8% p.a.). High growth in gas demand was due to a policy shift from kerosene to liquefied petroleum gas (LPG) subsidies in the residential sector. Coal consumption occurred solely in the industrial sector. The high growth in coal was due to the removal of the industrial diesel subsidy, resulting in industries switching from diesel to coal.

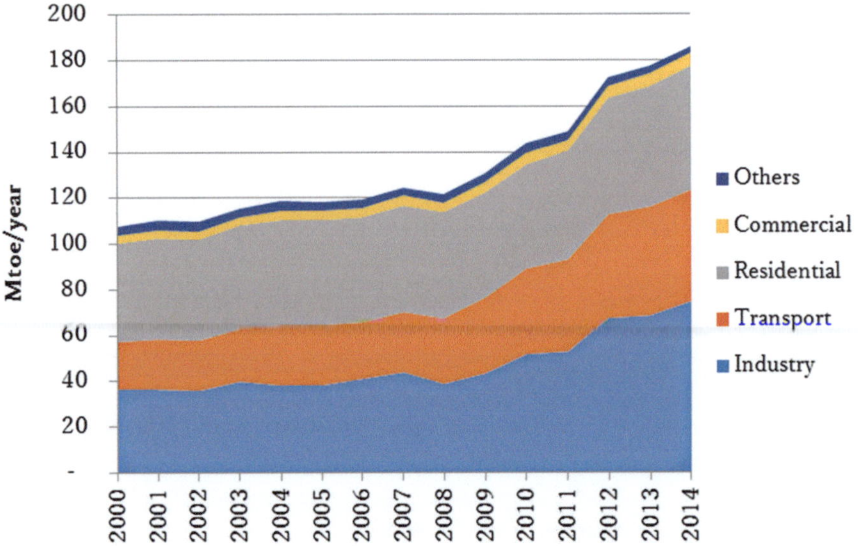

Fig. 8.3 Development of final energy demand by sector from 2000 to 2013

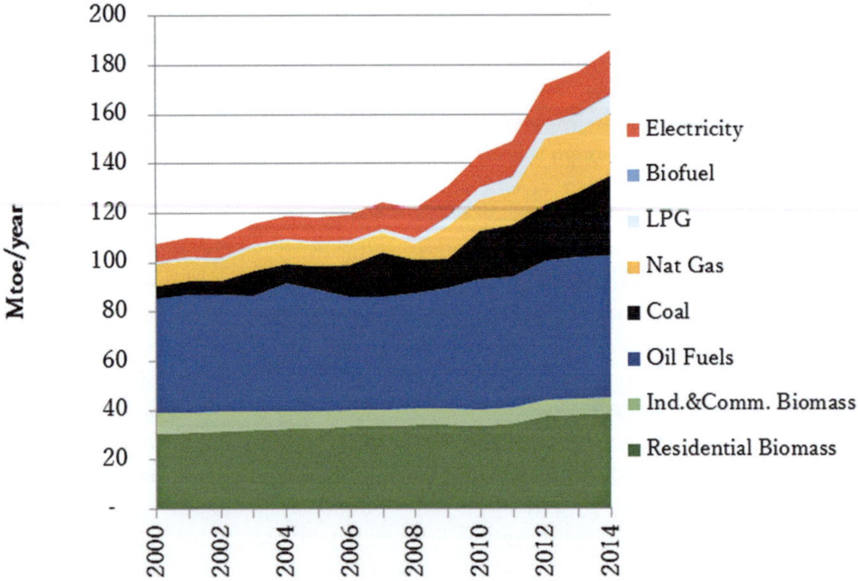

Fig. 8.4 Development of final energy demand by fuel type from 2000 to 2013

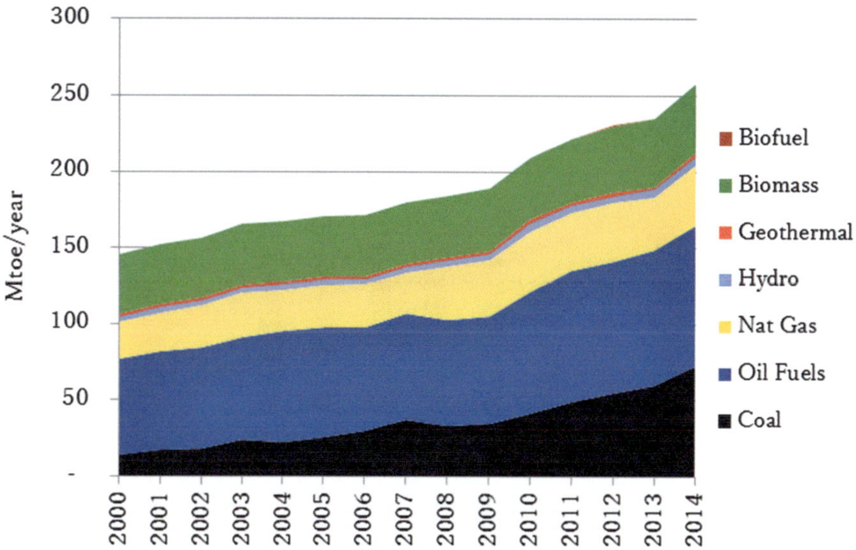

Fig. 8.5 Development of primary energy supply from 2000 to 2013

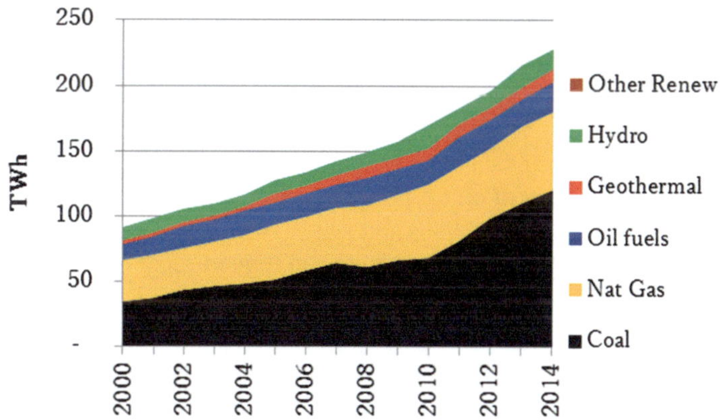

Fig. 8.6 Development of power generation mix from 2000 to 2013

Between 2000 and 2013, the primary energy supply grew 3.8% p.a., from 145 to 236 Mtoe. As shown in Fig. 8.5, the primary energy supply has been dominated by oil, followed by coal and natural gas. As the domestic oil production capacity continues to decline, the government has been attempting to move away from oil by promoting energy sources that are abundantly available in the country, for example, coal, natural gas, and renewables. These attempts have resulted in high growth in the coal supply (11.5% p.a.), which is significantly higher than those of oil (2.8% p.a.) and natural gas (3.9% p.a.). These growth rates have reduced oil's

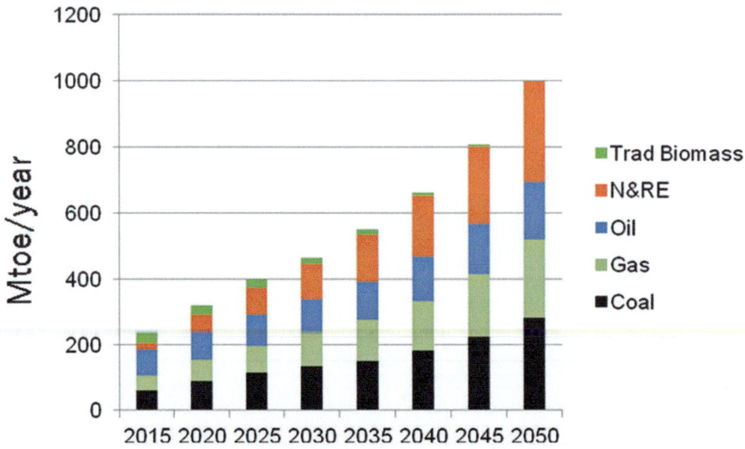

Fig. 8.7 National Energy Council's energy supply scenario (Nasional 2015); N&RE represents nuclear and renewable energy sources

share of the supply mix from 41% in 2000 to 38% in 2013 and increased that of coal from 9% in 2000 to 22% in 2013.

Around 10% of final energy consumption was in the form of electricity. The electricity demand was met by different types of power plant (coal, gas, hydropower, geothermal, and oil; Fig. 8.6). Over the past 13 years, geothermal and coal plants experienced high annual growth (10% and 9% p.a., respectively). The high growth rate of coal plants has increased the share of coal in the power generation mix from 37% in 2000 to 51% in 2013. Despite the high growth rate of geothermal energy, its relative share in the power generation mix remains small; it was 4.4% in 2013. The main consumers of oil for power generation are diesel plants installed to boost rural electrification in the 1980s and 1990s. Attempts to substitute these diesel plants with other types of power plant have had limited success, and therefore the share of electricity from oil remains high; it was 20% in 2013.

The future challenge for the energy sector is to meet rising energy demands of a growing economy and population by (i) improving energy efficiency to reduce energy demand growth and (ii) decarbonizing the energy supply. In 2014, the National Energy Council (DEN) endorsed the National Energy Policy to guide the development of Indonesia's energy. The main features of the energy policy are as follows:

- Strive for energy security: move away from oil by reducing it to 19.5% of the supply by 2050; promote abundantly available resources, such as natural gas and coal; and prioritize the use of domestic resources to satisfy the domestic market.
- Increase energy efficiency: target energy elasticity of <1 by 2025.
- Promote the development of renewable energy: target of 31% of the supply by 2050.

The Indonesia's energy policy heavily emphasizes energy security and independence as its ultimate objectives. Therefore, the Indonesian energy demand is to be satisfied using domestic resources as much as possible. With the current Indonesian policy framework, the Indonesian National Energy Council has developed a long-term energy supply scenario (Fig. 8.7). Figure 8.7 suggests that, despite high growth in new and renewable energy, Indonesia's primary energy supply will continue to be dominated by fossil energy for the next three decades.

The high share of fossil energy consumption in Indonesia and the coupling of energy and economic growth drive the energy sector's significant contributions to Indonesia's GHG emissions, which are second only to land use, land use change, and forestry (LULUCF). As will be shown in the next subsection, energy has been the fastest-growing GHG emitter among all sectors. To contribute to global climate change mitigation efforts, Indonesia is expected to decarbonize its future energy sector.

8.2.3 Current Agriculture and Land Use

Given its rapid income growth and steady population increase, Indonesia's crop production has increased over the last few decades. The major cereal produced in Indonesia is rice, which accounts for around 80% of production (Fig. 8.8a). Palm oil production has increased dramatically (Fig. 8.8b), which has resulted in considerable land use changes, including rain forest and peatland conversion. As a result, land use-related CO_2 emissions currently dominate Indonesia's GHG emissions. The main driver of the palm oil expansion is world demand for food and the oleochemical and energy industries (Wicke et al. 2011).

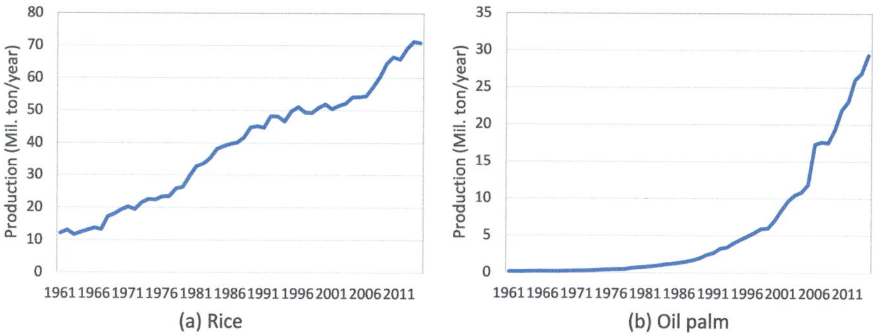

Fig. 8.8 (**a**) Rice and (**b**) oil palm production in Indonesia (FAO 2016)

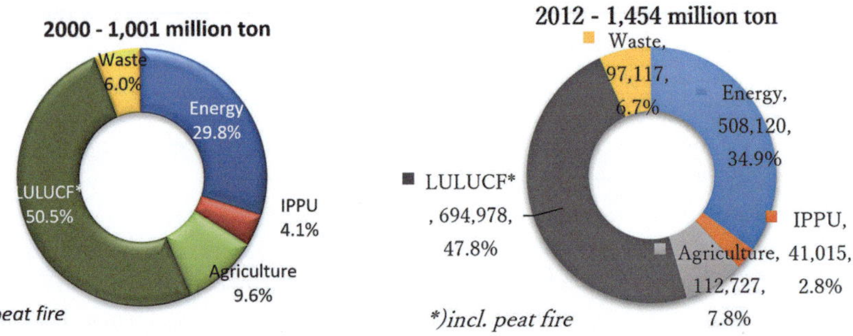

Fig. 8.9 Distribution of Indonesia GHG emissions in 2000 and 2012

8.2.4 *Current GHG Emissions*

In 2012, according to the latest Indonesia GHG inventory (Kehutanan 2015), Indonesia emitted around 1453 $MtCO_2e$, which was 453 $MtCO_2e$ higher than in 2000. The major contributors were LULUCF and peat fires, which together accounted for 48% of the emissions, and energy activities, which contributed 35% of the emissions (Fig. 8.9). The highest growth occurred in the energy sector (4.5% p.a.) followed by the waste sector (4% p.a.) and LULUCF (2.7% p.a.). At such rates, the energy sector is expected to become the largest GHG emitter in Indonesia.

8.3 Methods

8.3.1 *Model*

An AIM/CGE model was used for energy sector computations (Fujimori et al. 2017). For the agriculture and land sector simulations, we used a coupled national AIM/CGE and bottom-up type of AFOLU technology model after Hasegawa et al. (2016) (Fig. 8.10).

In the AIM/CGE national models (Fujimori et al. 2012), supply, demand, investment, and trade were described with individual behavioral functions that respond to changes in the price of production factors and commodities, as well as changes in technology and preference parameters based on assumed population, GDP, and consumer preferences.

The bottom-up type of AFOLU model was developed for assessing abatement costs and emission reduction in these sectors at a country-wide level (Hasegawa and Matsuoka 2015) and has been previously applied to other Asian countries (Hak et al. 2015; Jilani et al. 2015; Nguyen et al. 2014). The model calculates GHG emission reductions and costs of individual mitigation measures at cost minimization using a given carbon price and assumed future agricultural production and land

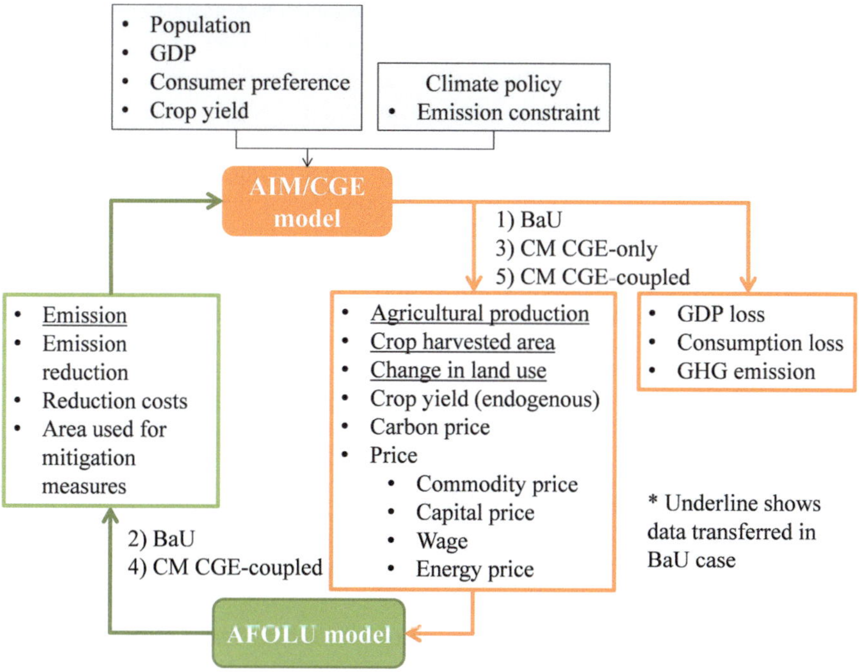

Fig. 8.10 Coupling scheme of the Asia-Pacific Integrated Model/Computable General Equilibrium (AIM/CGE) and agriculture, forestry, and other land use (AFOLU) models. Baseline, business-as-usual; *CM* countermeasure, *GHG* greenhouse gas, *GDP* gross domestic product (Hasegawa et al. 2016)

use changes. The model incorporates detailed information about mitigation measures using the following inputs: (1) future assumptions about agricultural production and area of land use change in a baseline case, where emission reduction is not strongly addressed; (2) information on mitigation measures (e.g., cost, reduction effects, and lifetime); and (3) carbon tax policy scenarios, which represent a willingness for emission reduction. Future production level was fixed, and cropland area and yields were calculated to meet the given agricultural production levels. To reflect long-term mitigation costs and effects, information about reduction effects, costs, and area used for the mitigation measures implemented within a given year were passed on to the next year (see (Hasegawa and Matsuoka 2015) for more details regarding the AFOLU model).

8.3.2 Scenario

To investigate the pathway to achieve INDC targets and the possibility of reaching deeper emission reductions, two scenarios were developed: "baseline" and "countermeasure" (CM). The baseline scenario projects Indonesia's current energy development and emissions assuming no additional climate change mitigation actions. The CM scenario models the developments required to achieve the INDC target, namely, 29% reduction by 2030. This target was applied only to the energy sector. The success of mitigation actions was assessed as the difference between the baseline and CM scenarios. We used the carbon price from this assessment in the AFOLU model and explored the mitigation potential of the AFOLU sector. The assumed economic (GDP) and population growth rates were 5% and 1.1%, respectively. The fuel share of the power sector was assumed to follow the utility's expansion plan (PLN 2016).

8.4 Results

8.4.1 Overall Emission Reduction

The simulation results under the baseline and CM scenarios are summarized in Fig. 8.11. In the baseline scenario during 2005–2030, the Indonesian population is projected to increase 1.3-fold, while GDP is projected to increase 4.3-fold. Over the

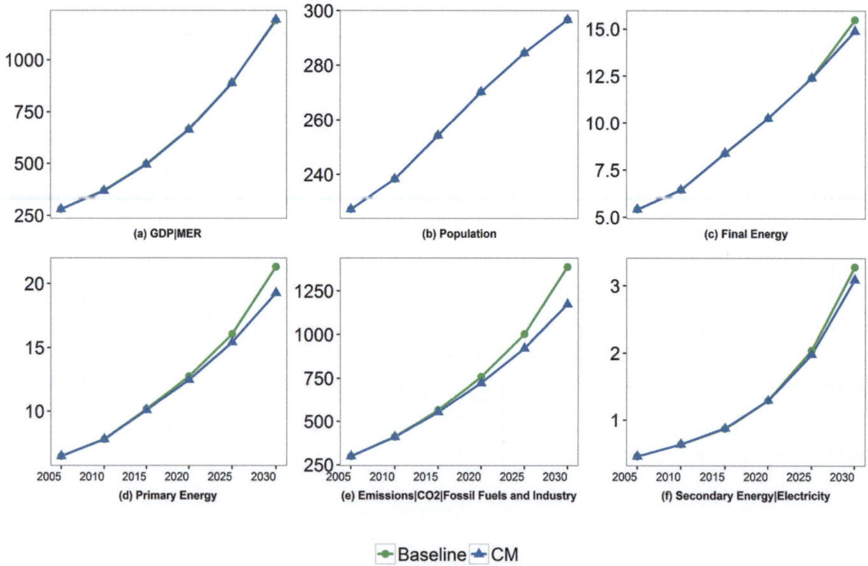

Fig. 8.11 Results for the baseline and countermeasure (CM) simulations from 2005 to 2030. (**a**) Projected GDP at market exchange rate (MER), in million USD; (**b**) population, in millions; (**c**) final energy consumption, in EJ/year; (**d**) primary energy supply, in EJ/year; (**e**) greenhouse gas emissions, in $MtCO_2$/year; and (**f**) power generation, in EJ/year

same period, energy demand is expected to increase 2.9-fold while energy supply is expected to increase 3.2-fold. Owing to expansion of the power supply infrastructure and the increase in wealth, electricity demand is expected to increase 7.2-fold. The increase in energy demand and supply and the type of fuels used in the energy system will result in a 4.5-fold increase in GHG emissions.

The CM scenario was similar to the baseline scenario in all indicators. Population was exactly the same between the two scenarios and GDP nearly so. The primary energy supply, electricity generation, and final energy demand were slightly lower than baseline levels due to the emission reduction.

8.4.2 Energy System

The projections for final energy demands under the baseline and CM scenarios are shown by sector in Fig. 8.12. As expected for a developing country, the fastest-growing energy consumers in the baseline scenario were the industry and transport sectors, which were projected to grow by 5.4% and 5% p.a., respectively. By 2030,

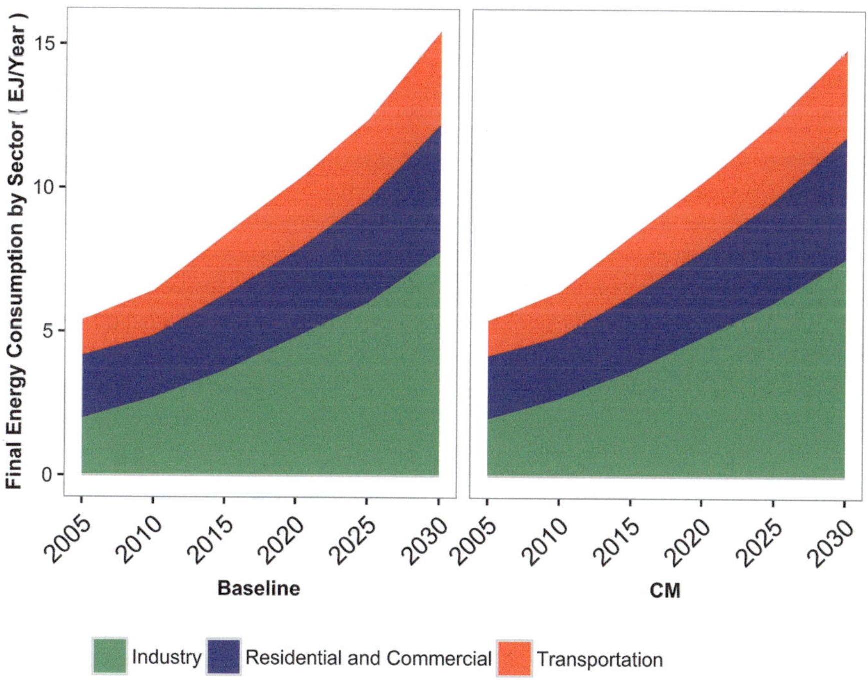

Fig. 8.12 Projected total energy demands by sector for the baseline and countermeasure (CM) scenarios from 2005 to 2030

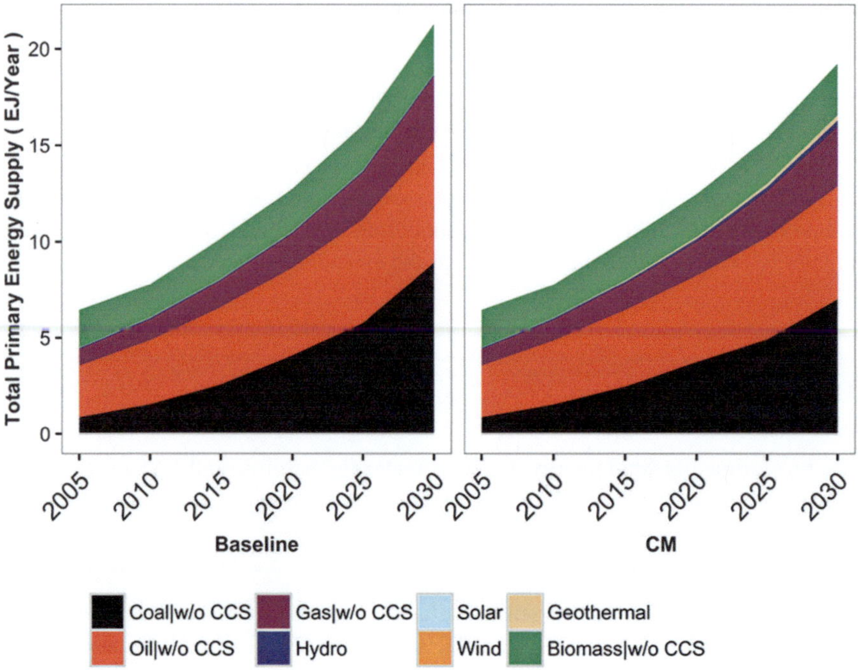

Fig. 8.13 Projected primary energy supply by resource in the baseline and countermeasure (CM) scenarios from 2005 to 2030

these two sectors will account for 70% of the total energy demand. The CM scenario showed similar results.

The projected development of primary energy under the baseline and CM scenarios is shown in Fig. 8.13. High supply growth is expected for coal and natural gas, which were predicted to increase by 8.7% and 5.5% p.a., respectively. Despite lower growth compared with coal and gas, the share of oil in the primary energy supply will remain high. By 2030, the share of oil in the primary energy mix was projected to be 36%. The growth of renewable energy development was projected to remain low. By 2030, the share of all hydro, geothermal, wind, and solar energies combined will be $< 2\%$ in terms of total primary energy. Biomass users in rural residential areas, however, were expected to remain significant. In 2030, the share of biomass in the total primary energy supply will be around 11%. The CM scenario predictions were almost identical but differed in a few key areas. The share of renewable energy increased while that of coal decreased. This was due mainly to electricity generation system changes (Fig. 8.14). Geothermal and hydropower in the CM scenario increased remarkably compared with those in the baseline scenario, while coal power decreased.

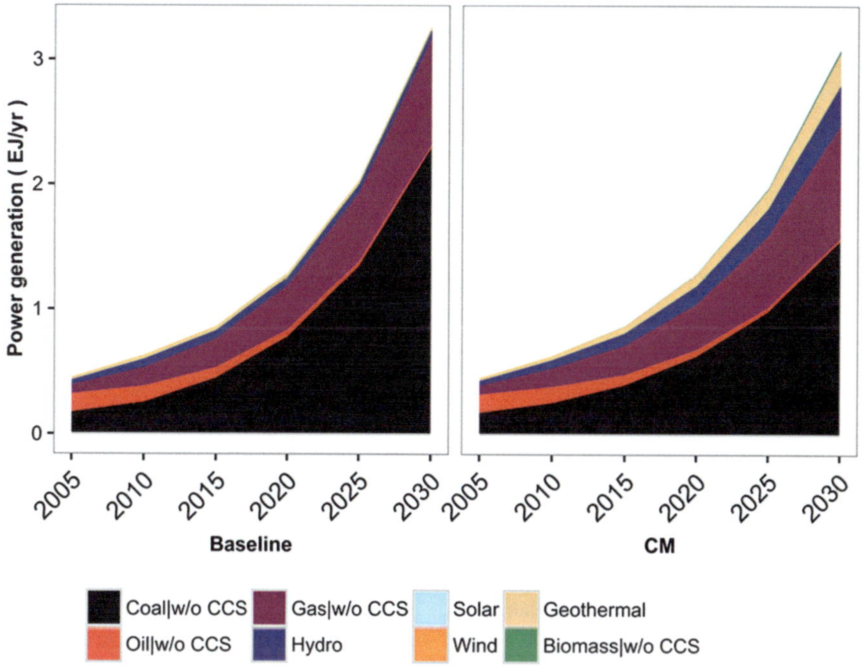

Fig. 8.14 Projected power generation in the baseline and countermeasure (CM) scenarios from 2005 to 2030

8.4.3 *Land Use Abatement*

Figure 8.15 shows the overall land use in the baseline and CM scenarios. The baseline scenario showed a gradual decrease in forest area due to the current rate of deforestation. Such deforestation ceased in the CM scenario, with afforestation and reforestation areas increasing in size. Figure 8.16 shows the predicted emission reductions and abatement costs by land use sector by 2030, with a breakdown of countermeasures calculated using the bottom-up model under a given reduction target. The reduction potential increased with the carbon price. The total emission reductions and abatement costs in the land use sector reached 480 $MtCO_2e$ p.a. and USD 3.2 billion p.a. by 2030 (~0.3% of Indonesia's GDP equivalent in 2030). FP, plantation and reforestation (RSS, RFS, PLR, and PSR; see Table 8.2 for details), and agroforestry contributed greatly to GHG emission reductions in the land use sector because they have relatively high economic efficiencies and large potential areas for their application.

Figure 8.17 shows the emission reductions in the agriculture and livestock sectors by 2030 broken down by each countermeasure. Total GHG reductions reached 47 $MtCO_2e$ p.a. in 2030, representing a much smaller reduction than that in the land use sector. High-efficiency fertilizer application (i.e., split fertilization)

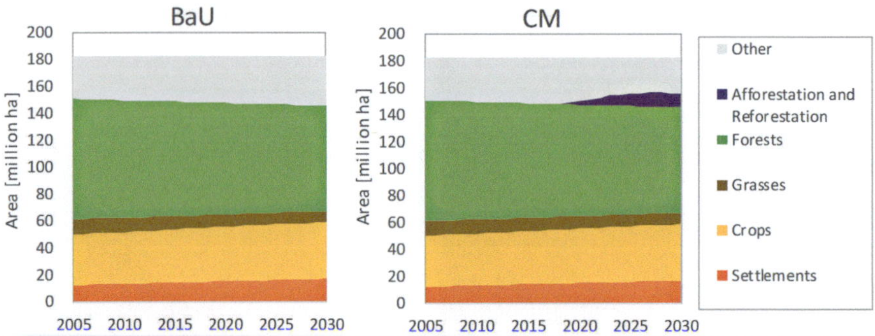

Fig. 8.15 Area of land use change under the baseline and countermeasure (CM) scenarios from 2005 to 2030 (Hasegawa et al. 2016)

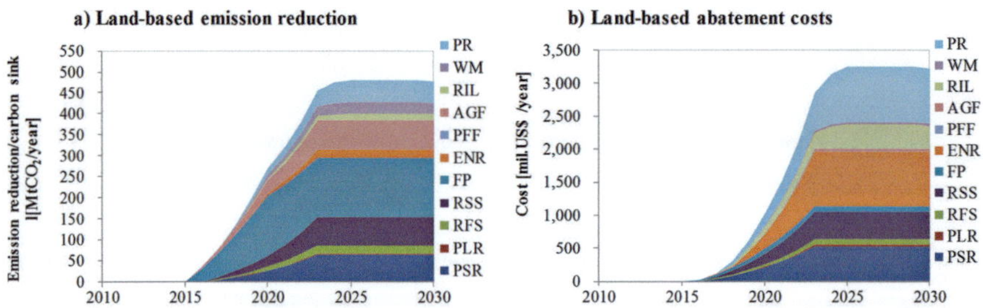

Fig. 8.16 (**a**) Emission reductions and (**b**) abatement costs in land use sectors arranged by mitigation measure, calculated using the bottom-up agriculture, forestry, and other land use (AFOLU) model. The mitigation measure codes are the same as those listed in Table 8.2 (Hasegawa et al. 2016)

Table 8.2 List of AFOLU technologies

Abbreviation	Explanation
PSR	Plantation – short rotation
PLR	Plantation – long rotation
RFS	Reforestation – fast-growing species
RSS	Reforestation – slow-growing species
FP	Forest protection
RIL	Reduced impact logging
ENR	Enhanced natural regeneration
WM	Improved water management on managed peatland
PR	Peatland restoration on unmanaged peatland
AGF	Agroforestry

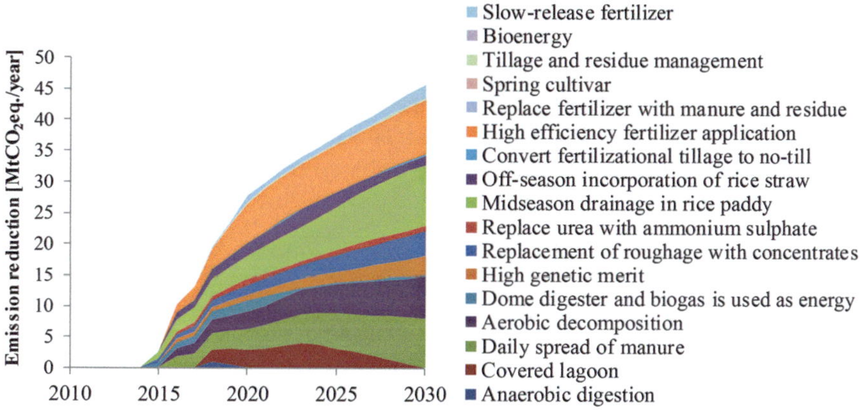

Fig. 8.17 Emission reductions in the agriculture and livestock sectors arranged by mitigation measure, calculated using the bottom-up AFOLU model (Hasegawa et al. 2016)

on cropland soils, midseason drainage in rice paddies, and the daily spread of manure for livestock management all contributed significantly to reducing emissions in the agriculture and livestock sectors. Increased crop and livestock production created additional opportunities to implement these effective mitigation measures.

8.5 Discussion and Policy Implications

There are two main policy implications of this study. First, the energy-related emissions grew rapidly while land use emissions were slightly increased, and these are comparable in 2030. Although AFOLU was the principal mitigation contributor (in terms of reduction amount), the role of the energy sectors should still grow. This implies that Indonesia's mitigation strategy should focus equally on the energy and AFOLU sectors. This is completely different from the previously pledged RAN/GRK (Indonesian National Mitigation Action Plan 2020), in which AFOLU was expected to be the major source of emissions and reduction.

Second, mitigation policy does not always harm economic development but can instead enhance it. The model results suggested that there may be a slight increase in GDP. This is due to coal expansion in the baseline scenario and the coal-related sector's low labor and capital productivity (particularly coalmining), whereas coal use was reduced in the CM scenario. Since we set the model to represent the use of coal in the baseline scenario, labor and capital shifted to more productive sectors in the CM scenario. These results depend on the base year social accounting matrix and future productivity assumption for individual sectors. However, as long as we do not assume any extraordinary events, the GDP loss would not be large, which is critically important for policymakers.

Although we derived meaningful policy implications from this modeling exercise, there are some important limitations. The model does not consider the spatial distribution of land, which may be crucial for AFOLU as well as energy grid connectivity. Ecosystem services maintained by forest protection were also not considered in the study. Furthermore, current land use regulations in Indonesia were not considered. Rather, we analyzed mitigation potential and options under the carbon-constrained scenario. CCS was not used in this exercise but may be a fundamental option to reduce CO_2 emissions.

References

Bappenas (2014) Rencana Pembangunan Jangka Menengah Nasional (RPJMN) 2015–2019

FAO (2016) FAOSTAT, Food and Agriculture Organization of the United Nations, Rome, Italy

Fujimori S, Masui T, Matsuoka Y (2012) AIM/CGE [basic] manual, Discussion paper series. Center for Social and Environmental Systems Research, NIES, Tsukuba

Fujimori S, Hasegawa T, Masui T, Takahashi K, Herran DS, Dai H, Hijioka Y, Kainuma M (2017) SSP3: AIM implementation of Shared Socioeconomic Pathways. Glob Environ Chang, 42:268–283

Hak M, Hasegawa T, Matsuoka Y (2015) An assessment of GHG emissions and mitigation potential from agriculture, forestry and other land-use in Cambodia. J Glob Environ Eng 71:12

Hasegawa T, Matsuoka Y (2015) Climate change mitigation strategies in agriculture and land use in Indonesia. Mitig Adapt Strateg Glob Chang 20:409–424

Hasegawa T, Fujimori S, Masui T, Matsuoka Y (2016) Introducing detailed land-based mitigation measures into a computable general equilibrium model. J Clean Prod 114:233–242

Hasegawa T, Fujimori S, Masui T (2016) Key mitigation options in agriculture and land use to achieve mid-term carbon reduction targets in Indonesia. Sustainabilities, 8(12):1283

Indonesia RO (2016) Intended Nationally Determined Contributions (INDC)

Jilani T, Hasegawa T, Matsuoka Y (2015) The future role of agriculture and land use change for climate change mitigation in Bangladesh. Mitig Adapt Strateg Glob Chang 20:1289–1304

Kehutanan KLHD (2015) Draft-Indonesian First Biennial Update Report (BUR)

Nasional DE (2015) Outlook Energi Indonesia 2015. Dewan Energi Nasional

Negara AHHPB (2015) Statistics Indonesia 1995–2015

Nguyen TH, Hasegawa T, Matsuoka Y (2014) Climate change mitigation strategies in agriculture, forestry and other land use sectors in Vietnam. Mitig Adapt Strateg Glob Chang 19:15–32

PLN (2016) Electricity supply business plan PT Perusahaan Listrik Negara (PLN) (Persero) 2016–2025, Jakarta

Pusdatin-ESDM (2013) Handbook of energy & economics statistics of Indonesia 2013

Wicke B, Sikkema R, Dornburg V, Faaij A (2011) Exploring land use changes and the role of palm oil production in Indonesia and Malaysia. Land Use Policy 28:193–206

Chapter 9
Quantitative Analysis of Japan's 2030 Target Based on AIM/CGE and AIM/Enduse

Ken Oshiro, Toshihiko Masui, and Mikiko Kainuma

Abstract Japan submitted its (Intended) Nationally Determined Contributions ((I)NDC) that is 26% reduction of GHG in 2030 compared with the 2013 level on July 2015. In this analysis, Japan's NDC is assessed using the technology selection model (AIM/Enduse) and the computable general equilibrium model (AIM/CGE) developed by AIM (Asia-Pacific Integrated Model) Project Team. In addition, consistency between the NDC and the 80% reduction target by 2050 is also assessed using AIM/Enduse. The analysis by 2030 suggests that Japan's NDC can be achieved with improvement of energy efficiency in the end-use sectors and deployment of low-carbon energies in the supply side. The marginal abatement cost in 2030 accounts for about 200 US$/t-$CO_2$, but the severe economic impact can be avoided by introducing the appropriate GHG mitigation options. Moreover, the 2050 target which is to reduce GHG by 80% is also technically feasible, although rapid and large-scale transformation of energy system is required between 2030 and 2050, including electrification in the end-use sectors.

Keywords Nationally determined contributions • Climate change mitigation • Top-down model • Bottom-up model

K. Oshiro (✉)
Mizuho Information & Research Institute, Inc., Tokyo, Japan
e-mail: ken.oshiro@mizuho-ir.co.jp

T. Masui
Center for Social and Environmental Systems Research, National Institute for Environmental Studies (NIES), 16–2 Onogawa, Tsukuba, Ibaraki 305–8506, Japan

M. Kainuma
Institute for Global Environmental Strategies (IGES), 2108–11 Kamiyamaguchi, Hayama, Kanagawa 240–0115, Japan

Center for Social and Environmental Systems Research, National Institute for Environmental Studies (NIES), 16–2 Onogawa, Tsukuba, Ibaraki 305–8506, Japan

S. Fujimori et al. (eds.), *Post-2020 Climate Action*,
DOI 10.1007/978-981-10-3869-3_9

Table 9.1 Estimated GHG emissions in 2030 in NDC

	Estimated emissions in FY 2030	Emissions in FY 2005	Emissions in FY 2013
GHG emissions (million ton CO_2)			
Energy-originated CO_2	927	1219	1235
Industry	401	457	429
Commercial and other	168	239	279
Residential	122	180	201
Transport	163	240	225
Energy conversion	73	104	101
Non-energy-originated CO_2	70.8	85.4	75.9
CH_4	31.6	39.0	36.0
N_2O	21.1	25.5	22.5
HFCs	21.6	12.7	31.8
PFCs	4.2	8.6	3.3
SF_6	2.7	5.1	2.2
NF_3	0.5	1.2	1.4
Forest carbon sinks measure reduction	−27.8		
Cropland management, grazing land management, and revegetation	−9.1		
Total	1042	1397	1408

Source: Government of Japan (2015)

9.1 Introduction

Prior to the twenty-first session of the Conference of the Parties (COP21) to the United Nations Framework Convention on Climate Change (UNFCCC) held in Paris in 2015, each country would submit the Intended Nationally Determined Contribution (INDC). In the case of Japan, INDC was submitted on July 17, 2015. According to the government of Japan (2015), the contents of Japan's INDC are the greenhouse gas mitigation target in 2030, 26% of reduction compared to the level in 2013. Tables 9.1 and 9.2 show the detailed target and the assumptions of socioeconomic conditions such as GDP and population.

From autumn in 2014, the mitigation target of GHG emissions in Japan was discussed at the Joint Experts' Meeting of the Central Environment Council (Subcommittee on Global Warming Measurement after 2020, Global Environment Committee) and the Industrial Structure Council (INDC WG, Global Environment Subcommittee, Committee on Industrial Science and Technology Policy and Environment). Japan's INDC presents contributions of various mitigation measures in detail based on a bottom-up calculation included in the Long-term Energy Supply and Demand Outlook (METI 2015), although the discussion on energy mix in 2030 was convened separated from that of INDC at the Advisory Committee for Natural Resources and Energy. In the energy mix in 2030, the share of renewable energies and nuclear power is assumed to range between 22–24% and 20–22% of electricity supply, respectively. Final energy consumption is assumed to decrease by around 10% by 2030 below the 2013 level.

Table 9.2 Assumptions on socioeconomic conditions in NDC

	FY2030	FY2005	FY2013
Real GDP (trillion JPY at 2005 price)	711	507	531
Population (million person)	117	128	127
Crude steel production (million ton)	120	110	110
Ethylene production (million ton)	5.7	7.5	6.8
Cement production (million ton)	56	70	59
Paper and paperboard production (million ton)	27	31	27
Floor space for commercial use (billion m^2)	1.97	1.76	1.85
Passenger transportation demand (trillion person km)	1.42	1.41	1.46
Freight transportation demand (billion ton km)	520	460	420

Source: METI (2015)

Since 1997, in order to discuss the GHG mitigation target in Japan domestically, the Asia-Pacific Integrated Model (AIM) has provided the quantitative results (e.g., Kainuma et al. (2000); Masui (2005); Oshiro et al. (2016)). Especially, in 2009, in order to assess the mitigation option in 2020, the AIM/Enduse [Global] to assess the differences among Japan, the USA, and EU, the AIM/Enduse [Japan] to assess the detailed mitigation options and the additional investment costs to achieve the target, and the AIM/CGE [Japan] to assess the economic impact to achieve the target were utilized. And after 2009, in order to show the road map to achieve the 25% reduction target by 2020, the AIM/Enduse [Japan] and AIM/CGE [Japan] had provided the numerical results. After East Japan Great Earthquake in 2011, these models presented the revised analysis considering uncertainty of availability of nuclear power. On the other hand, at the discussion on the INDC, the integrated assessment models including AIM were not invited.

In the following section, the results of the AIM/Enduse [Japan] and AIM/CGE [Japan] are explained based on the represented mitigation target. The study conducted by using AIM/Enduse [Japan] is written based on RITE and NIES (2015).

9.2 Models to Assess Japan's NDC

After the announcement of Japan's NDC, the AIM team assessed the proposed target using the AIM/Enduse [Japan] and AIM/CGE [Japan]. In this section, the structures of these two models are explained.

9.2.1 Structure of AIM/Enduse [Japan]

The AIM/Enduse [Japan] is a partial equilibrium model on energy with recursive dynamics and holds the very detailed energy technologies in each sector. The model covers both end-use sectors (industrial, commercial, residential, and transport) and the energy conversion sector. In addition, it includes non-CO_2 emissions

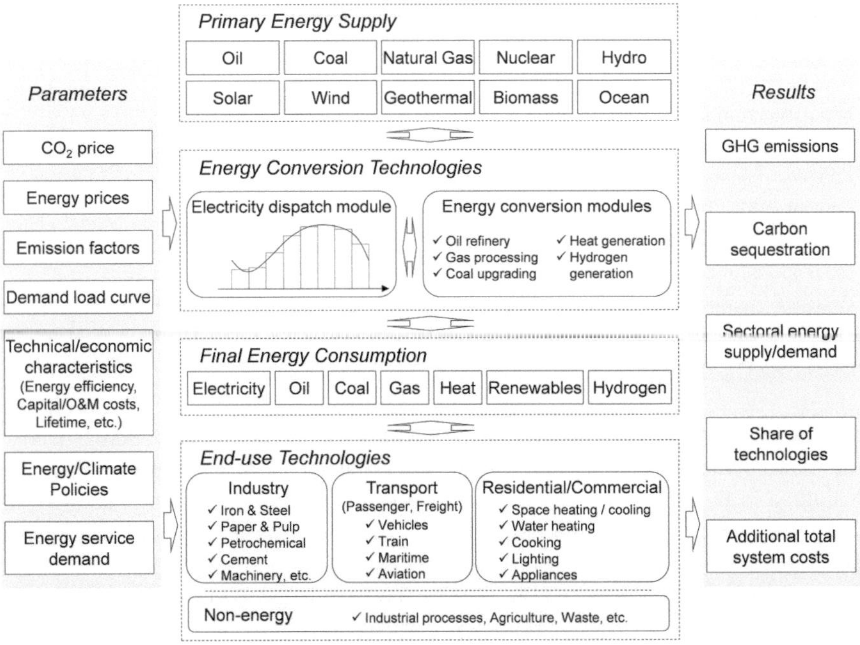

Fig. 9.1 Overview of AIM/Enduse [Japan]

(CH_4, N_2O, hydrofluorocarbon (HFC), perfluorocarbon (PFC), SF_6, and NF_3) from non-energy sectors such as industrial process and waste management as well as emissions from energy use. Non-CO_2 emissions are converted into CO_2 equivalents using GWP_{100} factors taken from the IPCC AR4. However, emissions from LULUCF are excluded for this analysis. In each sector, the least-cost technologies are selected to meet the exogenously given energy service demand. The technology cost consists of the annualized initial cost and operation and management costs. The operation cost includes the carbon cost. Consequently, key endogenous variables, such as final energy consumption, primary energy supply, GHG emissions, and marginal abatement cost, are calculated. Details of the modeling structure, including the list of parameters, variables, and equations, can be found in Kainuma et al. (2003).

The model used for this analysis is a multiregional version of AIM/Enduse [Japan] (Oshiro and Masui 2015). This model explicitly distinguishes ten regions which coincide with the business areas of ten power supply firms to assess the regional differences in renewable energy potentials and energy demand characteristics. The overview of the model is shown in Fig. 9.1.

Although AIM/Enduse [Japan] broadly takes into account the mitigation options included in the government's NDC by 2030, some options which contribute to cutting service demands, such as energy management in factories, promotion of Cool Biz/Warm Biz, and promotion of public transport and modal shift to railways, are not considered as mitigation options. In addition, as the technology selection is

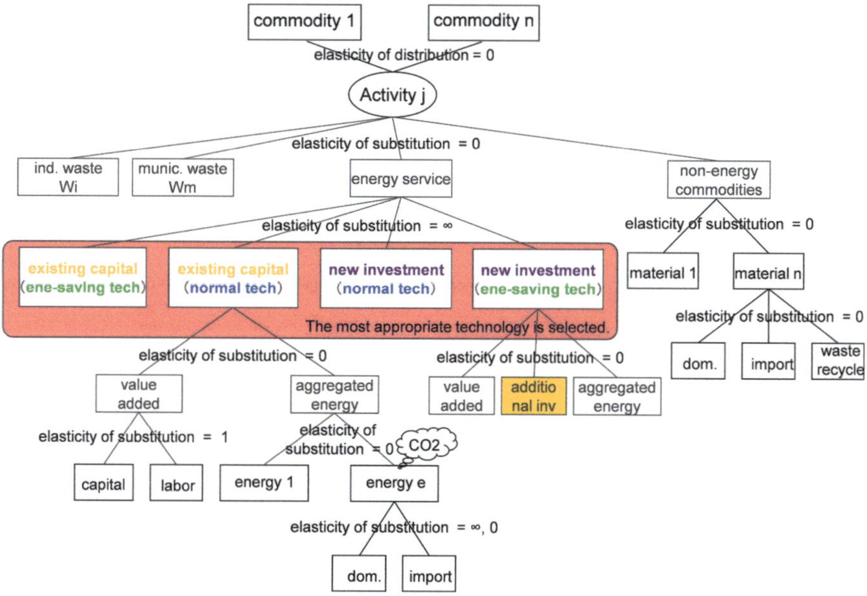

Fig. 9.2 Production structure of AIM/CGE [Japan]

determined based on cost minimization, the diffusion rate of the mitigation options is not precisely consistent with those of the government's NDC.

9.2.2 Structure of AIM/CGE [Japan]

The AIM/CGE [Japan] is based on the computable general equilibrium model with recursive dynamics. The base year of this model is 2000, and the input-output table in Japan in 2000 is replicated. This model covers more than 100 sectors and commodities including the solar power and wind power. Each sector maximizes its profit under a certain level of production function. On the other hand, one representative household sector maximizes its utility under the income constraint. In this model, at all markets of commodities and production factors, supply and demand can be equilibrium through the price mechanism. The CO_2 emissions are calculated from the fossil fuel combustions.

In each sector, the activities using the existing capital stock and the newly installed capital stock are distinguished as shown in Fig. 9.2. The energy-saving technologies are available as the newly installed equipment also needed the additional investment costs. At the representative household sector, the energy use equipment which has been hold and the newly bought equipment are distinguished.

The energy-saving investment is paid from the revenues of carbon tax. About the detailed model description, please see Masui et al. (2014).

In order to assess Japan's NDC, the CO_2 emissions will be reduced, and the energy-saving technologies will be available after 2014. The mix of the power supply is defined in advance to meet the Long-term Energy Supply and Demand Outlook.

9.3 Results and Discussions

9.3.1 Results of AIM/Enduse [Japan]

9.3.1.1 Cases

For the analysis using AIM/Enduse [Japan], three different cases are developed in order to analyze feasibility and robustness of Japan's NDC and an opportunity for further emission reduction by 2030 toward the 2050 target. In the NDC case, the level of GHG emissions is consistent with the government's NDC in 2030 and with long-term reduction target in 2050. In the enhanced action case, an opportunity for further reduction in 2030 is assessed. In the low-nuclear case, uncertainty of the availability of nuclear power is considered.

Regarding the long-term GHG emission reduction target, it is described in the Plan for Global Warming Countermeasures which was published on May 2016 that Japan aims to reduce GHG emissions by 80% by 2050 as its long-term goal, although the base year is not mentioned (MOE 2016). Considering this long-term goal, this analysis sets the 2050 target to reduce GHG emission by 80% compared to the 1990 level.

For all cases, carbon prices are set as a driver for introduction of mitigation options. It should be noted that other implemented or planned policies such as feed-in tariff for renewable energies, subsidies for low-carbon technologies, and fuel economy target for vehicles are not taken into account.

The following are the detailed descriptions of each case:

- NDC case – This case sets the target of reducing GHG emissions by 2030 by the same level with the government's NDC (25.4% reduction with respect to the 2005 level, including reduction in LULUCF) and achieving the 80% reduction target by 2050 thereafter. Excluding emissions from LULUCF, the NDC target of 25.4% reduction by 2030 corresponds to approximately 22.7% with respect to the 2005 level.

  In this case, availability of nuclear plant is given based on the New Policies Scenario developed by IEA's World Energy Outlook 2014 (IEA 2014); a lifetime of plants built since the mid-1980s is extended to 60 years and limited to 40 years for all others. Complying with this assumption, electricity generation from nuclear plant in 2030 accounts for approximately 232 TWh. It is broadly

Table 9.3 Assumptions on socioeconomic indicators

	2010	2030	2050
Population (million)	128	117	97
Real GDP (billion $US\$_{2005}$)	4619	6406	8285
Crude steel production (Mt)	111	120	85
Cement production (Mt)	56	56	47
Ethylene production (Mt)	7.0	5.7	4.3
Paper and pulp production (Mt)	27	27	18
Commercial floor space (billion m^2)	1.83	1.97	1.91
Passenger transport (billion p-km)	1366	1410	1166
Freight transport (billion t-km)	451	520	505

Sources: IIASA (2015); IPSS (2012); METI (2015)

consistent with the description in the Long-term Energy Supply and Demand Outlook for 2030 which projects between 217 TWh and 232 TWh in 2030 (METI 2015). Extrapolating this assumption toward 2050 without any new construction or replacement of nuclear plant, the electricity supply from nuclear power will be approximately 184 TWh in 2050.

- Enhanced action case – This case illustrates the opportunity for further decarbonization by 2030 from the level of the government's NDC. In this case, carbon prices are increased linearly from 2016 to 2050 in order to achieve 80% reduction target by 2050. This means a carbon price in 2030 reaches roughly a half of the 2050 level. Assumptions on availability of nuclear power are the same as the NDC case.
- Low-nuclear case – This case considers neither extension of lifetime nor new construction of nuclear plants. Therefore, electricity supply from nuclear power is entirely phased out by 2050. Assumptions on GHG emissions constraint are the same as the NDC case.

Assumptions on population and economic growth by 2030 are consistent with the government's NDC and the Long-term Energy Supply and Demand Outlook. From 2030 to 2050, economic growth is estimated using GDP growth per capita of SSP5, conventional development scenario of Shared Socioeconomic Pathways (IIASA 2015). Population prospects are taken from the perspectives by the National Institute of Population and Social Security Research (IPSS 2012). Assumptions on socioeconomic indicators up to 2050 are summarized in Table 9.3.

9.3.1.2 Assessment of Feasibility and Robustness of the Government's NDC

As summarized in Fig. 9.3, all cases in this analysis meet the 2030 target, and carbon prices in 2030 are estimated approximately as 187 US\$/t-CO_2 and 236 US\$/t-CO_2 in the NDC case and low-nuclear case, respectively. In the enhanced action case, GHG emissions are reduced furthermore, 25.3% with respect to the 2005 level, and the carbon price in 2030 rises to approximately 220 US\$/t-CO_2 (Fig. 9.3

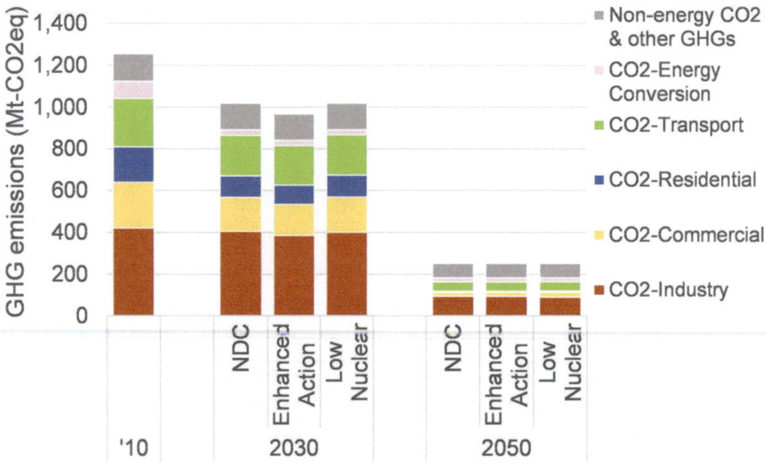

Fig. 9.3 GHG emissions estimated by AIM/Enduse [Japan]

Table 9.4 Carbon prices estimated by AIM/Enduse [Japan]

	2030	2050
NDC case	187	523
Enhanced action case	220	514
Low-nuclear case	236	631

Unit: US$/t-$CO_2$

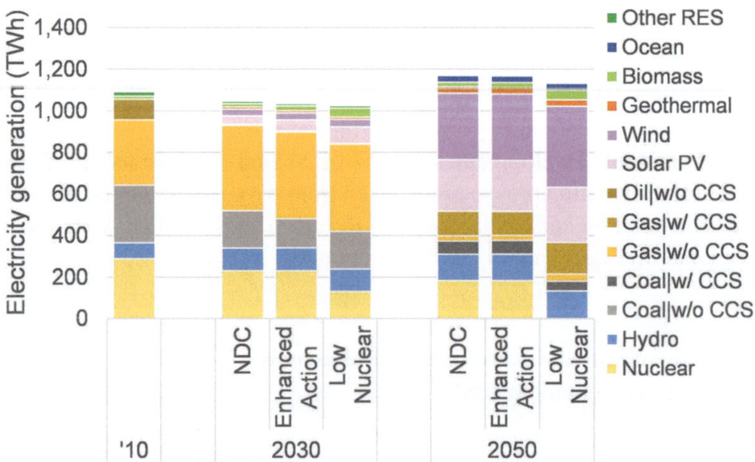

Fig. 9.4 Electricity generation estimated by AIM/Enduse [Japan]

and Table 9.4). The carbon price in this analysis is directly linked to the marginal abatement cost to achieve the NDC target; however, it should be noted that additional policies such as the recycle of carbon tax revenue could help to lower the actual carbon prices.

Further reduction in the enhanced action case by 2030 compared to the NDC case is due mainly to a switch from coal to natural gas and an additional deployment of renewable energies particularly in the power sector. As shown in Fig. 9.4, in the enhanced action case, power generation from gas-fired plants and renewable energies, such as solar PV and biomass, is increased, and a share of coal-fired plant falls to approximately 14%, while it accounts for 17% in the NDC case and the low-nuclear case, respectively. However, a share of coal-fired plants falls to lower level compared to the 2010 level even in the NDC case because of high carbon prices, while government's NDC requires a share of coal, where it is approximately 26% in 2030.

In the NDC case, a share of renewable energies rises to around 21% in 2030 which is slightly lower than the government's NDC (22–24%). The share of renewable energies accounts for approximately 23% and 28% in the enhanced action case and the low-nuclear case, respectively, where the coal-fired plants and nuclear power plants are replaced by renewable energies. A share of renewable energies stays below 30% in 2030 even in the low-nuclear case despite the high carbon prices due mainly to the fragmented potential of renewable energies with narrow interconnection capacity between the ten regions in Japan. Additionally, it should be noted that the additional policies, such as feed in tariff and subsidies for renewable energies, are considered; a share of renewable energies could be increased.

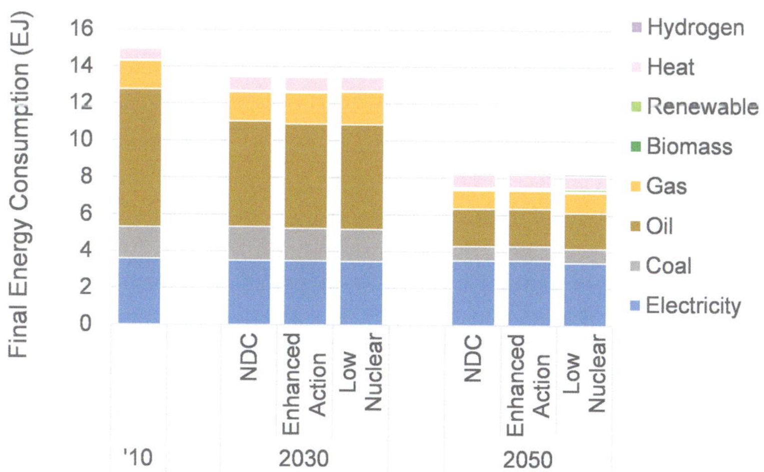

Fig. 9.5 Final energy consumption estimated by AIM/Enduse [Japan]

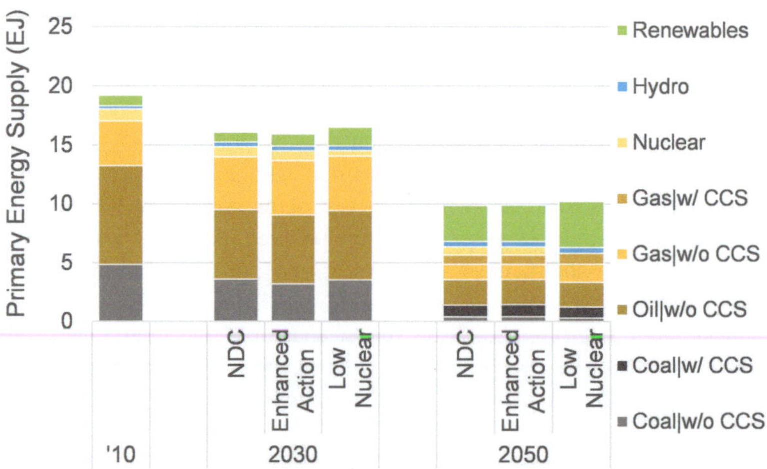

Fig. 9.6 Primary energy supply estimated by AIM/Enduse [Japan]

With respect to the 2013 level, the total electricity generation in 2030 is almost the same level because of economic growth and electrification in the demand side. This trend is roughly consistent with the government's NDC.

As shown in Fig. 9.5, final energy consumption in the NDC case is decreased by around 9% with respect to the 2010 level. The level of energy efficiency in the model used for this analysis is moderate compared to the government's NDC which accounts for 14% due mainly to the absence of some mitigation options which cut service demand, as it is difficult for technology selection model to take into account the effect of these abatement options.

Even in the enhanced case and the low-nuclear case, which are characterized by higher carbon prices, final energy consumption in 2030 stays the same level as the NDC case. It implies most of cost-effective energy efficiency option is introduced even in NDC case, and the remaining potential to additional reduction of energy use is insufficient excluding the options which cut service demand. Therefore, it seems that the level of reduction of final energy demand in government's NDC is appeared to be reasonable considering its high economic growth.

Figure 9.6 summarizes primary energy supply based on direct equivalent accounting method estimated by AIM/Enduse [Japan]. In 2030, a share of low-carbon energies is increased to 8–12% from 6% in 2010, although energy supply from nuclear power is declined in this period. Consequently, renewable energies act as the main driver of lowering emission intensity of energy by 2030, given that introduction of CCS in the power sector is not considered in this analysis.

Figure 9.7 shows the improvements of energy and carbon intensity in the final energy sectors under the NDC case. It can be seen that the industry, buildings, and passenger transport sectors require significant improvements in energy intensity by 2030 with respect to the 2010 level of approximately 27%, 18%, and 27%, respectively, while those in freight transport sector stay at around 7% in 2030.

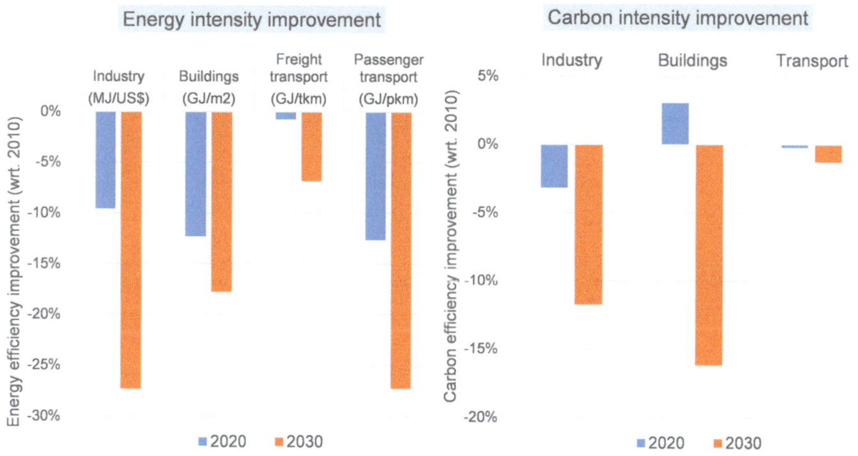

Fig. 9.7 Improvement of energy and carbon intensity by sector estimated by AIM/Enduse [Japan]

Likewise the carbon intensity (including direct and indirect emissions) of energy demand in the industry and building sectors improves by 12% and 16% between 2010 and 2030, due mainly to electrification and gasification. In building sector, it can be seen that carbon intensity by 2020 increased by 3% with respect to the 2010 level because of a temporal rise of carbon intensity of electricity due mainly to reduction of nuclear power. By contrast, the carbon intensity of transport fuel declines only by 1% by 2030, reflecting limited electrification and biofuel penetration.

9.3.1.3 Implications for the Long-Term Decarbonization Pathways

This section mainly focuses on the implications for the long-term target by 2050 which is to reduce GHG emission by 80% with respect to the 1990 level.

In the NDC case, GHG emission is reduced by 80% in 2050 due mainly to improvement in energy efficiency, electrification, and decarbonization of electricity. Final energy consumption in 2050 is nearly halved compared to the 2005 level, and electrification in demand side rises to more than 40% from 22% in 2005 (Fig. 9.5). In addition, electricity is almost decarbonized because of both a large-scale deployment of solar PV and wind power and a substitution from unabated coal-/gas-fired plant to CCS-equipped plant (Fig. 9.4).

By 2050, a carbon price in the NDC case rises to approximately 523 US$/t-$CO_2$. As it is not so far from the level of the enhanced action case, in which a carbon price in 2050 accounts for approximately 514 US$/t-$CO_2$, it is implied that the 80% reduction target is still technically feasible even if GHG emission reduction in 2030 stays the same level as the government's NDC.

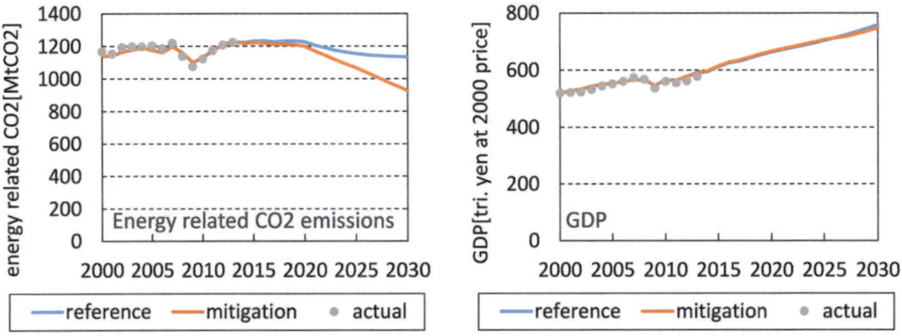

Fig. 9.8 GDP and energy-related CO_2 emission results from AIM/CGE [Japan]

In all cases assessed in this analysis, power generation from unabated coal-fired plant is almost substituted by low-carbon sources, while unabated gas-fired plants still play an important role as flexible resources in order to integrate VREs in the long term. Though the switch from coal to gas in early stage entails the challenge associated with energy security, it could contribute to effective mitigation in the long run because they avoid lock-in of high carbon intensity plant.

The 80% reduction target in 2050 is also technically feasible with the entire phaseout of nuclear power, with the deployment of CCS-equipped as well as renewable energies, though carbon price reaches approximately 631 US\$/t-CO_2 in 2050 in order to promote additional deployment of renewables, particularly variable renewable energies (VREs) and CCS-equipped natural gas plant as alternative low-carbon energies.

9.3.2 Results of AIM/CGE [Japan]

Based on the analysis using AIM/CGE [Japan], reference scenario (reference) and INDC scenario (mitigation) are assessed as shown in Fig. 9.8. In INDC scenario, the carbon price in 2030 will be around 26,300 JPY/t-CO_2. If the available mitigation measures change from the low level to the medium level, the carbon price in 2030 will be 22,900 JPY/t-CO_2. This means that to keep the low cost measures becomes important to reduce the carbon price.

In 2030, GDP in the NDC case is reduced by 1.4% compared with that in reference case, although the annual GDP growth rate between 2014 and 2030 becomes 1.6%. If the medium level of mitigations will be available, the GDP loss in 2030 will be 1.1% compared with that in the reference case. On the other hand, the tax revenue of carbon tax will be used to the energy-saving investment; the GDP loss becomes 4.7% compared with those in reference case.

9.4 Conclusions

In this study, Japan's NDC which aims to reduce GHG emissions by 26.0% below the 2013 level in 2030 is assessed by using both AIM/CGE [Japan] and AIM/Enduse [Japan]. These analyses highlight that emission reduction target of Japan's NDC could be achieved with several technological options such as improvement of energy efficiency in the demand sectors and deployment of low-carbon energies. However, additional mitigation effort beyond the current level which represents about 200 US$/t-$CO_2$ of carbon prices would be required by 2030. According to the analysis using AIM/CGE [Japan], it is estimated that the GDP in NDC case is reduced by 1.1–1.4% in 2030 compared with that in reference case, although in the NDC case, the annual GDP growth rate between 2014 and 2030 becomes 1.6%.

Moreover, the analysis using AIM/Enduse [Japan] suggests that the 2050 goal which is to reduce GHG emissions by 80% is also technically feasible, although acceleration of introduction of mitigation options is required after 2030. In particular, electrification in the end-use sectors would be one of the key options as well as successive improvement of energy efficiency and decarbonization of electricity.

Although this study presents the assessment of Japan's NDC and the long-term target mainly focusing on both technological and economic aspects, the effect of social changes including energy service demand reduction is not considered enough. As a future work, it would be useful to assess the possibility of deeper emission reduction by social changes as well as those of the technological options.

Acknowledgments The research in this chapter is supported by the Environmental Research and Technology Development Fund by the Ministry of Environment, Japan (2–1402).

References

Government of Japan (2015) Submission of Japan's Intended Nationally Determined Contribution (INDC)

IEA (2014) World energy outlook 2014. OECD/IEA, Paris

IIASA (2015) SSP database (Shared socioeconomic pathways) – Version 1.0. https://secure.iiasa.ac.at/web-apps/ene/SspDb/

IPSS (2012) Population projections for Japan (January 2012) -2011 to 2060. http://www.ipss.go.jp/site-ad/index_english/esuikei/gh2401e.asp

Kainuma M, Matsuoka Y, Morita T (2000) The AIM/end-use model and its application to forecast Japanese carbon dioxide emissions. Eur J Oper Res 122(2):416–425. doi:10.1016/S0377–2217(99)00243-X

Kainuma M, Matsuoka Y, Morita T (2003) Climate policy assessment: Asia-Pacific integrated modeling. Springer, Tokyo/New York

Masui T (2005) Policy evaluations under environmental constraints using a computable general equilibrium model. Eur J Oper Res 166(3):843–855. doi:10.1016/j.ejor.2004.07.002

Masui T, Hibino G, Oshiro K (2014) Assessment of Greenhouse gas emission reduction in Japan using computable general equilibrium model with simplified technology selection. J Jpn Soc Civil Eng, Ser. G (Environ Res) 70(5):I_43–I_51. doi:10.2208/jscejer.70.I_43. (in Japanese)
Ministry of Economy, Trade and Industry (METI) (2015) Long-term energy supply and demand outlook. http://www.meti.go.jp/english/press/2015/pdf/0716_01a.pdf
Ministry of Environment (MOE) (2016) Overview of the plan for Global Warming countermeasures – Cabinet decision on May 13, 2016. https://www.env.go.jp/press/files/en/676.pdf
Oshiro K, Masui T (2015) Diffusion of low emission vehicles and their impact on CO2 emission reduction in Japan. Energy Policy 81:215–225. doi:10.1016/j.enpol.2014.09.010
Oshiro K, Kainuma M, Masui T (2016) Assessing decarbonization pathways and their implications for energy security policies in Japan. Climate Policy 16(sup1):S63–S77. doi:10.1080/14693062.2016.1155042
RITE, NIES (2015) MILES (Modelling and Informing Low Emissions strategies) project – Japan policy paper: a joint analysis of Japan's INDC. http://www.iddri.org/Publications/Collections/Analyses/MILES_Japan%20Policy%20Paper.pdf

Chapter 10
Asian INDC Assessments: The Case of Thailand

Bundit Limmeechokchai, Puttipong Chunark, Shinichiro Fujimori, and Toshihiko Masui

Abstract On 1st October 2015, Thailand had submitted its Intended Nationally Determined Contributions (INDCs) and stated that by 2030 GHG emissions will be reduced by 20–25% when compared to the business-as-usual (BAU) scenario. The Paris Agreement was adopted on 12 December 2015 at the twenty-first session of the Conference of the Parties (COP21) to the United Nations Framework Convention on Climate Change (UNFCCC) held in Paris. In addition, Thailand provided signature at the United Nations in New York on 22 April 2016 and ratified the Paris Agreement on 21 September 2016. This paper assesses the impacts of GHG emission reduction targets in Thailand's INDC by using the Asia-Pacific Integrated Model/Computable General Equilibrium (AIM/CGE). Four scenarios are established by the given GHG emission constraints and the renewable power generation target. Results show that, under the Power Development Plan in 2015, the INDC target is achievable. As a result, macroeconomic loss is low in low reduction target, but it will be high in the high reduction target. In addition, it needs more renewable energy push to realize stringent climate policy. Thus, the availability of land for deploying the renewable energy technologies such as solar, wind, and biomass needs to be evaluated to meet higher GHG emission levels. Furthermore, the stringent GHG emission levels also induce the reduction of other air pollutants. Finally, the result of this study has been used in the design of

B. Limmeechokchai (✉) • P. Chunark
Sustainable Energy and Low Carbon Research Unit, School of Manufacturing System and Mechanical Engineering, Sirindhorn International Institute of Technology, Thammasat University, Bangkok, Pathum Thaini, Thailand
e-mail: bundit@siit.tu.ac.th

S. Fujimori
Center for Social and Environmental Systems Research, National Institute for Environmental Studies (NIES), 16–2 Onogawa, Tsukuba, Ibaraki 305–8506, Japan

International Institute for Applied Systems Analysis, Schlossplatz-1, Laxenburg 2361, Austria
e-mail: fujimori.shinichiro@nies.go.jp

T. Masui
Center for Social and Environmental Systems Research, National Institute for Environmental Studies (NIES), 16–2 Onogawa, Tsukuba, Ibaraki 305–8506, Japan

S. Fujimori et al. (eds.), *Post-2020 Climate Action*,
DOI 10.1007/978-981-10-3869-3_10

roadmap for GHG reduction targets in 2030, and Thailand has more confidence on the achievement of the Paris Agreement.

Keywords Greenhouse gas mitigation • Computable general equilibrium model • INDCs (Intended Nationally Determined Contributions) • Renewable energy • Thailand

10.1 Introduction

Thus, the climate change issue becomes a general consensus and a common issue. The Intergovernmental Panel on Climate Change (IPCC) Fifth Assessment Report (AR5) concluded that human activities are the main sources of GHG emission that induced the current climate change. The current emission reduction reveals the significant gap bringing about the global mean temperature rises by 3.7–4.8 °C by the end of the twenty-first century (IPCC 2014). Therefore, the AR5 remarkably proposed that the global carbon emission pathway stabilized the global mean temperature to be less than 2 °C compared to preindustrial level and GHG emissions need to peak before 2030 (IPCC 2014). The GHG emission should decrease to net zero emission at the end of the twenty-first century. However, such targets will take plenty of time for developing countries (IPCC 2014; Tran et al. 2016).

In order to lessen the GHG emissions while preserving both the economic growth and the social development, the United Nations Framework Convention on Climate Change (UNFCCC) established an international climate agreement during the Conference of the Parties (COP21) in December 2015 (den Elzen et al. 2016; Iyer et al. 2015). The parties have agreed to diminish the effect of climate change toward the low-carbon development and climate-resilient development by preparing the post-2020 climate actions, so-called Intended Nationally Determined Contributions (INDCs) (den Elzen et al. 2016; Iyer et al. 2015; Zhang and Pan 2016; Yamide Dagnet et al. 2016; Jennifer Morgan et al. 2015). The INDCs obviously outline the intended climate actions particularly the climate policies related to the cooperation between the government and policy makers and the infrastructure development. The agreement also stated that the adaptation plans are also engaged. Moreover, the implementation of INDC not only guarantees the countries' commitment but also provides the insight of climate action ambition and financial support (Zhang and Pan 2016). Thus, INDCs can become key points for improving the energy production system, preventing the damage of environment by the implementation of the ambitious climate policies, and providing a mechanism for the low-carbon development. As of May 2016, 162 INDCs have already been submitted to the UNFCCC representing 189 countries (WRI 2016). Thailand has submitted its INDCs to UNFCCC in which the GHG emission will be reduced by 20–25% by 2030 when compared to the business-as-usual (BAU) scenario (UNFCCC 2016).

There are several studies that focused on the achievement of climate change issues and INDC through the economic development by the implementation of renewable energy. China has studies about the economic aspect for achieving its INDC targets (Wu et al. 2016; Wang 2015; Qi and Weng 2016). Mittal et al. suggested that the role of renewable energy can reduce the economic loss, and also the introduction of carbon capture and storage can be another significant technology to control the GHG emission level (Mittal et al. 2016). Furthermore, Sundriyal and Dhyani (Sundriyal and Dhyani 2015) suggested that, to achieve the 40% nonfossil fuel in the energy system by 2030, India will need 200 GW of renewable energy power plants by 2030. In Thailand, there are several studies that investigate several climate policies under green growth economy by the role of renewable energy (Chunark et al. 2015; Winyuchakrit et al. 2016; Thepkhun et al. 2013; Jayatilaka and Limmeechokchai 2015; Kusumadewi and Limmeechokchai 2015; Chunark and Limmeechokchai 2015; Selvakkumaran and Limmeechokchai 2015; Selvakkumaran et al. 2015). Thepkhun et al. assessed Thailand's Nationally Appropriate Mitigation Action (NAMA) in the energy sector under the emission trading scheme (ETS), and they suggested that the ETS plays a vital role to reduce such GHG emissions by the energy efficiency improvement and the implementation of renewable energy (Thepkhun et al. 2013). They concluded that both ETS and CCS technologies also play a significant role for improving such corresponding indicators (Thepkhun et al. 2013). Moreover, Selvakkumaran et al. (2015) assessed CO_2 reduction potential together with energy security, other air pollutants, and marginal abatement cost through the low-carbon pathway of Thailand.

Many studies have presented assessment of global and national mitigation measures with several low-carbon measures. However, to facilitate a successful global climate agreement in COP21, the ambitious and stringent actions should be performed. Therefore, this paper aims to analyze three research questions: firstly, the capability of GHG emission reduction scenarios by the role of renewable energy in Thailand's INDC; secondly, the economic impact from GHG emission reduction targets; and thirdly, co-benefits of GHG mitigation from decreasing other air pollutants. In this paper, the AIM/CGE is used for the assessment. The AIM/CGE is a top-down computable general equilibrium model which is vastly used for assessing the macroeconomics together with environmental policies (Mittal et al. 2016; Dai et al. 2012, 2016; Fujimori et al. 2012, 2014, 2015; Fujimori and Matsuoka 2011; Hasegawa et al. 2014, 2015, 2016).

10.2 Background of Thailand and Energy System

10.2.1 About Thailand

Thailand is located in the heart of Southeast Asia. Its shape and geography are divided into four natural regions: the mountains and forests of the North, the vast rice fields of the Central Plains, the semi-arid farmlands of the Northeast plateau, and the tropical islands and long coastline of the peninsular South. It is bordered to the north by Myanmar and Laos, to the east by Laos and Cambodia, to the south by the Gulf of Thailand and Malaysia, and to the west by the Andaman Sea and the southern extremity of Myanmar. Its maritime boundaries include Vietnam in the Gulf of Thailand to the southeast and Indonesia and India in the Andaman Sea to the southwest.

The country comprises 76 provinces. The largest city in Thailand is Bangkok, the capital, which is also the country's center of political, commercial, industrial, and cultural activities. Thailand has a surface area of approximately 513,000 km^2, with approximately 64 million people.

In 2010, there were 65,981,659 persons residing in Thailand. One important demographic change in Thailand is the rising aging population. The number of people aged 60 and over in the country has continued to grow. Only 5% of the population were aged 60 and above in 1950. This figure reached 10% in 2005. The 2015 forecasts by Thailand's government suggest that it will increase up to a peak at 69 million by 2035 and steadily decline thereafter. The average population growth rate during 2015–2035 will be 0.05%. This trend of population is employed in the AIM/CGE modeling.

Thailand's GDP increased from 131 billion USD in 1990 to 170 billion USD in 2005 and accounted for 1.8% p.a. This low GDP growth rate was due to the Asian economic crisis during 1997–1999. In 1990, the per capita GDP for Thailand was about 2328 USD. In 2005, it increased to 2721 USD. The country is also the second largest economy among the ASEAN countries with a total gross domestic product (GDP) of about US$178 billion in 2005 and GDP per capita of US$ 2732. The 2015 forecasts by Thailand's government reveal that the economy is expected to grow at an annual average growth rate of 3.94% over the short-term period of 2015–2035.

10.2.2 Energy Situation in Thailand

The total primary energy supply (TPES) was 127 million tons of oil equivalent (Mtoe) in 2010. It corresponds to only 0.9% compared to the world TPES and 7.5% that of Asia excluding China. Oil has been the most dominant fuel comprising almost half of the energy supply in the past several decades. In 2010, oil accounted for almost 38% as the single largest fuel, and it was mainly used in the transport sector. Natural gas and biomass accounted for 34% and 17%, respectively. Currently, coal use is largely restricted due to environmental concerns. It accounted only 12% in 2010.

The share of fossil fuels has been increasing rapidly over the years. The share of fossil fuels accounted for 71% of the total TPES in 2010. Biomass and hydro accounted almost all new and renewable energy consumption. Biomass accounted for almost 98% of the total renewable energy use in 2010. This biomass came as means of firewood, agricultural residue, bagasse, etc. The contribution of wind, solar, and geothermal energy to renewable sources is almost negligible. Electricity imports to the country amounted to almost 621 ktoe comprising 0.8% of the TPES in 2010.

In 2011, the total energy demand of Thailand amounted to 70,562 ktoe, which rose to 0.4% from the previous year, while the economy expanded by 0.1%. Modern or commercial energy shared 57,424 ktoe or 81.4% of the total energy demand, whereas traditional renewable energy was consumed at 13,138 ktoe or 18.6%. The total energy supply was 128,092 ktoe and rose to 3.0% from 2010, with the net import of 54,949 ktoe, i.e., 42.9% of the total energy supply, while the domestic production was 73,143 ktoe (i.e., 57.1%).

In 2011, biofuel production comprises ethanol blended with gasoline (gasohol E20, E10 octane number 91 and 95) and biodiesel blended with diesel (high-speed diesel B2 and B5 and palm diesel) which totaled to 3729 ktoe. However, in 2010, share of renewable electricity in total generation was still very low, about 4%, and it mainly came from traditional renewable energy resources such as hydropower (storage type).

10.2.3 Renewable Energy Resources in Thailand

Hydro and biomass are the two main sources currently used at a greater extent. Country has already exploited almost all economically viable hydro for electricity generation. The maximum potential of solar and wind is estimated at 6000 MW and 3000 MW, respectively (Ministry of Energy (MOE) 2015). The maximum exploitable potential of agricultural residues is considered at 15 Mtoe. Likewise, plantation-based biomass is estimated to supply an installed capacity up to 10,000 MW. In addition, the total availability of municipal solid waste (MSW) is estimated at 2064 ktoe, and biogas is estimated at 570 ktoe. The use of solar, wind, plantation-based biomass, MSW, and biogas is considered for power generation, while agricultural residues are considered for cogeneration and residential cooking in the future. It is assumed that biomass-based energy resources used for power generation will be supplied by energy plantations grown on a sustainable basis; hence the CO_2 emissions for biomass are assumed zero.

10.2.4 Power Generation in Thailand

The national grid generation was 149,320 gigawatt-hour (GWh) with an installed capacity of 31,485 MW in 2010. The electricity consumption per capita was 2338 kWh. The peak generation was 25,094 MW. The line losses in the system are not significant at 6.3%. The national grid electricity generation is predominantly fossil fuel based. Out of the total electricity, almost three quarters is generated from

Table 10.1 Composition of Thailand power system (as of December 2014)

Thailand power system	Installed capacity (MW)	Ratio of installed capacity (%)
Classified by technology		
Combined cycle	21,145	56.2
Thermal	7538	20.0
Renewable	8476	22.5
Gas turbine/diesel generator	153	0.5
EGAT-TNB linkage	300	0.8
Total	**37,612**	**100.0**
Classified by power producer		
EGAT	15,482	41.2
Independent power producers (IPPs)	13,167	35.0
Small power producers (SPPs)	4530	12.0
Very small power producers (VSPPs)	2029	5.4
Power imports	2404	6.4
Total	**37,612**	**100.0**

natural gas in 2010. In April 2014, the maximum power generation of EGAT system reached 26,942.1 MW relating to 344.0 MW or 1.29% higher than 2013.

As of December 2014, the total contract capacity was 37,612 MW comprising 34,668 MW from firmed power plants EGAT, IPPs, and firmed SPPs, 915 MW from non-firmed SPPs, and 2029 MW from VSPPs. The details of contract capacity of Thailand power system are shown in Table 10.1. Figure 10.1 shows fuel mix in total generation in 2014.

10.2.5 Thailand's Power Development Plan

In 2015, Thailand launched the Power Development Plan (PDP) to respond to changes in economic and infrastructure development and the ASEAN Economic Community (AEC). Therefore, Thailand developed five integration master plans as follows: (1) Thailand Power Development Plan (PDP), (2) Energy Efficiency Development Plan (EEDP), (3) Alternative Energy Development Plan (AEDP), (4) Natural Gas Supply Plan, and (5) Petroleum Management Plan. The new PDP called "Thailand Power Development Plan 2015–2036 (PDP 2015)" focuses on (1) energy security (coping with the increasing power demand to correspond to the National Economic and Social Development Plan and taking into account fuel diversification), (2) economy (maintaining an appropriate cost of power generation for long-term economic competitiveness), and (3) ecology (lessening carbon dioxide intensity of power generation). The PDP 2015 was formulated in line with social and economic development direction addressed by the Office of the National

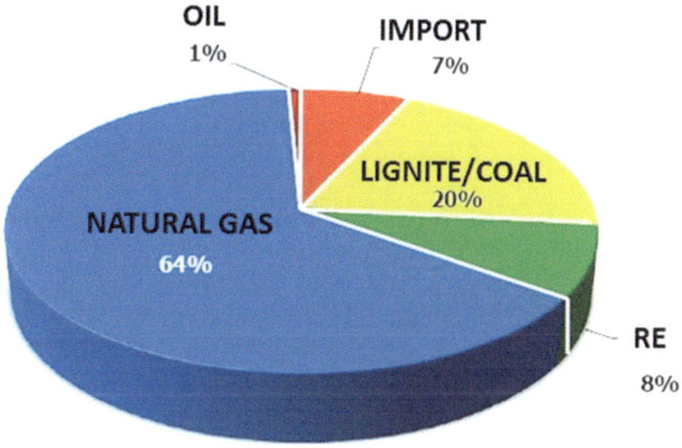

Fig. 10.1 Electricity generation by fuel types in 2014

Economic and Social Development Board (NESDB). The average growth of projected long-term Thai gross domestic product (GDP) estimated by the NESDB was 3.94%. With the integration of the PDP 2015 and the EEDP to foster energy efficiency, the expected energy savings would be 89,672 GWh in year 2036. Moreover, renewable energy, for instance, municipal waste, biomass, biogas, wind, and solar power generation, will be encouraged according to the AEDP. Investments in transmission and distribution system will accommodate renewable energy and smart grid development.

10.3 Thailand's INDC Commitments Under the Paris Agreement

On 1 October 2015, Thailand communicated its INDC to UNFCCC, and on 21 September 2016, Thailand ratified the Paris Agreement. The important messages in the pledged INDC are as follows (United Nations Framework Convention on Climate Change (UNFCCC) 2016):

> Thailand intends to reduce its greenhouse gas emissions by 20 percent from the projected business-as-usual (BAU) level by 2030.
>
> The level of contribution could increase up to 25 percent, subject to adequate and enhanced access to technology development and transfer, financial resources, and capacity building support through a balanced and ambitious global agreement under the United Nations Framework Convention on Climate Change (UNFCCC).

Table 10.2 shows the accompanying information in the submitted INDC.

Table 10.2 Accompanying information on the submitted Thailand's INDC

Baseline	BAU projection from 2005 in the absence of major climate change policies (BAU2030: approximately 555 $MtCO_2e$)
Time frame	2021–2030
Coverage	Economy-wide (inclusion of LULUCF will be decided later)
Gases	Carbon dioxide (CO_2), methane (CH_4), nitrous oxide (N_2O), hydrofluorocarbons (HFCs), perfluorocarbons (PFCs), and sulfur hexafluoride (SF_6)
Assumptions and methodological approaches	Global warming potential on a 100-year timescale in accordance with the Intergovernmental Panel on Climate Change (IPCC) Fourth Assessment Report
	National statistics, including sector activity and socioeconomic forecasts
Planning processes	Thailand's INDC was developed through participatory process.
	Stakeholder consultations were conducted through the establishment of an interministerial working group and steering committee comprising representatives from relevant sectoral agencies, academia, and private sector. In addition, three national consultations were held during the technical analysis phase. Thailand's INDC was formulated based on the national plans approved by the Cabinet:
	National Economic and Social Development Plans
	Climate Change Master Plan B.E. 2558–2593 (2015–2050)
	Power Development Plan B.E. 2558–2579 (2015–2036)
	Thailand Smart Grid Development Master Plan B.E. 2558–2579 (2015–2036)
	Energy Efficiency Plan B.E. 2558–2579 (2015–2036)
	Alternative Energy Development Plan B.E. 2558–2579 (2015–2036)
	Transport Infrastructure Master Plan B.E. 2556–2573 (2013–2030)
	National Industrial Development Master Plan B.E. 2555–2574 (2012–2031)
	Waste Management Roadmap
International market mechanism	Thailand recognizes the important role of market-based mechanisms to enhance the cost-effectiveness of mitigation actions and therefore will continue to explore the potentials of bilateral, regional, and international market mechanisms as well as various approaches that can facilitate, expedite, and enhance technology development and transfer, capacity building, and access to financial resources that support Thailand's efforts toward achieving low-carbon and climate-resilient growth, as appropriate.

10.4 Methodology

10.4.1 AIM/CGE Model

The AIM/CGE (Asia-Pacific Integrated Model/Computable General Equilibrium) model has been widely used for global assessing climate change mitigation and adaptation policies. It is a recursive-dynamic general equilibrium model, extended from the "Standard CGE model." Details of the model structure and mathematical formulas are described in Fujimori et al. (2014) and Chaps. 12 and 13.

Similar to most energy technology-rich CGE models, input parameters such as population, gross domestic product (GDP), energy service demand, the extraction cost of fossil fuels, and availability and cost of renewable energy are exogenously inputted into the model. The model reports energy demand and supply structure, GHG emissions, and prices for energy and carbon emissions. More technically, the production sector assumes maximized profits subject to multi-nested constant elasticity substitution (CES) functions and relative prices of inputs. Household expenditures on each commodity are described by a linear expenditure system (LES) function. The savings is the source for domestic and foreign direct investment. They are exogenously inputted as a proportion of GDP change relative to base year. To balance the savings account, the savings rate is treated as an endogenous variable. Moreover, the capital formation of each good is determined by a fixed coefficient of total investment. The Armington assumption is used for international trade, and the current account is assumed to be balanced. In addition to energy-related CO_2 emissions, CO_2 from other sources, methane (CH_4), and nitrous oxide (N_2O) (including changes resulting from land use and non-energy-related emissions) are also taken into consideration.

The constraint on the GHG emissions was specified based on the mitigation target. Once the emission constraint is added, the carbon tax becomes a complementary variable to the emission constraint, and it determines the marginal mitigation cost. In an emission-constrained scenario, the carbon tax increases the price of the fossil fuel-consuming goods which results in energy savings and substitution of fossil fuels with lower emission energy sources. The carbon tax also acts as an incentive to reduce non-energy-related emissions. GHGs other than CO_2 are weighted by their global warming potential and summed to estimate total GHG emissions in terms of CO_2 equivalent. Households are assumed to receive the revenue from the carbon tax.

The costs of the wind and solar technology were assumed as in the IEA energy technology perspective report (IEA 2012). We changed the input coefficients in the corresponding production function. Because the output prices of these technologies were endogenously determined, they were not exactly the same as those reported by the IEA (2012), but they were close enough.

10.4.2 Input Data

CGE models generally use a Social Accounting Matrix (SAM) to calibrate the model parameters. To assess energy flow and GHG emissions more precisely and more realistically, the CGE model should account not only for the original SAM but also for energy statistics. The Global Trade Analysis Project (GTAP) (Dimaranan 2006) and energy balance tables (International Energy Agency, (IEA) 2013a, b) were used as a basis for the SAM and energy balance table, and data were reconciled with other international statistics such as national account statistics (United Nations (UN) 2013). The concept behind the reconciliation method is described by Fujimori and Matsuoka (2011). GHG emissions and other air pollutant emissions were calibrated to EDGAR 4.2 (EC-JRC/PBL 2012). For the land use and agriculture sectors, agricultural statistics (Food and Agriculture Organization of the United Nations (FAO) 2013), land use RCP data (Hurtt et al. 2011), and GTAP data (Avetisyan et al. 2011) were used for physical data. The 2005 data are utilized as the base year for model calibration.

10.4.3 Scenario Description

In COP21, Thailand has announced the INDCs obligation to the UNFCCC until 2030. To align with such an obligation, the time horizon of this study is arranged in 2030 with Thailand's INDC. The socioeconomic indicators including gross domestic products (GDPs) and population growth are taken from the Thailand's Power Development Plan 2015 (PDP2015) (Ministry of Energy (MOE) 2015). The Office of the National Economic and Social Development Board (NESDB) published the GDP growth and the population growth during year 2014–2036 including outcomes from the master plan for sustainable transport system and mitigation of climate change impacts (Office of Transport and Traffic Policy and Planning (OTP) 2016). The averaged GDP growth and the population growth are expected to increase about 3.94% and 0.03% annually, respectively. The electricity generation assumption in 2030 in both the business-as-usual (BAU) scenario and the GHG emission reduction scenario is shown in Table 10.3. These scenarios are designed based on the stringent of GHG emission reduction level. However, the targets of these renewable energies are checked and complied with PDP 2015 of Thailand. Four scenarios are formulated in this study, namely, RED1, RED2, RED3, and RED4 scenarios. The RED1 and RED2 scenarios are designed to be similar to Thailand's INDC commitment. The RED3 and RED4 scenarios are considered alternative options to achieve the stringent GHG mitigation and effects on Thailand's economy. However, carbon capture and storage technologies and nuclear power plants are excluded from this study.

Table 10.3 Renewable electricity generation assumption in 2030

Fuel type	BAU (MW)	Emission constraint (MW)
Hydro power	1751	2950
Biomass	945	5760
Solar	779	4660
Wind	528	2317
Coal	9095	7760
Oil	60	1
Natural gas	54,485	17,627
Total	**67,643**	**41,075**

10.5 Results

10.5.1 The Future Trends of Socioeconomic Indicators

Owning to the economic development and the increment of income, GDP per capita level strongly increases in the BAU scenario without any climate policy interruption during 2005 and 2030. Thailand's per capita GDP will gradually grow by approximately 2.8 times when compared to the 2005 level in the BAU scenario (see Fig. 10.2).

10.5.2 Primary Energy Mix

Total primary energy supply (TPES) and total final energy consumption (TFC) will augment to 136.9 Mtoe and 104.4 Mtoe within 2030 or equivalent to an augmentation of 1.7 times and 1.6 times, respectively. Meanwhile, GHG emission will continue increasing from 383.2 million tons of carbon dioxide equivalent (Mt CO_2eq) to 561.8 Mt CO_2eq between 2005 and 2030 with an average increase by approximately 1.5% compound annual growth rate (CAGR).

The economic development together with the rise of income reveals the increase of TPES. The BAU scenario shows the highest amount of TPES in 2030 (137 Mtoe). Figure 10.2 shows that TPES will increase in all scenarios by 2030 when compared to 2005. The GHG reduction measures are introduced to the economy which enforce the descent of TPES under RED1, RED2, RED3, and RED4 scenarios compared to BAU scenario. TPES in RED1, RED2, RED3, and RED4 scenarios are 126 Mtoe, 122 Mtoe, 117 Mtoe, and 105 Mtoe, respectively. The decrease in TPES under RED1, RED2, RED3, and RED4 scenarios will be 11 Mtoe, 14 Mtoe, 20 Mtoe, and 32 Mtoe, respectively. RED4 scenario shows the lowest level of TPES due to the stringent GHG reduction which encourages the rise of the energy price. The RED4 scenario can be saved TPES by 30% when compared to the BAU scenario in 2030.

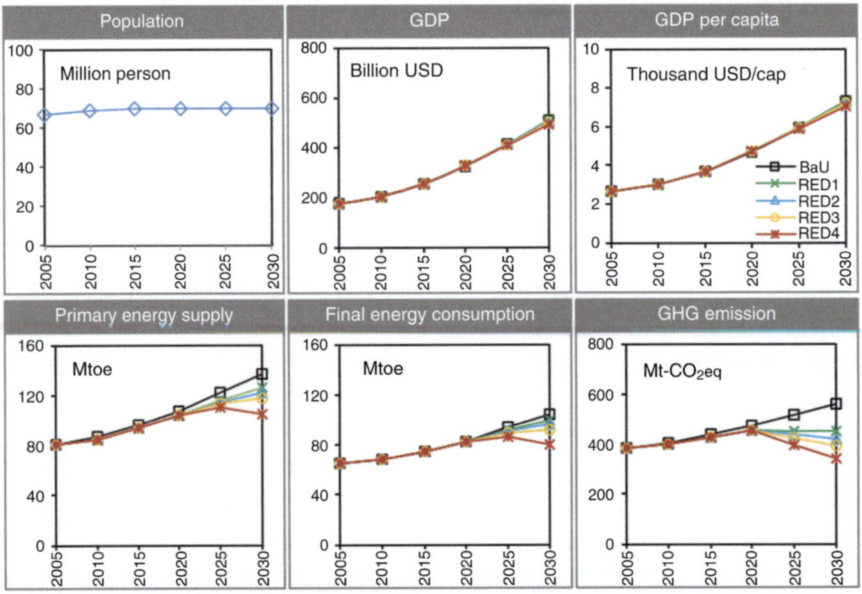

Fig. 10.2 Thailand's socioeconomic indicators, energy and emission trajectory

Figure 10.3 shows the energy intensity and GHG intensity under the GHG emission constraint scenarios. The energy intensity described in terms of TPES per GDP will gradually decrease. The GHG intensity represented as a ratio between GHG emission and GDP will slightly drop between 2.2 t CO_2eq and 1.1 t CO_2eq during 2005–2030 in the BAU scenario.

The share of fossil fuel, particularly share of coal consumption, will increase without any climate change policy intervention in the BAU scenario. The share of fossil fuels will diminish in the GHG reduction scenarios (see Fig. 10.4). The share of fossil fuels (coal, crude oil, and natural gas) will be reduced by 16%, 20%, 26%, and 39% under the RED1, RED2, RED3, and RED4 scenarios when compared to the 2030 BAU scenario, respectively. By contrast, the share of renewable energy will gradually drop during the study time frame in the BAU scenario as shown in Fig. 10.5. As a result, the share of renewable energy will decrease by 10% approximately in 2030. However, climate policy intervention will strongly affect the energy diversification. The share of renewable energy will moderately increase by 16.5%, 17.1%, and 18.1% under the RED1, RED2, and RED3 scenarios, respectively. Moreover, the RED4 scenario shows that the highest share of renewable energy will be 21% in 2030. Note that the share of renewable energy indicated in this section includes solar, wind, hydro, and biomass. Owing to Thailand being an agriculturally based country, thus, the share of biomass particularly bagasse and rice husks takes the highest share of renewable energy. In addition, the promotion of power plants integrated with biomass co-firing, together with feed-in tariff system, greatly persuades the spending for generated electricity by investors.

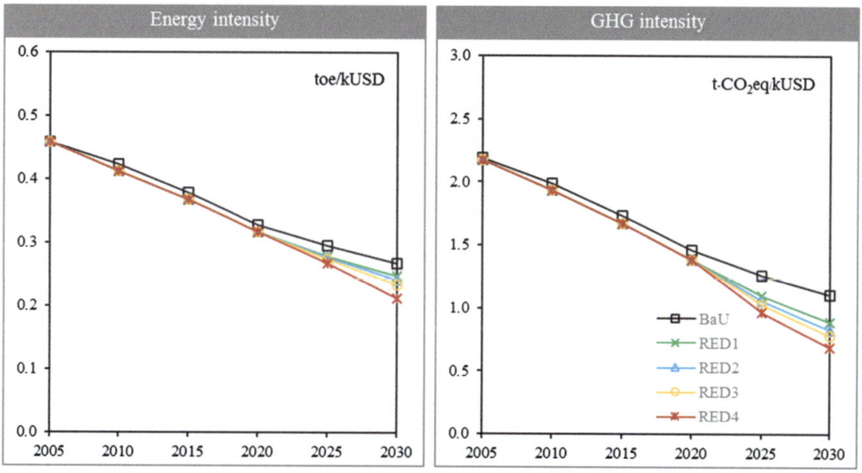

Fig. 10.3 Energy and emission intensity

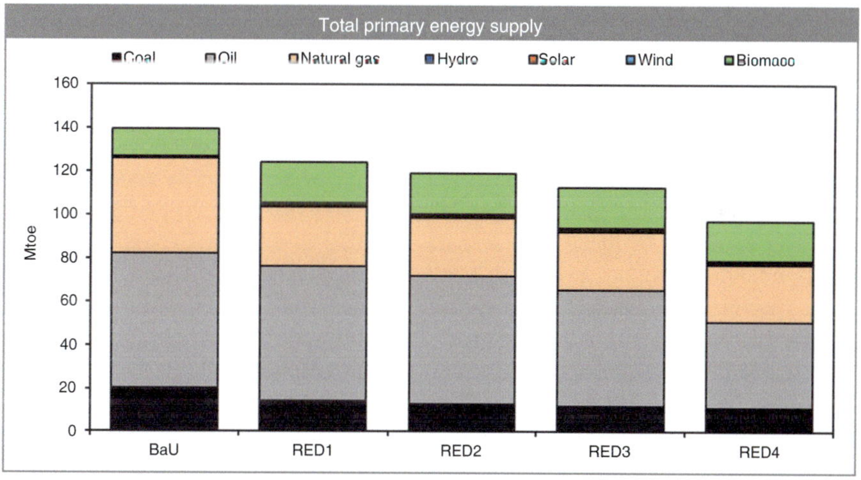

Fig. 10.4 Primary energy mix in 2030

10.5.3 GHG Emissions

According to fossil fuel-based combustion, total GHG emissions are forecasted to moderately increase to about 561 Mt CO_2eq in the BAU scenario in 2030 level (Fig. 10.6). The GHG emissions mostly rely on the human behaviors and industrial structural changes. The GHG emission constraints are externally given. The GHG emission pathway shows the descending trend starting from 2020 in the GHG emission reduction scenarios. The model projections show that Thailand's GHG

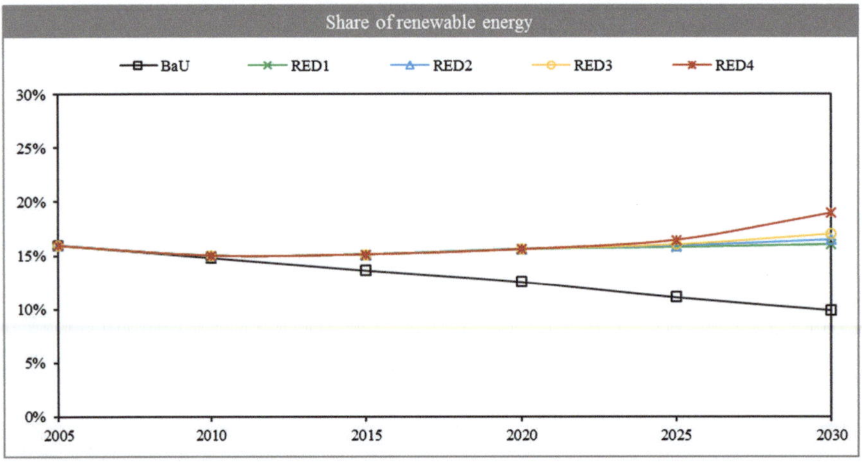

Fig. 10.5 Share of the renewable energy

emission will peak at 2020 (see Fig. 10.2). The amount of GHG emission in the 2020 is 456 Mt CO_2eq in the GHG emission reduction scenarios. The RED1 scenario shows the lowest GHG emission reduction. The GHG emission can be reduced by 20% when compared to the BAU scenario in 2030 level. The RED1 scenario is already aligned with Thailand's INDC commitment. As Thailand already pledged the INDC commitment, Thailand will reduce its economy-wide GHG emissions by 20% in 2030 when compared to its BAU in 2030. Furthermore, the GHG emission of the RED2 scenario in 2030 is 421 Mt CO_2eq. The GHG emission could be reduced by 25% when compared to the BAU scenario in 2030 level. The corresponding commitment further mentions that GHG emissions could be reduced by 25% with the international support and the technology knowledge transfer. Meanwhile, the RED3 and RED4 scenarios substantially reduce the GHG emissions. Therefore, the GHG emissions will be reduced by almost 30% and 40% in RED3 and RED4 scenarios, respectively.

10.5.4 Economic Implications

Another purpose of this study is to examine the mitigation cost resulting from the GHG emission constraint scenarios. Thus, GHG price, GDP loss, and welfare loss are presented in this section.

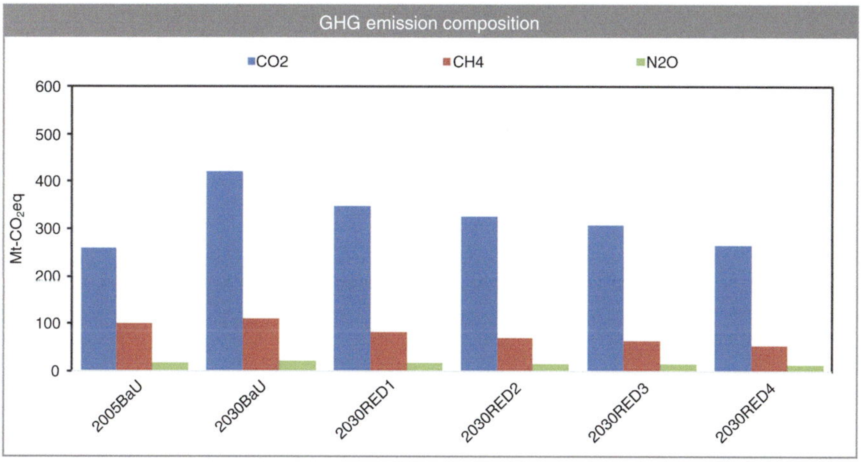

Fig. 10.6 GHG emission composition in 2030

10.5.4.1 GHG Price

Figure 10.7 depicts the GHG price trajectory affected by the GHG emission reduction scenario. The GHG prices are endogenously calculated, while GHG emission constraints are given exogenously. The induced emission price is directly related to the carbon-intensive sectors. The levels of the emission prices reveal the amount that should be paid for responding to the emission activities. The emission prices not only stimulate the GHG emissions reduction activities but also encourage the investment in the clean technology and the low-carbon pathway. The emission price apparently related the emission reduction between the BAU scenario and the GHG emission constraint scenarios. In order to investigate the transformation from high-carbon-intensive economy to low-carbon-intensive economy, therefore, it is reasonable for considering the emission prices within the economy.

The aforementioned results disclose that the industrial sector will significantly reduce the GHG emission. The emission price will obviously start to rise from 2021 which GHG emission reduction targets introduce. The emission prices gradually escalate through 2030. It can be seen that the emission price will be induced with the stringent emission reduction level. The induced emission prices start from US $1/t-$CO_2$eq in 2021 (see Fig. 10.7). The emission price in 2030 ranges from US$6/t-$CO_2$eq to US$16/t-CO_2eq in the RED1 and RED2 scenarios. The emission price ranges from US$35/t-$CO_2$eq to US$110/t-CO_2eq in the RED3 and the RED4 scenarios. In conclusion, the CO_2 emissions in the power sector will keep constant throughout GHG emission constraint scenarios following the PDP2015. Hence, the emission price will hurt the carbon-intensive sectors particularly in the industrial sector.

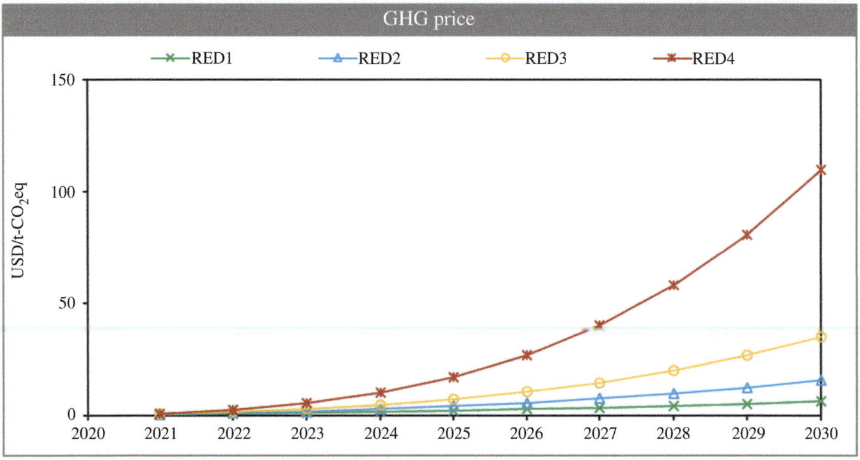

Fig. 10.7 GHG price trajectory

10.5.4.2 GDP Loss and Welfare Loss

Obviously, the emission prices stimulate the carbon-intensive sectors for reducing their fossil fuel combustion activities. Such emission prices directly have the adverse impacts on the economy. Consequently, GDP loss and welfare loss substantially increase, while the investment of clean technologies together with low-carbon societies is gradually focused. The GDP loss and welfare loss in this study are measured as relative change between the GHG emission constraint scenarios and the BAU scenario. The GDP loss substantially increases throughout the RED1 to RED4 scenarios (see Fig. 10.8). The GDP loss ranges from 0.2% in the RED1 scenario to 3.1% in the RED4 scenario in 2030. Moreover, welfare loss can be apparently investigated by the ratio between the household expenditure and government consumption in the GHG emission constraint scenarios and the BAU scenario. The import and export are balanced in the AIM/CGE model. Hence, the dissimilarity of GDP change in each scenario absolutely depends on the household expenditure. The welfare loss illustrates the surpassing amounts when compared to the GDP loss under the same GHG emission constraint scenario. Hence, welfare loss would be from 0.21% to 4.2% in 2030 under the RED1 to RED4 scenarios. In conclusion, the GDP loss and welfare loss apparently imply that Thailand will achieve a better living standard under the RED1 to RED4 scenarios. Both GDP loss and welfare loss can also reveal that there is an improvement in the end-use fuel switching, the end-use structural change, the end-use efficient appliances, and the end-use behavior changes.

Fig. 10.8 GDP loss and welfare loss in 2030

10.5.4.3 Co-benefits of Other GHG Emissions and Other Air Pollutants

In this study, the co-benefits of other GHG emissions and other air pollutants are investigated in terms of CH_4, N_2O, sulfur dioxide (SO_2), nitrous oxide (NO_x), ammonia (NH_3), carbon monoxide (CO), and non-methane volatile organic compound (NMVOC). The GHG emissions including the CH_4 and N_2O are calculated based on the global warming potential from an Intergovernmental Panel on Climate Change (IPCC). Moreover, the emission factors of other air pollutants are taken from the IPCC guidelines for national greenhouse gas inventories. Since Thailand is an agriculturally based country, Fig. 10.9 shows that the agricultural sector will gradually reduce the CH_4, N_2O, and SO_2 emission ranging from 16% to 37%, 19% to 33%, and 1% to 10%, respectively. The industrial sector will distribute the highest SO_2 and NO_x emissions reduction owing to the fossil fuel switching from the coal combustion to the natural gas combustion. The NO_x emissions will be reduced in the transport sector according to the fuel switching from oil to alternative energy.

10.6 Discussion

10.6.1 Implication of the Modeling Results, Policy Implication, and Limitation

The aforementioned results illustrated in the previous section show the remarkable insight for achieving Thailand's INDC. Therefore, there are five key points that can be discussed from the modeling outputs as follows.

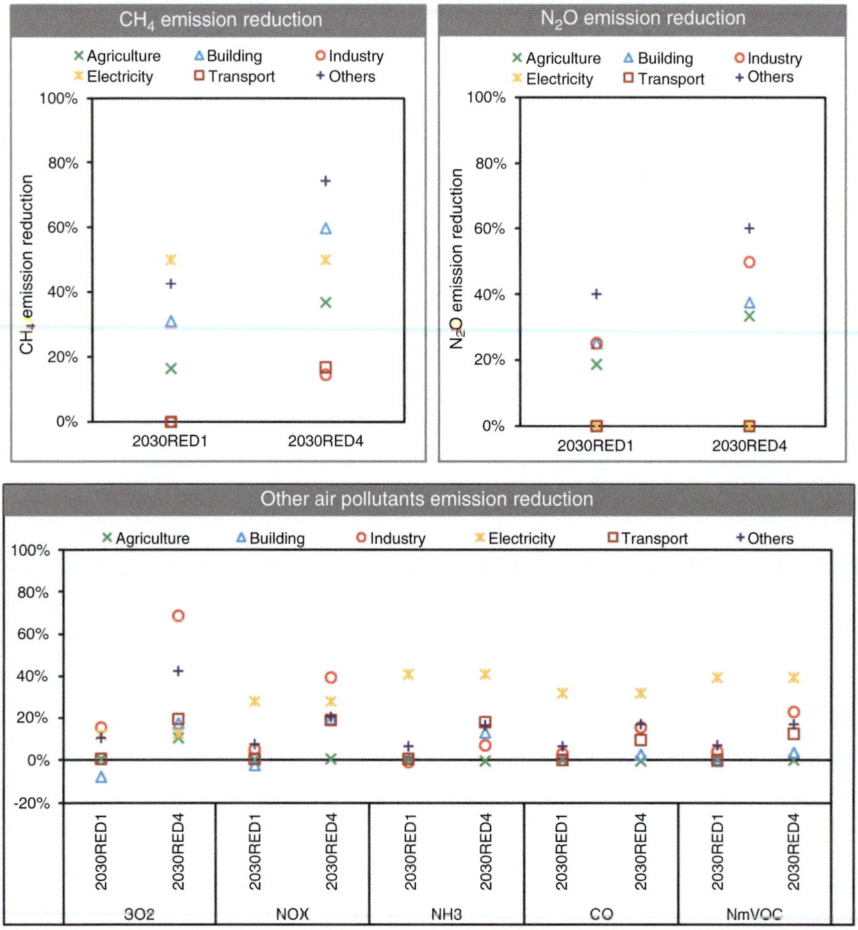

Fig. 10.9 Sectoral air pollutants emission reduction in 2030

First, the GDP loss and welfare loss will gradually increase. The RED1 scenario and the RED2 scenario imply that the renewable energy for the electricity generation sector in the PDP 2015 is appropriated for achieving Thailand's INDC target. Due to the fact that renewable energy can lessen the GDP loss and welfare loss, an additional renewable energy and the availability of land for deploying the renewable energy technologies such as solar, wind, and biomass need to be evaluated to meet the GHG emission levels. Thus, having plenty of renewable energy in the electricity generation sector not only reveals the possibility to achieve the stringent GHG emission reduction but also shows a cost-effective manner. Under the RED1 scenario and the RED2 scenario, the GHG prices of US$18 and US$37 per ton of GHG demonstrate that the renewable energy capacity is appropriately introduced for achieving Thailand's INDCs. However, the installed capacity of renewable energy in the PDP2015, which is designed for 20% renewable electricity, may not be sufficient to meet higher emission reduction targets. Thus, the government should provide not only the ambitious renewable energy target but also disclose the co-benefits of renewable energy to the community. Thus, it is recommended that policy makers should also present the investment cost, technological characteristics, and return of investment to the investors for their decision-making.

Second, Thailand has been upgraded from a lower-middle-income country to an upper-middle-income country in 2011. Moreover, Thailand has obviously switched from agriculture base to a major exporter in Southeast Asia alongside the substantial economic development in the last century (Asian Development Bank (ADB) 2015). Furthermore, the people earn more income and, thus, the capability of spending on high-quality goods with consumed less energy compared to the conventional one. Moreover, the stringent GHG emission reduction levels increase the price of fossil fuel in energy-related CO_2 industries; therefore, there is a shift from high-carbon-intensive commodities to low-carbon-intensive commodities. Such an impact can also induce the efficient technologies which will reduce the economic cost. However, these factors are an important aspect affecting the energy use and GHG emission. Furthermore, NESDB mentioned that Thailand will become an aging society in the future; therefore, aging people will expend more on health services thus diminishing the energy consumption and the GHG emission.

Third, a clear communication between government and private sectors needs to be discussed due to the rapid penetration of renewable energy that could reduce the mitigation cost and the macroeconomic loss. Thus, the renewable energy incentive policy should be aligned with the national climate policy. Nowadays, the government has already launched the incentive called the "feed-in tariff" mechanism. The mechanism particularly stimulates the private sector to invest in renewable energy including small hydropower, ground-mount solar farm, solar rooftop for residential sector, wind power, biomass power plants, and municipal solid waste power plants. Furthermore, we suggest that policy makers also consider the impacts of distance between renewable sites and an urban area. Therefore, the policy makers obviously provide the length of transmission lines and visibility restriction to the renewable energy sites.

Fourth, the development of infrastructure including smart grid and energy storage technologies is another mechanism to stimulate the penetration of renewable energy. Currently, the Thailand's smart grid policy plan and roadmap have been publicly disclosed. There are three stages of implementation: stage 1 (planning and pilot projects including microgrid and other related systems and equipment from 2012 to 2016), stage 2 (expanded the pilot projects into larger facilities covering major cities and developed efficient large-scale renewable energy and energy storage from 2017 to 2021, and stage 3 (enabled nationwide smart grid and applied "two-way power supply of electric vehicles"). However, smart grid and energy storage would be implemented successfully; Thailand would not only become a region hub for distributing the large-scale renewable energy and energy storage but also encourage the renewable energy industry to establish the factories in Thailand.

Finally, this study also investigates the reduction of other air pollutants under the GHG emission reduction levels. The study implies that all pollutants would be reduced economy-wide excluding the electricity generation sector. Therefore, the effect of GHG emission reduction under the GHG emission reduction levels not only gives the sustainable development insight but also reveals the co-benefits of human health.

10.7 Conclusions

We can conclude that the renewable energy plan published in PDP 2015 can achieve Thailand's INDC. GDP loss and welfare loss will increase from 0.2% to 3.1% and 0.21% to 4.2%, respectively. Moreover, the mitigation cost also increases under the corresponding targets. Thus, to achieve the stringent GHG emission reduction conditions, the government needs to revise and harmonize the availability of renewable energy and the available land with the national climate policy. In addition, the role of renewable energy combined with the GHG emission reduction targets also provides the reduction of other air pollutants thus improving the quality of health.

The future works will include nuclear power in the analysis since the Thai government plans to add nuclear power plants after 2035. Moreover, the carbon capture and storage obviously show the tremendous emission reduction potential. Therefore, both technologies would play a vital role in GHG mitigation. Further studies would be covering the impacts of smart grid on renewable energy deployment and estimate the role of energy storage. The economic implication of electric vehicles requires another further analysis.

Acknowledgments The authors would like to acknowledge the support from Sirindhorn International Institute of Technology (SIIT), Thammasat University, Thailand, and the National Institute for Environmental Studies (NIES), Japan, which was made possible by JSPS KAKENHI Grant Number JP16K18177 and the Global Environmental Research Fund 2–1402 of the Ministry of Environment of Japan. The authors are most grateful to the generosity of these funds.

References

Asian Development Bank (ADB) (2015) Thailand: industrialization and economic catch-up. Asian Development Bank, Mandaluyong City

Avetisyan M, Baldos U, Hertel TW (2011) Development of the GTAP Version 7 Land use data base. GTAP Research Memorandum

Chunark P, Limmeechokchai B (2015) Energy saving potential and CO_2 mitigation assessment using the Asia-Pacific integrated model/enduse in Thailand energy sectors. Energy Procedia 79:871–878

Chunark P, Thepkhun P, Promjiraprawat K, Winyuchakrit P, Limmeechokchai B (2015) Low carbon transportation in Thailand: CO_2 mitigation strategy in 2050. SpringerPlus 4:1

Dai H, Masui T, Matsuoka Y, Fujimori S (2012) The impacts of China's household consumption expenditure patterns on energy demand and carbon emissions towards 2050. Energy Policy 50:736–750

Dai H, Silva Herran D, Fujimori S, Masui T (2016) Key factors affecting long-term penetration of global onshore wind energy integrating top-down and bottom-up approaches. Renew Energy 85:19–30

den Elzen M, Admiraal A, Roelfsema M, van Soest H, Hof AF, Forsell N (2016) Contribution of the G20 economies to the global impact of the Paris agreement climate proposals. Clim Chang 137:1–11

Dimaranan BV (2006) Global trade, assistance, and production: the GTAP 6 data base. Center for Global Trade Analysis, Purdue University

EC-JRC/PBL (2012) Emission database for global atmospheric research (EDGAR), release version 4.2. Retrieved from http://edgar.jrc.ec.europa.eu. Accessed May 2012

Food and Agriculture Organization of the United Nations (FAO) (2013) FAOSTAT. FAO, Rome

Fujimori S, Matsuoka Y (2011) Development of method for estimation of world industrial energy consumption and its application. Energy Econ 33:461–473

Fujimori S, Masui T, Matsuoka Y (2012) AIM/CGE [basic] manual. Discussion paper series. Center for Social and Environmental Systems Research, National Institute Environmental Studies, pp 1–87

Fujimori S, Masui T, Matsuoka Y (2014) Development of a global computable general equilibrium model coupled with detailed energy end-use technology. Appl Energy 128:296–306

Fujimori S, Masui T, Matsuoka Y (2015) Gains from emission trading under multiple stabilization targets and technological constraints. Energy Econ 48:306–315

Hasegawa T, Fujimori S, Shin Y, Takahashi K, Masui T, Tanaka A (2014) Climate change impact and adaptation assessment on food consumption utilizing a new scenario framework. Environ Sci Technol 48:438–445

Hasegawa T, Fujimori S, Shin Y, Tanaka A, Takahashi K, Masui T (2015) Consequence of climate mitigation on the risk of hunger. Environ Sci Technol 49:7245–7253

Hasegawa T, Fujimori S, Masui T, Matsuoka Y (2016) Introducing detailed land-based mitigation measures into a computable general equilibrium model. J Clean Prod 114:233–242

Hurtt GC, Chini LP, Frolking S, Betts RA, Feddema J, Fischer G et al (2011) Harmonization of land-use scenarios for the period 1500–2100: 600 years of global gridded annual land-use transitions, wood harvest, and resulting secondary lands. Clim Chang 109:117–161

IEA (2012) Energy technology perspectives 2012. International Energy Agency (IEA)

Intergovernmental Panel on Climate Change (IPCC) (2014) Summary for policymakers. Climate change 2014: mitigation of climate change. Cambridge University Press, Cambridge/New York

International Energy Agency (IEA) (2013a) Energy balances for OECD countries. OECD/IEA, Paris

International Energy Agency (IEA) (2013b) Energy balances for non-OECD countries. OECD/IEA, Paris

Iyer GC, Edmonds JA, Fawcett AA, Hultman NE, Alsalam J, Asrar GR et al (2015) The contribution of Paris to limit global warming to 2 °C. Environ Res Lett 10:125002

Jayatilaka PR, Limmeechokchai B (2015) Scenario based assessment of CO_2 mitigation pathways: a case study in Thai transport sector. Energy Procedia 79:969–975

Jennifer Morgan, Yamide Dagnet, Niklas HÖhne, Sebastian OberthÜr, Lina Li (2015) Race to the top: driving ambition in the post-2020 International Climate Agreement. World Resources Institute

Kusumadewi TV, Limmeechokchai B (2015) Energy efficiency improvement and CO_2 mitigation in residential sector: comparison between Indonesia and Thailand. Energy Procedia 79:994–1000

Ministry of Energy (MOE) (2015) Thailand Power Development Plan 2015–2036 (PDP2015). Energy Policy and Planing Office, (EPPO)

Mittal S, Dai H, Fujimori S, Masui T (2016) Bridging greenhouse gas emissions and renewable energy deployment target: comparative assessment of China and India. Appl Energy 166:301–313

Office of Transport and Traffic Policy and Planning (OTP) (2016) The master plan for sustainable transport system and mitigation of climate change impacts. Retrieved from http://www.otp.go.th/index.php/project-plan/2010-07-06-06-31-21/4745-2012-10-09-09-14-54.html. Accessed 10 Aug 2016

Qi T, Weng Y (2016) Economic impacts of an international carbon market in achieving the INDC targets. Energy 109:886–893

Selvakkumaran S, Limmeechokchai B (2015) Low carbon society scenario analysis of transport sector of an emerging economy – the AIM/Enduse modelling approach. Energy Policy 81:199–214

Selvakkumaran S, Limmeechokchai B, Masui T, Hanaoka T, Matsuoka Y (2015) A quantitative analysis of low carbon society (LCS) measures in Thai industrial sector. Renew Sust Energ Rev 43:178–195

Sundriyal R, Dhyani P (2015) Significance of India's INDC and climate justice: an appraisal. Indian Acad Sciences CV Raman Avenue, Sadashivanagar, PB# 8005, Bangalore 560080, India

Thepkhun P, Limmeechokchai B, Fujimori S, Masui T, Shrestha RM (2013) Thailand's low-carbon scenario 2050: the AIM/CGE analyses of CO_2 mitigation measures. Energy Policy 62:561–572

Tran T, Fujimori S, Masui T (2016) Realizing the intended nationally determined contribution: the role of renewable energies in Vietnam. Energies 9:587

United Nations (UN) (2013) National accounts main aggregates database. United Nations, New York, USA

United Nations Framework Convention on Climate Change (UNFCCC) (2016) Thailand intended nationally determined contributions. Retrieved from http://www4.unfccc.int/submissions/INDC/Published%20Documents/Thailand/1/Thailand_INDC.pdf. Accessed May 2016

Wang X, Zhang S (2015) Exploring linkages among China's 2030 climate targets. Clim Pol:1–12

Winyuchakrit P, Limmeechokchai B, Matsuoka Y, Gomi K, Kainuma M, Fujino J et al (2016) CO_2 mitigation in Thailand's low-carbon society: the potential of renewable energy. Energy Source Part B Econ Plan Pol 11:553–561

WRI (2016) CAIT climate data explorer. Retrieved from http://cait.wri.org/indc/. Accessed 8 May 2016

Wu R, Dai H, Geng Y, Xie Y, Masui T, Tian X (2016) Achieving China's INDC through carbon cap-and-trade: insights from Shanghai. Appl Energy 184:1114–1122

Yamide Dagnet, David Waskow, Cynthia Elliott, Eliza Northrop, Joe Thwaites, Kathleen Mogelgaard, Melisa Krnjaic, Kelly Levin, Heather McGray (2016) Staying on track from Paris: advancing the key elements of the Paris Agreement. World Resources Institute

Zhang W, Pan X (2016) Study on the demand of climate finance for developing countries based on submitted INDC. Advances in Climate Change Research

Chapter 11
Realizing the Intended Nationally Determined Contribution: The Role of Renewable Energies in Vietnam

Thanh Tu Tran, Shinichiro Fujimori, and Toshihiko Masui

Abstract This study contributes to the realization of intended nationally determined contributions (INDCs) by analyzing their implications for the energy production system and the economy and determines the role of renewable energies (RE) in reducing the challenge of committing to the INDCs. The Asia-Pacific Integrated Model/Computable General Equilibrium (AIM/CGE) model was used to assess seven scenarios having the same socioeconomic development but different shares of RE in power generation. By comparing different relative reductions caused by the emission constraints vis-a-vis the business-as-usual (BaU) scenario, the mitigation costs can be estimated. Results show that the economic impact could be reduced by around 55% in terms of welfare loss (from 6.0 to 2.7%) and by around 36% in terms of gross domestic product (GDP) loss (from 3.4 to 2.1%) through the incorporation of high levels of renewable energy. Furthermore, the additional double deployment of wind and SPV to 5.4% and 12.0%, respectively, which currently comprise 43.1% of the renewable energies used in electricity generation, could reduce the GDP loss from 2.1 to 1.9% and reduce the welfare loss from 2.7 to 1.5% in order to achieve a 25.0% GHG emissions reduction. These losses are less than those in the pricing-only scenario (2.1% and 2.3%, respectively).

T.T. Tran (✉)
Regional Centre of Expertise on Education for Sustainable Development in Southern Vietnam, International University—Vietnam National University Ho Chi Minh City, Quarter 6, Linh Trung Ward, Thu Duc District, Ho Chi Minh City 700000, Vietnam
e-mail: tttu@hcmiu.edu.vn

S. Fujimori
Center for Social and Environmental Systems Research, National Institute for Environmental Studies (NIES), 16–2 Onogawa, Tsukuba, Ibaraki 305–8506, Japan

International Institute for Applied Systems Analysis, Schlossplatz-1, Laxenburg 2361, Austria
e-mail: fujimori.shinichiro@nies.go.jp

T. Masui
Center for Social and Environmental Systems Research, National Institute for Environmental Studies (NIES), 16–2 Onogawa, Tsukuba, Ibaraki 305–8506, Japan
e-mail: masui@nies.go.jp

S. Fujimori et al. (eds.), *Post-2020 Climate Action*,
DOI 10.1007/978-981-10-3869-3_11

Keywords Energy outlook of Vietnam • Intended nationally determined contribution • AIM/CGE model • Power development plan • Renewable energies • Emissions gap

11.1 Introduction

On 12 December 2015, the Conference of the Parties 21 (COP21) in Paris achieved a so-called "historic" or "universal" agreement to combat climate change and unleash actions and investment toward a low-carbon, resilient, and sustainable future. Under this agreement, 195 countries were brought together for the first time to tackle a cause based on their historic, current, and future responsibilities. The main aim was to keep the global temperature rise in this century well below 2 °C and aim for an even more ambitious limit of 1.5 °C. The intended nationally determined contributions (INDCs) is the latest pledge for the climate nehiion (Jacoby and Chen 2014; Levin et al. 2015) that brings several concerns (Höhne et al. 2014). First, most of the discussions focus on the political negotiation and financial supports, rather than the possibility of bridging the gap between current emissions pathways and the 2 °C target (United Nations Environment Programme (UNEP) 2015). Second, targeting greenhouse gas (GHG) emissions reduction may compromise national energy security (Turton and Barreto 2006), especially in developing countries where conventional energies are still the main sources (Tran et al. 2010).

Thus, it is not the issue of one single country to determine, between INDCs target and energy security, which one is more important. In developing countries, the priority would be on energy security in which the country uses the existing domestic conventional energy sources to ensure the economic development. As a result, it is necessary to have an integrated assessment of national climate change mitigation target and energy policy, especially renewables (Mathiesen et al. 2011). Particularly, the controversial concerns across nations are on the renewable energy's technology (Ludig et al. 2011; Luderer et al. 2014) and investment (Fuss et al. 2012). With regard to climate mitigation in Vietnam, in September 2012, the Vietnamese government approved the national green growth strategy (NGGS) (Vietnam-Prime Minister 2012). The GHG emissions reduction target in NGGS was 10–20% of the projected 2030 emissions under the business-as-usual (BaU) scenario. In September 2015, the Vietnamese government submitted their INDCs to the United Nations Framework Convention on Climate Change (UNFCCC), in which the target was to reduce GHG emissions in 2030 by 8–25% relative to the BaU scenario (Vietnam-Prime Minister 2015). The former target is unconditional, while the latter is conditional on receiving international support.

The INDCs' target is more ambitious than the NGGS target and thus more risky to the nation's energy security (Karki et al. 2005), with the country still being very dependent on traditional fossil fuels (England and Kammen 1993). Therefore, to support the achievement of the INDCs' target, in November 2015, the Vietnamese government approved the renewable energy development strategy (REDS) (Vietnam-Prime Minister 2015). Compared to previous energy development plans such as the national energy development strategy (NEDS) (Vietnam-Prime Minister 2007) and power development plan 7 (PDP7) (Vietnam-Prime Minister 2011a), published in 2007 and 2012, respectively, the contribution of renewable energies (RE) in the REDS was increased and included more deployment of biomass, solar photovoltaic (SPV), and wind power. Previous studies have also identified the potential of renewable energies in Vietnam, either for final consumption or for power generation (Universit et al. 2005), highlighting the economic benefit (Nguyen and Ha-Duong 2009) of sustainable development (Hai and Lien 2009). The role of renewable energies in climate change mitigation has been assessed, especially biomass (Kumar et al. 2003), SPV (Polo et al. 2015; Dung 1997), and wind (Nguyen 2007a, b). However, GHG emissions reduction generally has a cost and sometimes is not small enough to ignore, especially in developing country like Vietnam (Tran et al. 2011). For example, Mittal et al. (2016) reported that the cost involved depends on the strength of the emissions reduction and renewable energy policy.

Considering the above backgrounds, this study contributes to the realization of Vietnam's INDCs by analyzing its implication to energy production system and economy and determined the role of renewable energies in reducing the challenge of committing to the INDCs. To achieve this goal, the Asia-Pacific Integrated Model/Computable General Equilibrium (AIM/CGE) model was used, and seven scenarios (BaU and six mitigation scenarios with different emissions targets and renewable energy policies) were assessed. The remaining part of this paper is structured as follows: Sect. 11.2 describes the methodology and the design of scenarios. Sect. 11.3 analyzes the main results, with a focus on the energy supply and demand changes. Some additional issues related to this study are discussed in Sect. 11.4. Finally, the main findings of this study are provided in Sect. 11.5.

11.2 Materials and Methods

11.2.1 AIM/CGE Model

The AIM/CGE model has been widely used for assessing global climate change mitigation and adaptation policies (Fujimori et al. 2014a, b, 2015; Hasegawa et al. 2014, 2015;Dai et al. 2016). It is a recursive-dynamic general equilibrium model, extended from the "standard CGE model" (Lofgren et al. 2002), in which industry is classified into 42 sectors (shown in the Table 13.10 in Chap. 13). Details of the model structure and mathematical formulas are described in Fujimori et al. (2012). The model is specified and solved as a pure mixed complementary problem (MCP). This model has a national version, which was applied in previous studies (Mittal et al. 2016; Namazu et al. 2013a, b; Hasegawa et al. 2016), and thus we used the national version in this study.

As with most energy technology-rich CGE models, parameters such as population, gross domestic product (GDP), energy efficiency improvement, the extraction cost of fossil fuels, and the availability and cost of renewable energy were exogenously input into the model. The model outputs were energy demand and supply structure, GHG emissions, and prices for energy and carbon emissions. More technically, the production sectors were assumed to maximize profits subject to multi-nested constant elasticity substitution (CES) functions and the relative prices of inputs. Household expenditures on each commodity were described by a linear expenditure system (LES) function. The savings were the source of domestic and foreign direct investment. They were exogenously inputted as a proportion of the change in GDP relative to the base year. To balance the saving account, the saving rate was treated as an endogenous variable. Moreover, the capital formation of each good was determined by a fixed coefficient of total investment. The Armington assumption was used for international trade, and the current account was assumed to be balanced. In addition to energy-related CO_2 emissions, CO_2 from other sources, CH_4, and N_2O (including changes resulting from land use and non-energy-related emissions) were also considered.

The constraint on the GHG emissions was specified based on the mitigation target. Once the emission constraint was added, the carbon tax became a complementary variable to the emission constraint and determined the marginal mitigation cost. In this emission-constrained scenario, the carbon tax increased the price of the fossil-fuel-consuming goods, which resulted in an energy saving and substitution of fossil fuels with lower emission energy sources. The carbon tax also acted as an incentive to reduce non-energy-related emissions. GHGs other than CO_2 were weighted by their global warming potential and summed to estimate total GHG emissions in terms of CO_2 equivalent. Households were assumed to receive the revenue from the carbon tax. In the power sector, the fired power plants had an option to see carbon capture and storage (CCS). We assumed a maximum annual penetration rate for CCS technology. The maximum annual penetration rate differed over time. More specifically, in the early stages, new installations were limited (if the share of CCS was under 10%, 2% per annum was the maximum); but as the technology matured, the rate of penetration increased (if the share of CCS was over 10%, 4% per annum was the maximum). Although the model had this CCS option, in this analysis, it was found that CCS did not penetrate much because the carbon price did not reach a critical level.

The costs of the wind and solar technology were assumed as in the IEA energy technology perspective report (International Energy Agency 2012). We changed the input coefficients in the corresponding production function. Because the output prices of these technologies were endogenously determined, they were not exactly the same as those reported by the IEA (International Energy Agency 2012), but they were close enough. As a reference, the solar and wind power output prices for the baseline scenario are shown in Fig. 11.1.

The energy source share is determined by logit function with two parameters, one determines base share and the other represents price elasticity (Fujimori et al.

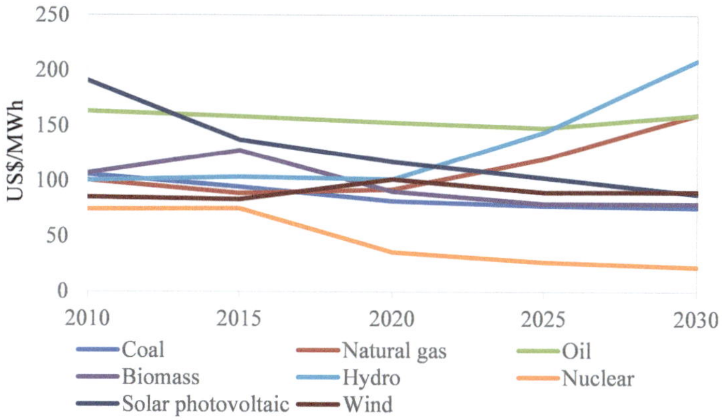

Fig. 11.1 Solar and wind power price

2012). Price includes tax and carbon emission tax which is formulated as multiplying energy use ratio and emission coefficient associated with carbon tax rate.

11.2.2 Scenario Settings

Seven scenarios were developed, with the detailed assumptions shown in Table 11.1. All scenarios had the same socioeconomic assumption, in which we adjusted the GDP growth rate targeted in the national socioeconomic development plan (SEDP) (Vietnam-Prime Minister 2011b) based on the actual development status of Vietnam, and the subsequent population was estimated using this plan. Specifically, the annual average growth rates (AAGRs) during the 2005–2030 period of population growth and GDP were 0.9% and 6.2%, respectively.

The BaU and low renewable (LowRE; LowRE-LowINDC and LowRE-HighINDC) scenarios had the same assumed power generation mix. The power generation assumption was based on the NEDS (Vietnam-Prime Minister 2007) and PDP7 (Vietnam-Prime Minister 2011a), which were formerly disclosed. In addition to fossil fuels, nuclear contributed 10.1% of the total power generation in 2030, followed by hydro (9.3%), SPV (2.5%), wind (2.4%), and biomass (1.1%). On the other hand, in the HighRE scenarios (HighRE-LowINDC and HighRE-HighINDC), the shares of biomass, SPV, and wind increased to 6.3%, 6.0%, and 2.7%, respectively, while the shares of nuclear and hydro were the same as in the previous three scenarios. This assumption was based on the REDS (Vietnam-Prime Minister 2015), which is more renewable aggressive than the NEDS (Vietnam-Prime Minister 2007) and PDP7 (Vietnam-Prime Minister 2011a); therefore, we referred to these scenarios as HighRE. Moreover, the additional double deployment of wind and SPV was also assessed to determine the role of renewable energies in reducing

Table 11.1 Main assumptions and greenhouse gas (GHG) emissions reduction targets

Indicators	BaU	LowRE-LowINDC	LowRE-HighINDC	HighRE-LowINDC	HighRE-HighINDC	AddRE-HighINDC	FreeRE-HighINDC
Socioeconomic development (%/year) (Source: SEDP (Vietnam-Prime Minister 2011b))	Period		Population growth		GDP growth		
	2006–2019		1.1		7.0		
	2010–2020		1.0		6.0		
	2021–2030		0.6		6.0		
	2005–2030		0.9		6.2		
Share of renewables in power generation (%)	**Energy**	2020	2030	2020	2030	2030	No exogenous setting of RE share
	Nuclear	2.1	9.6	2.1	10.1	10.1	
	Hydro	18.3	8.8	19.6	9.3	9.3	
	Biomass	0.6	1.0	3.0	6.3	6.3	
	SPV	3.1	3.1	3.5	6.0	12.0	
	Wind	0.7	2.3	1.0	2.7	5.4	
	Total RE	24.8	24.8	29.2	34.4	43.1	
	(Source: NEDS (Vietnam-Prime Minister 2007), PDP7 (Vietnam-Prime Minister 2011a))			(Source: REDS (Vietnam-Prime Minister 2015))		(Source: REDS (Vietnam-Prime Minister 2015) and assumption)	
GHG emissions reduction in 2030 compared to BaU (%) (Source: Vietnam INDC (Vietnam-Prime Minister 2015))	None	At least 8	25	At least 8	25	25	25

Notes: *BaU* business-as-usual, *RE* renewable energies, *INDC* intended nationally determined contribution, *SEDP* socioeconomic development plan, *SPV* solar photovoltaic, *GDP* gross domestic product, *NEDS* national energy development strategy, *PDP7* power development plan 7, *REDS* renewable energy development strategy, *GHG* greenhouse gas the impacts of climate change policy (AddRE-HighINDC). The increased deployment of renewable technologies in HighRE scenarios and the AddRE scenario substituted for the decline of coal-fired power plants. These scenarios indicated how much the economic burden could be reduced if the government considered a more ambitious renewable target in its energy development plan. In contrast to the six scenarios above, where the RE share is fixed exogenously, we developed the FreeRE-HighINDC scenario, in which the share of RE in power generation is not fixed but rather is driven by the energy price. This was compared with other scenarios to clearly understand the impact of energy policy targets.

The BaU scenario has no GHG emissions reduction target, while the targets for the other scenarios were assumed to follow the INDCs (Vietnam-Prime Minister 2015). The LowRE-LowINDC and HighRE-LowINDC scenarios aimed to reduce at least 8.0% of 2030s BaU emissions as an unconditional target, while the LowRE-HighINDC, HighRE-HighINDC, AddRE-HighINDC, and FreeRE-HighINDC aimed to reduce 25.0% of 2030s BaU emissions as a conditional target.

In policy-based scenarios, the total amount of power generation is assumed following the national power development plan (PDP7), and the energy mix for power generation is assumed following the targeted shares in PDP7 and renewable energy policy. Particularly, the contribution amount of natural gas and nuclear is assumed to be the same in all scenarios. The share of other renewable energies (wind, biomass, and solar power) increases as assumed in the scenario settings, in corresponding with the reduction of hydro and coal.

The assumption on the solar and wind power generation cost is basically following the IEA energy technology perspective. The input coefficients are changed in order to represent the cost decrease. Since the current model assumption is that the power generation price is endogenously determined given the capital and intermediate inputs prices, the outcomes are not exactly same as the IEA, but we think that they are almost similar.

Figure 11.2 shows an outline of the seven scenarios according to the share of different renewable energies (%) in the power generation mix (X axis) and the GHG

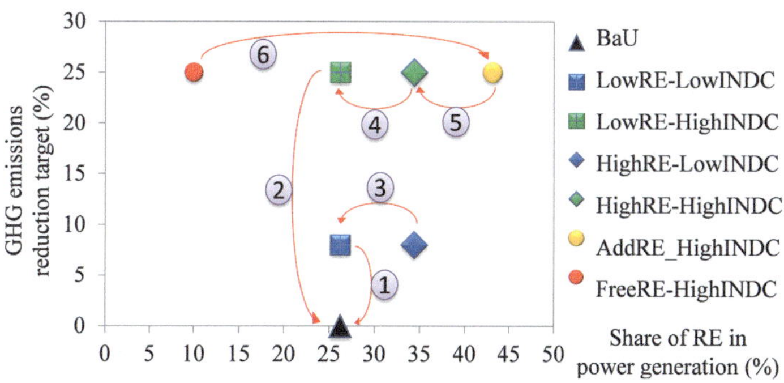

Fig. 11.2 Outline of scenario setting and the subsequent analysis. *GHG* greenhouse gas, *RE* renewable energies, *BaU* business-as-usual, *INDC* intended nationally determined contributions

emissions reduction target (Y axis), with six comparisons made. The first comparison (① in Fig. 11.2) was made between the LowRE-LowINDC and BaU scenarios, to determine the implication of an 8% reduction in total GHG emissions on the energy supply and demand in Vietnam. The second comparison (② in Fig. 11.2) was made between the LowRE-HighINDC and BaU scenarios and indicated that under a more stringent GHG emissions reduction target (25% of BaU), there was a greater challenge to the energy generation system and economy of Vietnam. In the third comparison (③ in Fig. 11.2), the HighRE-LowINDC and LowRE-LowINDC scenarios were compared to determine the contribution of renewable energies in achieving the 8% reduction target. The final three comparisons (④, ⑤, and ⑥ in Fig. 11.2), among the AddRE-HighINDC, HighRE-HighINDC, LowRE-HighINDC, and FreeRE-HighINDC scenarios, assessed the impact of different energy policy and pricing-only scenario, indicating that the role of renewable energies became even more important to achieve the most stringent target in Vietnam's INDCs, with less impact on the economy.

11.3 Results

11.3.1 Outlook for Vietnam in 2030: The Base Scenario

Following the assumed socioeconomic development shown in Table 11.1, the population and GDP of Vietnam increased by 1.2 and 4.5 times in 2030 compared to 2005, respectively. Thus, the population in 2030 was 102.9 million people, and the GDP was around 236.6 billion USD (2005 market exchange rate; MER), resulting in the GDP per capita increasing by 3.6 times, from 638.0 USD in 2005 to nearly 2300.0 USD in 2030 (2005 MER). Under these driving forces, the final energy demand increased 1.8 times, with an AAGR of 2.0% per year, reaching 83.0 milliontonnes of oil equivalent (TOE). The primary energy supply increased 2.9 times (AAGR was 3.9% per year) to 146.0 millionTOE; and there was a rapid increase in power generation of 9.0 times (AAGR was 9.2% per year) to 42.2 millionTOE. The increasing supply and demand of energy resulted in an increase in GHG emissions of 2.4 times, with an AAGR of 3.3% per year (Fig. 11.3).

11.3.1.1 Primary Energy Supply

The primary energy supply increased from 50.7 millionTOE in 2005 to 160.0 millionTOE in 2030, which was a bit higher than the projection in the Asia Pacific Energy Research Centre (APERC) report (150.0 millionTOE in 2030) (Vietnam-Prime Minister 2011b). Currently, Vietnam is a coal exporter; however, the actual trend and even the PDP7 show that Vietnam will become an energy

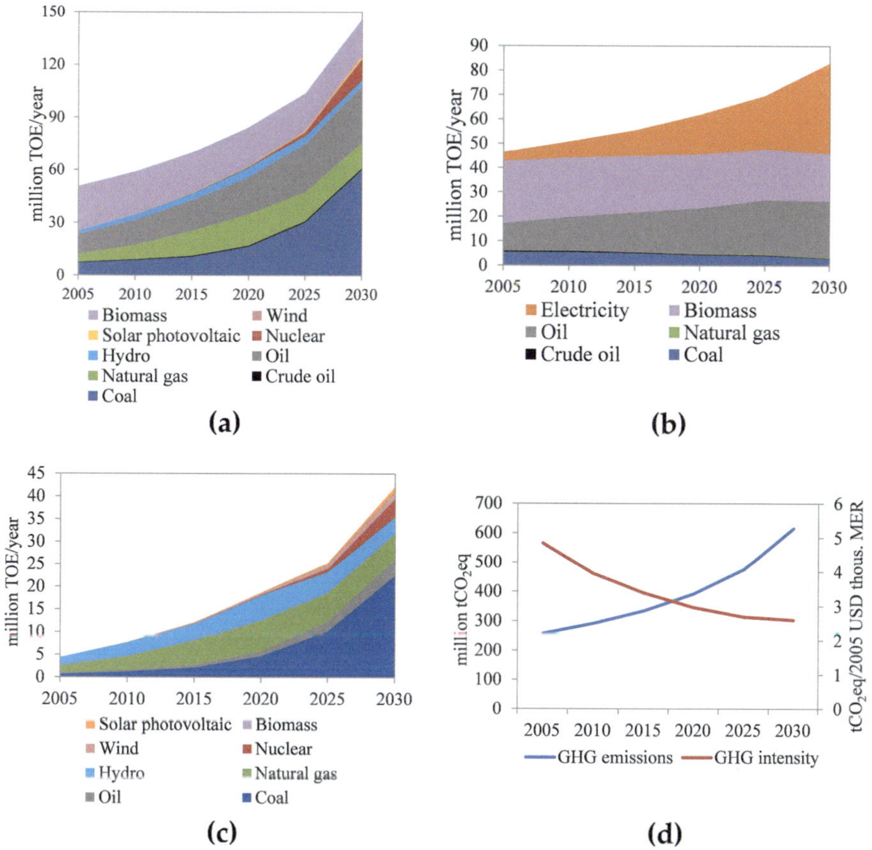

Fig. 11.3 Future development of Vietnam in the BaU scenario. *GHG* greenhouse gas, *BaU* business-as-usual. (**a**) Primary energy supply; (**b**) final energy demand; (**c**) electricity generation; (**d**) GHG emissions and GHG intensity

importer in the next few years, as a result of its high growth in energy demand and the limitations on available energy resources (particularly coal, oil, and electricity). Among the energy carriers, coal and nuclear have experienced a rapid increase due to the demand for these energy types in power generation since 2020. In 2030, the expected coal supply contributed 41.2% of total primary energy supply, followed by oil (21.8%), biomass (13.9%), natural gas (10.1%), nuclear (8.3%), and other renewable energies (hydro, wind, and solar – 4.1%) (Fig. 11.3a). The primary energy intensity was projected to be reduced from 937.8 TOE per unit of GDP to 585.6 TOE per unit of GDP (37.6% reduction). This projection is consistent with the APERC report, where the primary energy intensity is projected to reduce from 286.0 TOE per 2005 USD million purchasing power parity (PPP) in 2005 to 164.0 TOE per 2005 USD million PPP in 2035 (Vietnam-Prime Minister 2011b).

11.3.1.2 Final Energy Demand

The final energy demand increased from 46.5 million TOE in 2005 to 83.0 million TOE in 2030, with the main consumers being the industry (40.6%) and residential sector (36.5%). This projection was less than in the APERC report (nearly 120.0 million TOE in 2030) (APERC 2013). The total demand for oil and biomass would be the same (51.1%) as in the base year, in which their shares were 27.8% and 23.3%, respectively (Fig. 11.3b). Meanwhile, the demand for electricity increased nearly 9.9 times, contributing to 44.9% of the total final energy demand. The final energy intensity was expected to decline by about 60.0% between 2005 and 2030 – from 876.1 TOE per unit of GDP (in 2005 USD million MER) to 350.9 TOE per unit of GDP. This reduction was slightly higher than the 52.0% reduction projected in the APERC report (APERC 2013).

In the baseline scenario, industry is the main energy consumer (37.7 million TOE), with a rapid increase in electricity consumption (72.7%), followed by oil (18.4%), coal (7.5%), natural gas (0.9%), and crude oil (0.4%). The increasing consumption of electricity in industry is mainly due to the assumption that power generation follows the power development plan.

11.3.1.3 Electricity Generation

The total electricity generation and the electricity generation mix were assumed to follow the NEDS (Vietnam-Prime Minister 2007) and the PDP7 (Vietnam-Prime Minister 2011a), in which the total electricity generation was expected to increase rapidly from 4.4 million TOE in 2005 to 42.2 million TOE in 2030. The amount generated covered 97.0% of the total electricity demand, with the additional 3.0% being imported. Within the 42.2 million TOE generated, 53.3% was produced by coal-fired plants and 13.6% was gas-fired generation. Renewable energies contributed 24.8% of total electricity generation, in which nuclear accounted for 9.6%, followed by hydro, SPV, wind, and biomass with 8.8%, 3.1%, 2.3%, and 1.0%, respectively (Fig. 11.3c). This projection was very different from the APERC report (APERC 2013), where the shares of coal and natural gas were 39.0% and 30.0% respectively, followed by 15.0% from hydro, 14.0% from nuclear, and the remainder being imported.

11.3.1.4 GHG Emissions

The GHG emissions per capita of Vietnam in 2005 were about 3.1 tonnes of CO_2 equivalent (tCO_2eq), which is one of the lowest values in the Asia-Pacific region (APERC 2013). With the rapid increase of socioeconomic development and energy consumption, the total GHG emissions increased from 257.1 million tCO_2eq in 2005 to 614.9 million tCO_2eq in 2030. This projection is consistent with Vietnam's industrialization vision, in which the economy uses more carbon-intensive energy sources, and in particular is dependent on coal-fired power plants. The increase in GHG emissions in 2030 was caused by a threefold increase of CO_2 emissions due to energy-related activities, while the increases in N_2O and CH_4 emissions were relatively modest. The emission intensity was expected to be reduced by half, from 4.8 tCO_2eq per 2005 USD thousand MER in 2005 to 2.6 tCO_2eq per 2005 USD thousand MER in 2030 (Fig. 11.3d).

11.3.2 Implication of Vietnam's INDC and the Role of Renewable Energies

Figure 11.4 summarizes the socioeconomic, energy, and GHG emissions changes in the six climate mitigation scenarios compared to the BaU scenario in 2030. In LowRE scenarios where the share of renewable energies in electricity generation was around 24.8% (6.4% biomass-SPV-wind, 8.8% hydro, and 9.6% nuclear), the GHG emissions reduction potential of Vietnam in 2030 was 9.4%. This 9.4% reduction in GHG emissions was achievable through reductions of 1.0% in the final energy demand and 0.7% in the primary energy supply, as well as no effect on electricity generation, causing a welfare loss of 0.5% and a GDP loss of 0.3% (LowRE-LowINDC). For the same situation with regard to renewable energies, to achieve the 25.0% GHG emissions reduction, a much greater reduction in energy demand and supply was required.

In particular, the final energy demand was reduced by 11.7%, the primary energy supply was reduced by 7.9%, and the electricity generation was reduced by 1.2%.

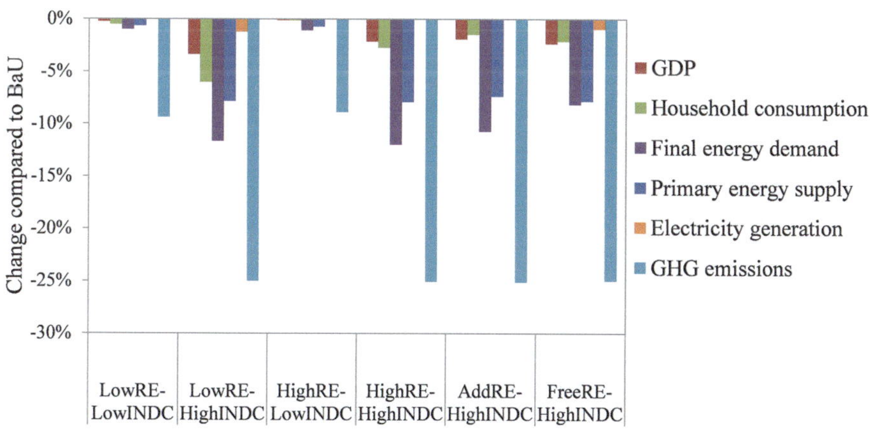

Fig. 11.4 Energy and economic changes under different INDC scenarios in 2030. *GHG* greenhouse gas, *RE* renewable energies, *BaU* business-as-usual, *INDC* intended nationally determined contribution

These reductions would affect the development of Vietnam's economy and were projected to cause a welfare loss of 6.0% as well as a GDP loss of 3.4% (LowRE-HighINDC). In HighRE scenarios, where the share of renewable energies in electricity generation increased to 34.4% (15.0% biomass-SPV-wind, 9.3% hydro, and 10.1% nuclear), it was possible to achieve the GHG emissions reduction target, with less effect on the energy system compared to the LowRE scenarios. In particular, the GHG emissions reduction potential of Vietnam was 8.9%, leading to a 1.1% reduction in final energy demand, a 0.7% reduction in primary energy supply, no effect to the electricity generation, and with a welfare loss of 0.2% and GDP loss of 0.1% (HighRE-LowINDC). Furthermore, the GHG emissions potential of Vietnam was increased to 25.0%, resulting in a welfare loss of 2.7% and GDP loss of 2.1%, in addition to a 12.0% reduction in final energy demand, a 7.4% reduction in primary energy supply, and no effect to the electricity generation (HighRE-HighINDC).

In the AddRE-HighINDC scenario, where the share of wind and SPV increased twofold compared to the HighRE scenarios, the contribution of renewable energies in the power generation mix increased to 43.1% (23.7% biomass-SPV-wind, 9.3% hydro, and 10.1% nuclear). An 8.7% increase from SPV and wind substituted for the same reduction in the amount of coal consumed for electricity generation. In this scenario, there was a 1.9% GDP loss and a 1.5% welfare loss, with the potential to reduce total GHG emissions by 25.0%. The effect on the energy system was a 10.8% reduction in final energy demand, a 7.4% reduction in primary energy supply, and no effect to the electricity generation (AddRE-HighINDC).

On the other hand, in the FreeRE-HighINDC scenario, where the RE share is driven by the energy price rather by than following the policy, the contribution of RE in the power generation mix was only 26.7% (9.6% biomass-SPV-wind, 16.4% hydro, and 0.7% nuclear). This low contribution caused an 8.2% reduction in the final energy demand, a 7.8% reduction in primary energy supply, and a 0.9% reduction in electricity generation, leading to a 2.1% welfare loss and a 2.4% GDP loss to achieve the 25.0% GHG emissions reduction target. The sub-sections below describe the effect of the different INDC scenarios on the energy system and economy of Vietnam in more detail.

11.3.2.1 Change in Primary Energy Supply

Similar to the final energy demand, Fig. 11.5 shows that a very small reduction (nearly 0.7%) in the primary energy supply was required to achieve the less stringent GHG emissions reduction target. However, a 7.9% reduction of primary energy supply, corresponding to a 12.0% reduction of final energy demand, was required to achieve a 25.0% reduction in GHG emissions.

The total primary energy supply was reduced from 146.0 million TOE in the BaU scenario to 135.0 million TOE in the LowRE-HighINDC scenario. The supply of other energies was retained, except for the reduction in the oil supply from 31.8 million TOE in BaU to 23.2 million TOE in LowRE-HighINDC. Coal was still

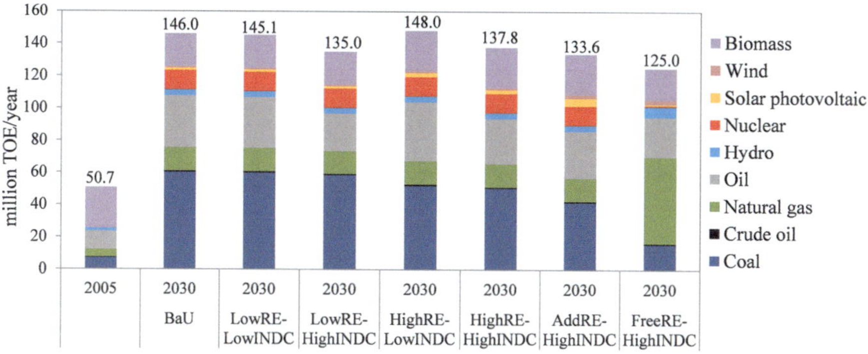

Fig. 11.5 Change in primary energy supply. *RE* renewable energies, *BaU* business-as-usual, *INDC* intended nationally determined contribution

the main primary energy source due to the dependence on coal-fired power generation in Vietnam. Under the low INDC scenarios, the increasing contribution of renewable energies helped to secure the total primary energy supply compared to the BaU scenario, from a 0.6% reduction in the LowRE-LowINDC scenario to a 1.4% increase in the HighRE-LowINDC scenario. Moreover, the supply of coal was reduced by around 14.3% and was replaced by renewable energies, especially the increase of biomass, SPV, and wind. Similarly, in high INDC scenarios, renewable energies would enable the energy system to become less dependent on coal.

Different from the results of policy-based scenario, the pricing-only scenario (FreeRE-HighINDC) showed the domination of natural gas in the total primary energy supply rather than the deployment of renewable energies. The total primary energy supply in this pricing-only scenario was reduced 14.4% compared to the BaU scenario that may challenge the energy security of the country.

11.3.2.2 Change in Final Energy Demand

As shown in Fig. 11.6, under the low GHG emissions reduction target, the total final energy demand was slightly reduced from 83.0 to 82.2 million TOE. This small reduction in the final energy demand in the LowRE-LowINDC scenario was due to a reduction in oil consumption, while other energy sources were the same as in the BaU scenario. To achieve a more stringent GHG emissions reduction target, the total final energy demand was more significantly reduced (by about 11.7%) to reach 73.3 million TOE in the LowRE-HighINDC scenario. This reduction was mainly due to the reduction of oil consumption, from 34.6 million TOE in the BaU scenario to 15.7 million TOE in the LowRE-HighINDC scenario. The reduction of oil consumption was distributed across all energy-related sectors in the economy, including residential and commercial, industry, and transport. It was expected that the consumption of electricity would dominate the final energy demand, due to the increasing use of electrical appliances and machines.

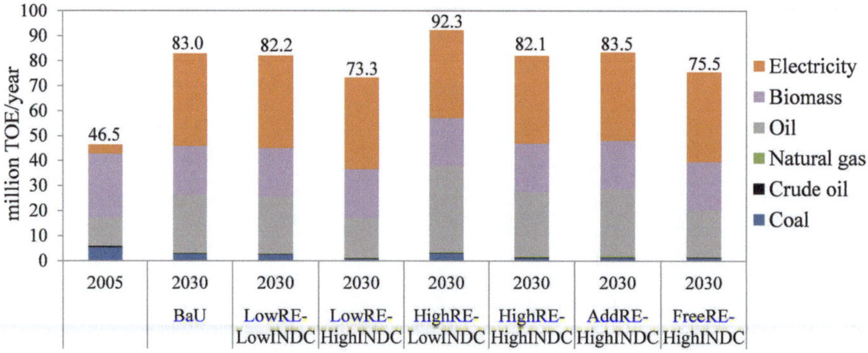

Fig. 11.6 Change in final energy demand. *RE* renewable energies, *BaU* business-as-usual, *INDC* intended nationally determined contribution

However, with a greater deployment of renewable energies, especially the double contribution of biomass, the pressure on energy consumption under the high INDC scenarios was reduced. Instead of reducing 11.7% of its final energy demand, Vietnam might not need to reduce its final energy consumption in order to achieve the 25.0% GHG emissions reduction target (HighRE-HighINDC).

Moreover, when the contribution of wind and SPV increased twofold, the final energy consumption was maintained at 83.5 million TOE (AddRE-HighINDC). This lesser reduction in the final energy demand, compared to BaU, would enable some industries to retain the use of their oil-consuming production machinery; thus, there was less pressure for them to fully switch to new or advanced technology in the medium term. On the other hand, in the FreeRE-HighINDC scenario, the final energy consumption was reduced to 75.7 million TOE, with the energy mix similar to LowRE-HighINDC scenario.

11.3.2.3 Change in Electricity Generation

In the LowRE scenarios, the total amount of electricity generation is not largely affected by the GHG emissions reduction targets, because the main primary source (coal-fired generation) was almost the same due to the contribution of CCS technology in power plants that consumed fossil fuels (Fig. 11.7). Compared to the LowRE scenarios, the total electricity generation in HighRE scenarios was reduced by 0.5%, from 42.2 to 39.9 million TOE. This reduction was due to the 14.5% decline in coal-fired power generation, which was substituted for by the increasing contribution (a nearly sixfold increase) of biomass generation and nearly twofold increase of SPV. In addition, there was also a small share of electricity generation using CCS technology in thermal power generation.

Moreover, in the AddRE scenario, the amount of wind and SPV was twice that in the HighRE scenarios, which helped to reduce the coal consumed in electricity generation from 22.5 to 15.4 million TOE. Similar to the change in

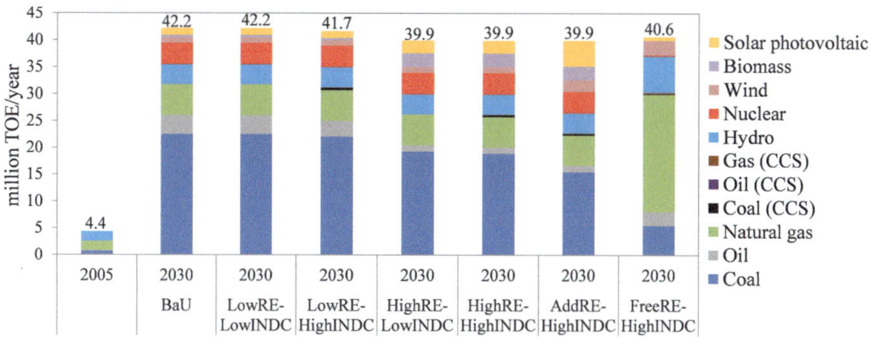

Fig. 11.7 Change in electricity generation. *CCS* carbon capture and storage, *RE* renewable energies, *BaU* business-as-usual, *INDC* intended nationally determined contribution

total primary energy supply, in FreeRE-HighINDC, the electricity generation is mainly from natural gas (52.9%), followed by hydro (16.4) due to the effect of energy price.

The share of RE (wind, biomass, and SPV) in power generation increases from 6.4% in BaU scenario to 6.5% in LowRE-HighINDC and reaches 15.0% and 23.4% in HighRE-HighINDC and AddRE-HighINDC, respectively. However, in the FreeRE-HighINDC, this share is only 8.2%.

11.3.2.4 Welfare Loss and GDP Loss

The economic implication of Vietnam's INDC was shown through the reduction of household consumption (welfare loss) and GDP (GDP loss). These losses are calculated by deriving the mitigation cost through the difference between mitigation scenarios and their respective BaU scenario, i.e., the respective scenario without emissions target but same electricity system configuration. Under the less stringent GHG emissions reduction target, the welfare loss was reduced from 0.5% in the LowRE-LowINDC scenario to 0.2% in the HighRE-LowINDC scenario. When the GHG emissions reduction target was more stringent, the welfare loss increased significantly to 6.0% in the LowRE-HighINDC scenario and reduced to 2.7% in the HighRE-HighINDC scenario when the installed RE increases. Moreover, the whole economy was also expected to experience a loss of GDP when GHG emissions reduction was considered. A similar situation was observed in the consumption reduction: Vietnam experienced the highest GDP loss of 3.4% in the LowRE-HighINDC scenario, which was reduced to 2.1% in the HighRE-HighINDC scenario, and reached 0.1% and 0.3% in the HighRE-LowINDC and LowRE-LowINDC scenarios, respectively. Additionally, the double deployment of biomass, wind, and SPV (AddRE-HighINDC) minimized the GDP loss to 1.9% when the most stringent target was achieved and 1.5% of welfare loss. Moreover, the pricing-only scenario with no fixed share of renewable energy (FreeRE-

HighINDC) showed a 2.1% welfare loss and a 2.4% GDP loss in order to achieve the most stringent target. This implies that the economy would be damaged when the INDC was committed to; however, the increasing deployment of renewable energies could reduce the losses.

11.3.2.5 Carbon Price

There were a wide range of GHG abatement costs in Vietnam among the different scenarios. In the LowINDC scenarios, there was a very small abatement cost of less than 5.0 USD per tCO_2eq (4.3 USD per tCO_2eq in the LowRE-LowINDC scenario and 3.6 USD per tCO_2eq in the HighRE-LowINDC scenario). The HighINDC target placed much more pressure on the economy to reduce the GHG emissions, causing the abatement costs to become 89.2 USD per tCO_2eq in the LowRE-HighINDC scenario and 77.6 USD per tCO_2eq in the HighRE-HighINDC scenario. In both INDC target scenarios, the higher installed RE capacity helped to reduce around 12.0% to 16.0% of the GHG abatement costs. Furthermore, the double deployment of wind and SPV reduced the GHG abatement cost to 65.8 USD per tCO_2eq in the AddRE-HighINDC scenario. Besides, the GHG abatement cost in FreeRE-HighINDC was around 47.5 USD per tCO_2eq.

11.4 Discussion

11.4.1 Implications

There are several implications and interpretations of these modeling results that are discussed in this section. First, the more stringent the GHG emissions reduction target that a country commits to, the more challenges will be faced by the energy system and economy of the country. In particular, the primary energy supply might be reduced, and thus final energy consumption and electricity consumption could decline, with subsequent effects on the development of the economy. However, an appropriate energy policy, with strong deployment of renewable energies, could lessen the challenges and make climate change policy more feasible. This was obvious in Vietnam when the AddRE-HighINDC, FreeRE-HighINDC, and LowRE-HighINDC scenarios were compared. The macroeconomic loss was almost completely diminished by the large renewable energy deployment. The price distortion caused by the carbon price can be substantially reduced and the energy demand would not be largely affected. Thus, we suggest that policymakers give more consideration to developing a national energy development plan, especially when it is necessary to reduce the dependence on coal-fired power plants.

Second, the rapid increase in electricity demand could be met when the contribution of renewable energies in the power generation mix was significantly increased. This could encourage both public and private sectors to invest more in the deployment of renewable energies, especially when the potential of biomass, wind, and solar photovoltaic is very large. The comparisons made in this study showed that the greater the deployment of renewable energies, the more likely it would be to achieve energy security and maintain the 7.0% economic growth target in the next period for Vietnam. It is not an obligation, but if Vietnam could start GHG emissions reduction as soon as possible after 2020 and could achieve its INDCs as considered in this study, then there is a strong possibility that the country could maintain the reduction trend and achieve the target of a 45.0% emissions reduction from energy production in 2050, as discussed in the latest REDS. For this to happen, the energy development plan and policy should be integrated with the actual socioeconomic situation and should consider environmental targets as well as a vision for sustainable development.

Third, it is obvious that there may be some reduction in economic development due to the reduction of GHG emissions. This cost should not be interpreted independently. This study did not take into account any climate impacts, such as increased flooding, agricultural damage, and other issues. Achieving the ultimate climate mitigation goal (2 °C target) reaffirmed in the Paris Agreement more or less avoids such climate change impacts. The mitigation cost should be compared with the impacts of such climate change. Therefore, there should be some benefits of avoiding climate change, such as a reduction in investment for climate change adaptation and resilience and the cost of resource losses, which are uncertain until the impacts actually happen. Nonetheless, a greater deployment of renewable energies and a more stringent GHG emissions reduction target can be achieved, with little economic loss or less mitigation efforts. This predictable economic loss could reassure policymakers and encourage them to implement climate change mitigation actions to manage controllable risks. Furthermore, the health benefits associated with the reduction in air pollution were ignored in this study. This is also one of the benefits of GHG emissions reduction and the deployment of renewable energies.

Finally, the price-based selection of an energy option might allow the energy system to choose the cheapest energy sources. However, it might not encourage the deployment of REs in the primary energy supply as well as the power generation option, because previous studies have shown the importance of early investment in low-carbon infrastructure to avoid carbon lock-in (Erickson et al. 2015).

11.4.2 Limitations

We considered the contribution of CCS technology in power generation. In reality, CCS is still at the pilot project stage, and its potential in Vietnam is still being evaluated due to political and financial barriers. On the other hand, the carbon price in this study was not high enough to implement large-scale CCS, and so this would not significantly affect the main implications of the study.

Another concern pertains to the carbon price, which was calculated in the model as a carbon tax to be borne by the economy. In reality, it might be impossible for Vietnamese policymakers to implement the carbon tax in the country, because there is no such market for carbon in Vietnam yet. Even in high-income countries (e.g., Japan), there have been ongoing discussions about the implementation of a carbon tax, and it is not likely to be easy to implement (Jenkins 2014).

In terms of economic development, it might be perceived that the final coal consumption for energy production is small. There are three reasons for this. First, the base year coal consumption for power generation is small and the CES function is calibrated from that data. The industrial sector remains that characteristics for the future scenario, in which the coal consumption is very small. The reason we kept the exogenous trajectories for the share or efficiency parameters is because in the model we are lacking of information to predict the change of these parameters in final energy demand for the case of Vietnam. Second, the power generation assumption shows a strong increase in the future, which would force the model to indicate a large final share of electricity consumption in the industrial sector. Third, large amounts of heavy industrial goods are imported in the base year and that tendency is also calibrated in the trade functions. We might be able to change the parameters arbitrarily to increase the coal consumption in industrial sectors, but this treatment requires many parameter adjustments. The development of more realistic economic development scenarios would be a very useful research area, but this was beyond the scope of this study. This topic should be investigated in future studies.

11.5 Conclusions

One discussion point in free power generation scenario (freeRE) is that the mitigation cost is high but high renewable energy is not selected. This is because logit is not a cost-minimizing function and parameters representing energy security concerns or preferences in power generation sources other than direct cost measures remain in the future simulation. Thus, the mitigation cost is not necessarily low in free power scenario. On the other hand, the primary goal of this paper is to clarify the mitigation cost of INDC emissions given the alternative renewable energy policies and assess the renewable policy consistency with the climate policy in Vietnam. In that sense, we have stylized the study that has almost fixed power generation scenarios. As a result, the most important implication was that high renewable energy can reduce the mitigation efforts in energy demand side and finally macroeconomic loss can be lower and logit formulation itself did not affect so much to the conclusion.

The primary goal of this paper is to clarify the mitigation cost of INDC emissions given the alternative renewable energy policies and assess the renewable policy consistency with the climate policy in Vietnam. In that sense, we have stylized the study that has almost fixed power generation scenarios. As a result, the most important implication was that high renewable energy can reduce the mitigation efforts in energy demand side and finally macroeconomic loss can be lower.

We assessed the Vietnamese 2030 emissions target set by the INDCs in conjunction with several renewable energy policy options (Vietnam-Prime Minister 2007) using the AIM/CGE model. The socioeconomic assumptions behind these assessments were based on the actual developmental situation of Vietnam, which enabled us to make appropriate simulations. This paper implies that if all nations are willing to put efforts and accept the predictable implications of achieving the most ambitious INDCs target, then it might be possible to bridge the gap between emissions pathways and the limitation of no more than 2 °C above preindustrial levels (Boyd et al. 2015). Particularly, developing countries should invest more on renewable energies and implement the renewable energy policy as early as possible to avoid a carbon-intensive lock-in that makes the 2050 target too expensive to reach (Bertram et al. 2015; Vogt-Schilb and Hallegatte 2014).

Overall, the higher contribution of renewable energies, especially biomass and SPV technologies in the Vietnam energy system, will help the country to achieve the more stringent GHG emissions reduction target with less impact on the economy and energy production. This higher level of deployment of renewable energies can be considered as an additional effort made by Vietnam to achieve its INDC's targets. In addition to the assessment of the latest REDS, this paper also proposes an additional scenario in which the deployment of wind and SPV is double that of the current strategy. The results of this scenario indicated that the mitigation costs were dramatically reduced and it would be less challenging for Vietnam to achieve the most stringent GHG emissions reduction target if the country started to utilize the highest potential of renewable energies as early as possible. In summary, the mitigation costs can be substantially reduced through a portfolio of energy policy instruments. However, mitigation costs may be still significant and the costs tend to rise with more stringent levels of atmospheric stabilization.

The development of power generation should be integrated with the actual energy demand of the economy (reasonable GDP growth), and the increase in energy production should be planned and implemented in advance (i.e., no sudden increase from previous levels). This applies for the case of the PDP7 in Vietnam, in which the target for power generation does not increase rapidly.

Following the finding in this paper, showing the 2050s low-carbon visions is expected as one of the extensions of the study in the future since the Vietnam's latest renewable energy development strategy (Vietnam-Prime Minister 2015) mentions the target of reducing 45% GHG emissions from energy sector in 2050 toward the sustainable development.

Acknowledgments This work was supported by JSPS KAKENHI Grant Number JP16K18177, Japan.

Author Contributions Thanh Tu Tran operated the model simulation and wrote the manuscript. Shinichiro Fujimori supported the modification and improvement of this paper. Toshihiko Masui provided the research funding.

Conflicts of Interest The authors declare no conflict of interest.

References

APERC (2013) APEC energy demand and supply outlook, 5th edn. Vietnam Energy Working Group, Singapore

Bertram C, Luderer G, Pietzcker RC, Schmid E, Kriegler E, Edenhofer O (2015) Complementing carbon prices with technology policies to keep climate targets within reach. Nat Clim Chang 5:235–239

Boyd R, Turner JC, Ward B (2015) Intended nationally determined contributions: what are the implications for greenhouse gas emissions in 2030, Policy paper October 2015 ESRC. Centre for Climate Change Economics and Policy, London

Dai H, Silva Herran D, Fujimori S, Masui T (2016) Key factors affecting long-term penetration of global onshore wind energy integrating top-down and bottom-up approaches. Renew Energy 85:19–30

Dung TQ (1997) P.V technology and success of solar electricity in Vietnam. In: Proceedings of the conference record of the twenty-sixth IEEE on photovoltaic specialists conference, Anaheim, CA, 29 Sept–3 Oct 1997, pp 1309–1312

England SB, Kammen DM (1993) Energy resources and development in Vietnam. Annu Rev Energy Environ 18:137–167

Erickson P, Kartha S, Lazarus M, Tempest K (2015) Assessing carbon lock-in. Environ Res Lett 10:84023

Fujimori S, Masui T, Matsuoka Y (2012) Center for Social and Environmental Systems Research; N.I.E.S.AIM/CGE [Basic] manual. National Institute Environmental Studies, Tsukuba, pp 1–87

Fujimori S, Masui T, Matsuoka Y (2014a) Development of a global computable general equilibrium model coupled with detailed energy end-use technology. Appl Energy 128:296–306

Fujimori S, Kainuma M, Masui T, Hasegawa T, Dai H (2014b) The effectiveness of energy service demand reduction: a scenario analysis of global climate change mitigation. Energy Policy 75:379–391

Fujimori S, Masui T, Matsuoka Y (2015) Gains from emission trading under multiple stabilization targets and technological constraints. Energy Econ 48:306–315

Fuss S, Szolgayová J, Khabarov N, Obersteiner M (2012) Renewables and climate change mitigation: irreversible energy investment under uncertainty and portfolio effects. Energy Policy 40:59–68

Hai LD, Lien NTH (2009) Renewable energy policies for sustainable development in Vietnam. J Sci Earth Sci 25:133–142

Hasegawa T, Fujimori S, Shin Y, Takahashi K, Masui T, Tanaka A (2014) Climate change impact and adaptation assessment on food consumption utilizing a new scenario framework. Environ Sci Technol 48:438–445

Hasegawa T, Fujimori S, Shin Y, Tanaka A, Takahashi K, Masui T (2015) Consequence of climate mitigation on the risk of hunger. Environ Sci Technol 49:7245–7253

Hasegawa T, Fujimori S, Masui T, Matsuoka Y (2016) Introducing detailed land-based mitigation measures into a computable general equilibrium model. J Clean Prod 114:233–242

Höhne N, Ellermann C, Li L (2014) Intended nationally determined contributions under the UNFCCC; discussion paper. Ecofys, Utrecht

International Energy Agency (2012) Energy technology perspectives 2012. International Energy Agency, Paris

Jacoby HD, Chen YH (2014) Expectations for a new climate agreement. MIT Joint Program for the Science and Policy of Global Change, Cambridge, MA

Jenkins JD (2014) Political economy constraints on carbon pricing policies: what are the implications for economic efficiency, environmental efficacy, and climate policy design? Energy Policy 69:467–477

Karki SK, Mann MD, Salehfar H (2005) Energy and environment in the ASEAN: challenges and opportunities. Energy Policy 33:499–509

Kumar A, Bhattacharya S, Pham H (2003) Greenhouse gas mitigation potential of biomass energy technologies in Vietnam using the long range energy alternative planning system model. Energy 28:627–654

Levin K, Rich D, Bonduki Y, Comstock M, Tirpak D, Mcgray H, Noble IAN, Mogelgaard K, Waskow D (2015) Designing and preparing the intended nationally determined contributions (INDCs). World Resources Institute, Washington, DC

Lofgren H, Harris RL, Robinson S, Institute, I.F.P.R (2002) A standard computable general equilibrium (CGE) model in gams. International Food Policy Research Institute, Washington, DC

Luderer G, Krey V, Calvin K, Merrick J, Mima S, Pietzcker R, Van Vliet J, Wada K (2014) The role of renewable energy in climate stabilization: results from the EMF27 scenarios. Clim Chang 123:427–441

Ludig S, Haller M, Schmid E, Bauer N (2011) Fluctuating renewables in a long-term climate change mitigation strategy. Energy 36:6674–6685

Mathiesen BV, Lund H, Karlsson K (2011) 100% renewable energy systems, climate mitigation and economic growth. Appl Energy 88:488–501

Mittal S, Dai H, Fujimori S, Masui T (2016) Bridging greenhouse gas emissions and renewable energy deployment target: comparative assessment of China and India. Appl Energy 166:301–313

Namazu M, Fujimori S, Shukla PR, Matsuoka Y (2013a) Two low-carbon development pathways in India. Glob Environ Res 17:119–128

Namazu M, Fujimori S, Jiang K, Matsuoka Y (2013b) Feasibility of low-carbon development in China. Glob Environ Res 17:109–118

Nguyen KQ (2007a) Impacts of wind power generation and CO_2 emission constraints on the future choice of fuels and technologies in the power sector of Vietnam. Energy Policy 35:2305–2312

Nguyen KQ (2007b) Wind energy in Vietnam: resource assessment, development status and future implications. Energy Policy 35:1405–1413

Nguyen NT, Ha-Duong M (2009) Economic potential of renewable energy in Vietnam's power sector. Energy Policy 37:1601–1613

Polo J, Bernardos A, Navarro AA, Fernandez-peruchena CM, Ramírez L, Guisado MV (2015) Solar resources and power potential mapping in Vietnam using satellite-derived and GIS-based information. Energy Convers Manag 98:348–358

Tran TT, Fujimori S, Matsuoka Y (2010) Changes of energy consumption and the CO_2 emissions structure in Vietnam from 1986 to 2005. Environ Syst Res (38):289–299

Tran TT, Fujimori S, Matsuoka Y (2011) Potential of GHG emission reduction in Vietnam and its implications. Glob Environ Eng Res 19:161–169

Turton H, Barreto L (2006) Long-term security of energy supply and climate change. Energy Policy 34:2232–2250

United Nations Environment Programme (UNEP) (2015) The Emissions Gap Report 2015 A UNEP Synthesis Report. United Nations Environment Programme (UNEP), Nairobi

Universit O, Erlangung Z, Nguyen QK, Korreferent WP, Welsch H (2005) Long term optimization of energy supply and demand in Vietnam with special reference to the potential of renewable energy. University of Oldenburg, Oldenburg

Vietnam-Prime Minister (2007) Decision No. 1855/QD-TTg of 27 December 2007, approving Vietnam's National Energy Development Strategy up to 2020, with 2050 vision. Vietnamese Government Publisher, Hanoi

Vietnam-Prime Minister (2011a) Decision No. 1208/QD-TTg dated 21 July 2011 on the approval of "national master plan on power development (PDP7) for the 2011–2020 period with the vision to 2030". Vietnamese Government Publisher, Hanoi

Vietnam-Prime Minister (2011b) Vietnam's socio-economic development strategy for the period of 2011–2020. Vietnamese Government Publisher, Hanoi

Vietnam-Prime Minister (2012) Decision No. 1393/QD-TTg dated 25 September 2012 on the approval of the National Green Growth Strategy for Vietnam. Vietnamese Government Publisher, Hanoi

Vietnam-Prime Minister (2015a) Intended nationally determined contribution of VietNam. Vietnamese Government Publisher, Hanoi, pp 1–11

Vietnam-Prime Minister (2015b) Decision No. 2068/QD-TTg dated 25 November 2015 on the approval of "renewable energy development strategy of Vietnam up to 2030 with vision to 2050". Vietnamese Government Publisher, Hanoi

Vogt-Schilb A, Hallegatte S (2014) Marginal abatement cost curves and the optimal timing of mitigation measures. Energy Policy 66:645–653

Chapter 12
AIM/CGE V2.0 Model Formula

Shinichiro Fujimori, Toshihiko Masui, and Yuzuru Matsuoka

Abstract This chapter describes the (1) model structure of AIM/CGE (Asia-Pacific Integrated Model/Computable General Equilibrium), (2) data structure (social accounting matrix (SAM)), and (3) formula and list of sets, parameters, equations, and variables. The aim of this chapter is to present all equations written in the model which is used in the other chapter's analysis and make the analysis made in this book transparent.

Keywords CGE • Global model • Climate change • Energy • Land use

12.1 Introduction

CGE model is one of the most useful tools for empirical economic analysis. During the last few decades, the CGE (computable general equilibrium) model has been applied to various countries and to the whole world for not only economic analysis but also energy consumption and greenhouse gas emission analysis.

We, the AIM (Asia-Pacific Integrated Model) team, have been investigating CGE model for more than 10 years (Masui et al. 2006). The CGE model is used for the analysis of global and country CO_2 emissions, mitigation costs, or carbon taxes. Recently, we rebuilt the CGE model called as "AIM/CGE 7.0" referring to

S. Fujimori (✉)
Center for Social and Environmental Systems Research, National Institute for Environmental Studies (NIES), 16–2 Onogawa, Tsukuba, Ibaraki 305–8506, Japan

International Institute for Applied Systems Analysis, Schlossplatz-1, Laxenburg 2361, Austria
e-mail: fujimori.shinichiro@nies.go.jp

T. Masui
Center for Social and Environmental Systems Research, National Institute for Environmental Studies (NIES), 16–2 Onogawa, Tsukuba, Ibaraki 305–8506, Japan

Y. Matsuoka
Department of Urban and Environmental Engineering, Kyoto University, 367, C1-3, Kyoto-daigaku Katsura campus, Nishikyo-ku, Kyoto City, Kyoto 606-8540, Japan

S. Fujimori et al. (eds.), *Post-2020 Climate Action*,
DOI 10.1007/978-981-10-3869-3_12

"a standard CGE model" which was developed by Lofgren et al. (2002). This document describes the new of CGE model. This document is organized as follows.

1. Social accounting matrix (used in the model)
2. Model structure
3. Mathematical statement of the model

12.2 Social Accounting Matrix (Used in the Model)

12.2.1 Social Accounting Matrix Framework

A social accounting matrix (SAM) is a comprehensive, economy-wide data framework, typically representing the economy of a nation. In this model we use global SAM so it covers not only nation but also some regions. A SAM is a square matrix in which each account is represented by a row and a column. Each cell shows the payment from the account of its column to the account of its row. Thus, the incomes of an account appear along its row and its expenditures down its column. The underlying principle of double-entry accounting requires that, for each account in the SAM, total revenue (row total) equals total expenditure (column total). Table 12.1 shows an aggregated SAM concept with verbal explanations in the cells instead of numbers. The payments are not permitted in the blank cells of Table 12.1.

The typical SAM distinguishes between accounts for "activities" (the entities that carry out production) and "commodities." They are definitely different from general input-output table framework. The receipts are valued at producer prices in the activity accounts and at market prices (including value-added commodity taxes which are a part of tax on products and production) in the commodity accounts.

Same as the IFPRI's framework[4], the domestic nongovernment institutions in the SAM consist of households and enterprises. The enterprises earn factor incomes (reflecting their ownership of capital and/or land). They may also receive transfers from other institutions. Their incomes are used for direct taxes, savings, and transfers to other institutions. As opposed to households, enterprises do not consume. Technically, the standard CGE model requires that the SAM have at least one household account; enterprise accounts are not necessary.

The property income and current transfer are accounted in this SAM as they may be different from other SAM frameworks. They represent the transaction among institution sectors. They are basically aggregated into one category but optionally can be differentiated as the objectives of the model simulation.

AIM/CGE couples multiple accounting frameworks for SAM. Concretely, typical SAM does not include physical volume transaction data and is described as only monetary unit, but this CGE model deals with energy commodity with physical unit volume. Therefore, the SAM attaches energy input-output table whose structure is same as the monetary unit SAM.

12.2.2 Sector Classification

The SAM sector classification is shown below. The AIM/CGE 7.0 default column shows the default classification and the original SAM column shows the classification of the original SAM data. The default classification has 14 nonenergy sectors and 26 energy sectors. The commodity is coded as "COM" sector name. The commodity classification is basically the same as industrial sectors, but there are a few exceptions. One is related to electricity all power supplying sectors which produce the same commodity as electricity "COM_ELY." The second is biofuel as "COM_BIO." The third is petroleum products and coal transformation products, namely, COM_P_P and COM_COP (Table 12.2).

12.3 Model Structure

The AIM/CGE model explains all of the payments recorded in the SAM. The energy commodity transaction and GHG emissions are particularly dealt with in detail. The model therefore follows the SAM disaggregation of factors, activities, commodities, and institutions. It is written as a set of simultaneous equations without any objective functions. In addition, there are complement variables to solve the problem like the carbon tax attached to carbon emissions constraint. Therefore, mixed complementarity problem (MCP) is used to solve them. The equations define the behavior of the different actors. In part, this behavior follows simple rules captured by fixed coefficients. For production and consumption decisions, behavior is captured by nonlinear, first-order optimality conditions – that is, production and consumption decisions are driven by the maximization of profits and utility, respectively. The equations also include a set of constraints that have to be satisfied by the system as a whole but are not necessarily considered by any individual actor, namely, macroeconomic balance and balance of payment and so on. An overview of this structure is shown in Fig. 12.1.

There are four blocks: production, income distribution, final consumption, and market. The first block, production, represents the structure of the production functions. We apply a nested CES function for production activities with multiple nested CES functions. Secondly, incomes are distributed to three institutional sectors: enterprises, government, and households. The government takes in income by collecting taxes. Thirdly, institutions consume goods as final consumption. Government expenditure and capital formation are defined as a constant coefficient function. The LES (linear expenditure system) or AIDADS (an implicit direct additive demand system) is used for household consumption. Lastly, the CES function is applied to the import of goods and the CET function is applied to the export of goods. A goods-consumption-and-supply equilibrium is achieved for each market.

Table 12.1 Standard social accounting matrix

		Industry			Products			Factor	
		Industry1	Industry2	Industry3	Products1	Products2	Products3	Labor	Operating surplus
Industry	Industry1				Production				
	Industry2								
	Industry3								
Products	Products1	Intermediate consumption							
	Products2								
	Products3								
Factor	Labor	Compensation of employees							
	Operating surplus	Operating Surplus and mixed income							
Corporation									Resource of operating surplus and mixed income
Household								Resource of compensation	
Government									
Saving and investment									
Property									
Current transfer									
Import tax					Import tax				
Export tax					Export tax				
Value-added tax					Value-added tax				
Taxes on production		Taxes on production							
Tax on income and wealth									
Trade margin					Trade margin				
Rest of the world					Imports of goods and services			Compensation to rest of the world	
Total									

Corporation	Household	Government	Saving and investment	Property	Current transfer	Import tax	Export tax	Value-added tax	Taxes on production	Tax on income and wealth	Rest of the world	Total
	Household consumption	Government consumption	Gross capital formation								Exports of goods and services	
											compensation from rest of the world	
				Resource of property income	Resource of current transfer							
						Receipt of import tax, export tax, value-added tax, taxes on production, and tax on income and wealth						
Gross saving												
Use of property income											Property income from rest of the world	
Use of current transfer											Current transfer from rest of the world	
Tax on income and wealth												
			Net foreign direct investment	Property income to rest of the world	Current transfer to rest of the world							

Table 12.2 Sector classification

CGE code	Original SAM		ISIC rev3 code	GTAP
	Code	Description		
PDR	PDR	Rice	01	PDR
WHT	WHT	Wheat		WHT
GRO	GRO	Other grains		GRO
OSD	OSD	Oilseed crops		OSD
C_B	C_B	Sugar crops		C_B
OTH_A	V_F, PFB, OCR	Vegetable, fruits and nuts, fiber crops, other crops		V_F, PFB, OCR
CTL	CTL	Ruminant livestock		CTL
RMK	RMK	Raw milk		RMK
OTH_L	OAP, WOL, FSH	Nonruminant livestock, other livestock, and fishery	05	OAP, WOL, FSH
FRS	FRS	Forestry	02	FRS
COA	COA	Coal mining	101, 102	COA
OIL	OIL	Oil mining	111, 112 (related to oil extraction), 103	OIL
GAS	GAS	Gas mining	111, 112 (related to gas extraction)	GAS
OMN	OMN	Mineral mining and other quarrying	12, 13, 14	OMN
FPR	OMT, VOL, MIL, SGR, OFD, B_T	Food products		CMT, OMT, VOL, MIL, SGR, OFD, B_T, PCR
LIN	TEX	Textiles and apparel and leather	17, 18, 19, 243	TEX, WAP, LEA
LIN	LUM	Wood products	20	LUM
PPP	PPP	Paper, paper products, and pulp	21, 2211, 2212, 2213, 2219, 222, 223	PPP
CRP	CRP	Chemical, plastic, and rubber products	241, 242, 25	CRP
PRF	P_C	Petroleum refinery	231, 232, 233	P_C
CTF		Coal transformation		
BTR		Biomass transformation (first generation)		
BTR2		Biomass transformation (second generation with residue)		

(continued)

Table 12.2 (continued)

CGE code	Original SAM		ISIC rev3 code	GTAP
	Code	Description		
BTR3		Biomass transformation (second generation with energy crop)		
NMM	NMM	Mineral products nec	25	NMM
I_S	I_S	Iron and steel	271, 2731	I_S
NFM	NFM	Non-ferrous products	272, 2732	NFM
OMF	FMP, OME, ELE, MVH, OTN, OMF	Other manufacturing	28, 29, 31, 33, 30, 32, 34, 35, 36, 37	FMP, OME, ELE, MVH, OTN, OMF
E_COL	E_COL	Coal-fired generation without CCS	Part of 401	ELY
E_OIL	E_OIL	Oil-fired generation without CCS		
E_GAS	E_GAS	Gas-fired generation without CCS		
E_NUC	E_NUC	Nuclear electric power generation		
E_HYD	E_HYD	Hydroelectric power generation		
E_GEO	E_GEO	Geothermal power generation		
E_SPV	E_SPV	Photovoltaic power generation		
E_WIN	E_WIN	Wind-power generation		
E_BIO	E_BIO	Waste biomass power generation		
E_ORN	E_ORN	Other renewable energy power generation		
E_BIN	E_BIO	Advanced biomass power generation		
GDT	GDT	Gas manufacture distribution	402, 403	GDT
CNS	CNS	Construction	45	CNS
TRS	TRS	Transport and communications	60, 61, 62, 63, 64	ATP, OTP, WTP, CMN
CSS	WTR, TRD, FIR, CSS	Other service sectors	41, 50, 51, 52, 55, 65, 66, 67, 70, 71, 72, 73, 74, 75, 80, 85, 90, 91, 92, 93, 94, 95, 99	ROS, OSG, WTR, TRD, OFI, ISR, OBS, DWE
CCSC		Industrial CCS service		
CCSO		Oil refinery CCS service		
CCSEC		Coal-fired CCS service		
CCSEO		Other CCS services		

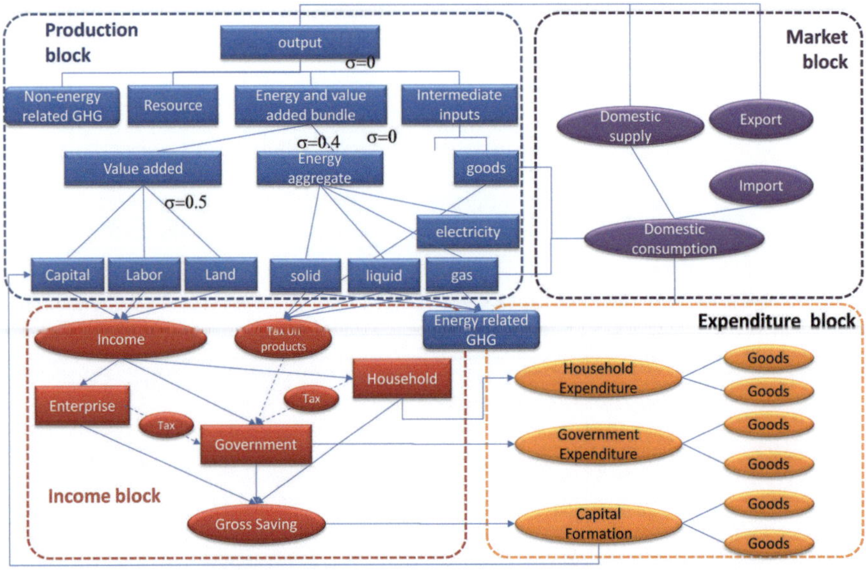

Fig. 12.1 Overview of AIM/CGE model structure (for one region) (The production function excludes emission gases in this figure

12.3.1 *Activity Production and Factor Markets*

Each producer (represented by an activity) is assumed to maximize profits, defined as the difference between revenue earned and the cost of factors and intermediate inputs. Profits are maximized subject to a production technology, the structure of which is shown in Fig. 12.2. At the top level, there is nonenergy-related GHG emission and conventional inputs. This GHG emission treatment is described in detailed by Hyman et al. (2003). Conventional inputs technology is specified by a Leontief function of the quantities of energy and value-added bundle, aggregate nonenergy intermediate input, and resource input. Energy and value-added bundle is nested by valued added and energy inputs. Value added is itself a constant elasticity of substitution (CES) function of primary factors. The aggregated energy inputs are specified by a Logit function. The aggregate intermediate input is a Leontief function of disaggregated intermediate inputs.

The energy transformation sectors such as the power and petroleum refinery sectors are assumed to be different production functions from the other sectors. The structure is drawn as below. Value-added aggregation and energy inputs are specified by Leontief (Fig. 12.4).

There is an option to specify each sector's energy consumption by using detailed end-use technology device information. Once the option turned on, the energy consumption is separated from value-added nest and determined by device selection Logit function and the devices input energy sources (Fig. 12.3).

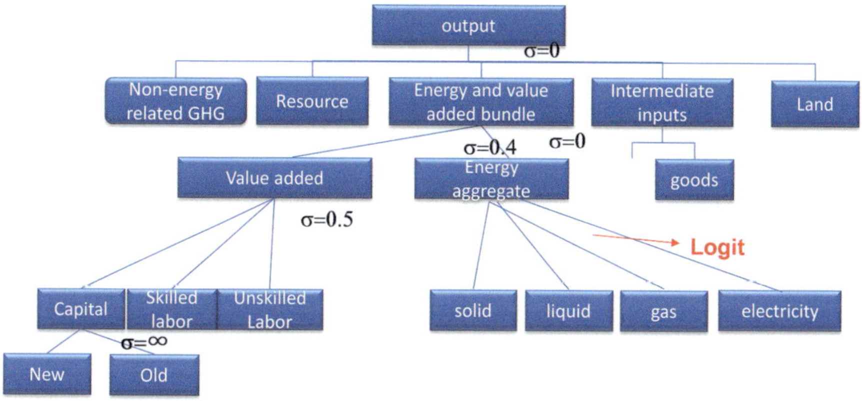

Fig. 12.2 Production structure (nonenergy transformation sectors)

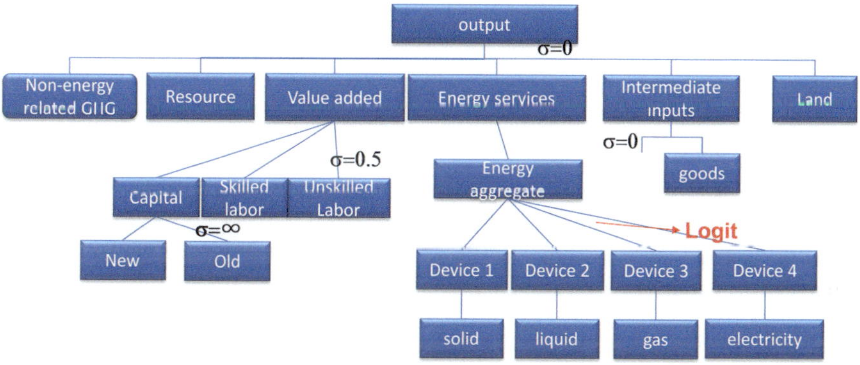

Fig. 12.3 Production structure (nonenergy transformation sectors) with end-use option

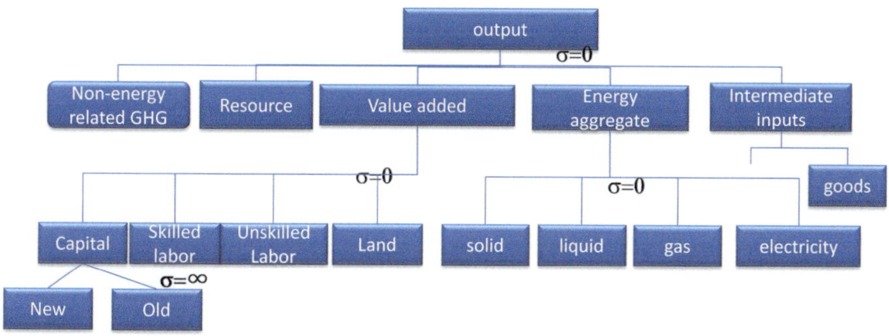

Fig. 12.4 Production structure (energy transformation sectors)

Each activity produces one or more commodities according to fixed yield coefficients. A commodity may be produced by more than one activity. In this model, the electricity is produced by many types of power supply technology. The revenue of the activity is defined by the level of the activity, yields, and commodity prices at the producer level.

As part of its profit-maximizing decision, each activity uses a set of factors up to the point where the marginal revenue product of each factor is equal to its factor price. The quantity supplied of each factor is fixed at a certain level. As for the capital, the capital accumulation is taken into account. This level is recursive dynamically decided. An economy-wide wage variable is free to vary to assure that the sum of demands from all activities equals the quantity supplied. New capital is allocated to anywhere. Old capital is assumed to be fixed to each sector, which is named the putty-clay assumption. The rate of return of capital is same for both of old and new capitals. However, if the new capital is not allocated, the price of old capital is lower than the new one and the operation ratio could be lower than 1.

12.3.2 Institutions

In the CGE model, institutions are represented by households, enterprises, the government, and the rest of the world.[1] The households (disaggregated as in the SAM) receive income from the factors of production (directly or indirectly via the enterprises) and transfers from other institutions. The households use their income to pay direct taxes, save, consume, and make transfers to other institutions. In the basic model version, direct taxes and transfers to other domestic institutions are defined as fixed shares of household income, whereas the savings share is flexible for selected households. The income that remains after taxes, savings, and transfers to other institutions is spent on consumption. When the economy has to implement carbon tax, all of the revenues are assumed to be receipt by the household.

Household consumption covers marketed commodities, purchased at market prices that include commodity taxes. Household consumption is allocated across different commodities according to linear expenditure system (LES) demand functions, derived from maximization of a Stone-Geary utility function.

Instead of being paid directly to the households, factor incomes may be paid to the enterprise. The enterprise may also receive transfers from other institutions. Enterprise incomes are allocated to direct taxes, savings, and transfers to other institutions. Enterprises do not consume. Apart from this, the payments to and from enterprises are modeled in the same way as the payments to and from households.

[1]This is the default version classification, but if the long-term simulation like till 2100 is the objectives, household government and enterprises are treated as one aggregated representative household.

The government collects taxes and receives transfers from other institutions. In the basic model version, all taxes are at fixed *ad valorem* rates.

The government uses this income to purchase commodities for its consumption and for transfers to other institutions. Government consumption is fixed in real (quantity) terms, whereas government transfers to domestic institutions (households and enterprises) are CPI-indexed. Government savings (the difference between government income and spending) is a flexible residual.

The saving rate made by all institution is fixed coefficient also.

The final institution is the rest of the world. As noted, transfer payments between the rest of the world and domestic institutions and factors are all fixed in foreign currency. Foreign savings (or the current account deficit) is the difference between foreign currency spending and receipts. Commodity trade with the rest of the world is discussed in the next section.

12.3.3 Commodity Market

All commodities (domestic output and imports) enter markets. Domestic output may be sold in the market. For marketed output, the first stage in the chain consists of generating aggregated domestic output from the output of different activities of a given commodity. These outputs are imperfectly substitutable as a result of, for example, differences in timing, quality, and distance between the locations of activities. A CES function is used as the aggregation function.[2] The demand for the output of each activity is derived from the problem of minimizing the cost of supplying a given quantity of aggregated output subject to this CES function. Activity-specific commodity prices serve to clear the implicit market for each disaggregated commodity.

In the next, aggregated domestic output is allocated between exports and domestic sales on the assumption that suppliers maximize sales revenue for any given aggregate output level, subject to imperfect transformability between exports and domestic sales, expressed by a constant elasticity of transformation (CET) function. In the international markets, export demands are infinitely elastic at given world prices. The price received by domestic suppliers for exports is expressed in domestic currency and adjusted for export taxes (if any). If the commodity is not exported, total output is passed to the domestic market.

Domestic demand is made up of the sum of demands for household consumption, government consumption, investment (the determination of which is discussed below), and intermediate inputs.

To the extent that a commodity is imported, all domestic market demands are for a composite commodity made up of imports and domestic output, the demands for which are derived on the assumption that domestic demanders minimize cost

[2]Energy commodity has specific treatment. It is discussed below.

subject to imperfect substitutability. This is also captured by a CES aggregation function which is so-called Armington assumption. Total market demand is directed to imports for commodities that lack domestic production and to domestic output for nonimported commodities. The derived demands for imported commodities are met by international supplies that are infinitely elastic at given world prices. The import prices paid by domestic demanders also include import tariffs (at fixed *ad valorem* rates) and the cost of a fixed quantity of transactions services per import unit, covering the cost of moving the commodity from the border to the demander. Similarly, the derived demand for domestic output is met by domestic suppliers.

The AIM/CGE model treats the volume of energy commodities[3] as energy units, though usual CGE model does not as well. In the calibration procedure, the energy unit transaction data is used.

Although energy commodities must satisfy the energy balance condition, CES and CET function make the balanced condition. Therefore, we apply specific formulas to the energy commodities market transaction, which is a sort of Logit function. We determine the share of imported and domestic consumption by the ratio of the current price to previous year's price. The share of exported and domestic consumption and the share of the same commodity production are determined as well as the import composition. The share is controlled by the ratio of previous year's and calculation year price with an elasticity parameter. For example, if import price is increased over that of domestic products, the share of import commodity would be decreased.

12.3.4 Land Allocation and Its Market

Land is disaggregated into AEZ (agroecological zone) categories, and default version assumes three kinds of land type (AEZ1 to AEZ3). The land-use sectors, cropping, livestock, and forestry, input these land categories. The farmers and forestry activities determine the share of the land input according to the land price.

On the other hand, the land owners decide the share of land use according to the land price, and there are multiple decisions to allocate the land to each activity. Firstly, agriculture and forestry land are identified. Then, forestry is disaggregated into primary and secondary forestry. The other blanch has grassland and cropland disaggregation. Grassland has unused primary area and grazing area. Cropland and grazing area have final disaggregation to each specific commodity production activity. Each decision is based on Logit function. This decision process is illustrated in Fig. 12.5.

[3]Coal (COA), crude oil (OIL), natural gas (GAS), petroleum products (P_C), town gas (GDT), and electricity (ELY).

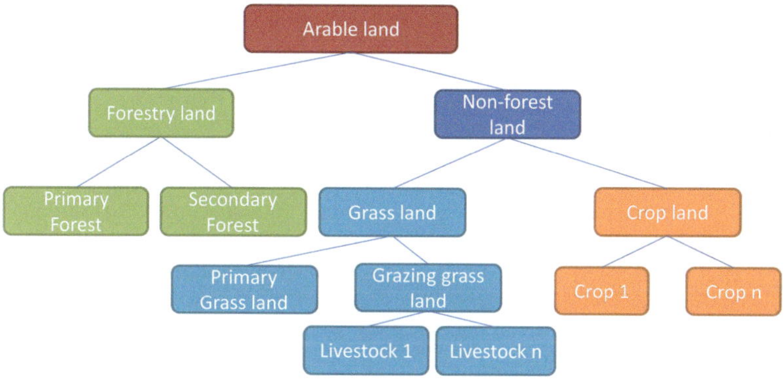

Fig. 12.5 Land allocation structure

12.3.5 *Macroeconomic Balances*

The CGE model includes three macroeconomic balances: the (current) government balance, the external balance (the current account of the balance of payments, which includes the trade balance), and the savings-investment balance.

For the government balance, the closure is that government savings (the difference between current government revenues and current government expenditures) is a flexible residual, while all tax rates are fixed.

For the external balance, which is expressed in foreign currency, the closure is that the real exchange rate is flexible, while foreign savings (the current account deficit) is fixed. Given that all other items are fixed in the external balance (transfers between the rest of the world and domestic institutions), the trade balance is also fixed.

For the savings-investment balance, the closure is investment-driven. Real investment quantities are fixed. In order to generate savings that equal the cost of the investment bundle, the base-year savings rates of selected nongovernment institutions are adjusted by the same number of percentage points. Implicitly, it is assumed that the government is able to implement policies that generate the necessary private savings to finance the fixed real investment quantities.

12.3.6 *Air Pollutants and GHG Emissions*

The CGE model includes air pollutants and GHG emissions module. The emission gases treated in this model are CO, NH_3, NMVOC, NOx, SO_2, BC, OC, CO_2 CH_4, and N_2O. The emission sources are classified into two groups; (1) One is related to fuel combustion and this kind of emission is proportional to the energy consumption. (2) The other is related to the activity level, e.g. CO_2 emission from cement production, and this kind of emission is proportional to the activity level. The emission sources and their groups are shown in Table 12.3. In addition, biomass consumption is also introduced in this model to estimate the air pollutants.

Table 12.3 Emission sources

Emission source related to activity level		Emission source related to fuel combustion	
IPCC category	Explanation	IPCC category	Explanation
2A	Production of minerals	1A1a	Public electricity and heat production
2A1	Cement production	1A1bc	Other energy industries
2A2	Lime production	1A2	Manufacturing industries and construction
2B	Production of chemicals	1A3a	Domestic aviation
2C	Production of metals	1A3b	Road transportation
2D	Production of pulp/paper/food/drink	1A3c	Rail transportation
2G	Nonenergy use of lubricants/waxes (CO_2)	1A3d	Inland navigation
3A	Solvent and other product uses: paint	1A3e	Other transportation
3B	Solvent and other product uses: degrease	1A4	Residential and other sectors
3C	Solvent and other product uses: chemicals		
3D	Solvent and other product uses: others		
4A	Enteric fermentation		
4B	Manure management		
4C	Rice cultivation		
4D1	Direct soil emissions		
4D2	Manure in pasture/range/paddock		
4D3	Indirect N_2O from agriculture		
4D4	Other direct soil emissions		
4E	Savanna burning		
4F	Agricultural waste burning		
5A	Forest fires		
5C	Grassland fires		
5F2	Forest fires: post-burn decay		
6A	Solid waste disposal on land		
6B	Wastewater handling		
6C	Waste incineration		
6D	Other waste handling		

Table 12.4 Notational principles

Item	Notation
Endogenous variables	Upper-case Latin letters without a bar
Exogenous variables	Upper-case Latin letters with a bar
Parameters	Lower-case Latin letters (with or without a bar) or lower-case Greek letters (with or without superscripts)
Set indices	Lower-case Latin letters as subscripts to variables and parameters

12.4 Mathematical Statement

This chapter presents the mathematical model statement equation by equation. In its mathematical form, the CGE model is a system of simultaneous, nonlinear equations. The model is square – that is, the number of equations is equal to the number of variables. In this class of models, this is a necessary (but not a sufficient) condition for the existence of a unique solution. The chapter divides the equations into six blocks: prices, production and trade, institutions, international trade,[4] energy and CO_2 emissions, and system constraints. New items (sets, parameters, and variables) are defined the first time that they appear in the equations. Table 12.4 summarizes the notational principles. Parameter and variable names are chosen to facilitate interpretation; most importantly, commodity and factor quantities start with q, commodity prices with p, and factor prices with w.

12.4.1 Price Block

The price system of the model is rich, primarily because of the assumed quality differences among commodities of different origins and destinations (exports, imports, and domestic outputs used domestically). The price block consists of equations in which endogenous model prices are linked to other prices (endogenous or exogenous) and to non-price model variables.

Import price:

$$PM_{r,c} = PWM_c \cdot dis\_imp_{r,c} \cdot (1 + tm_{r,c}) \cdot EXR_r, \quad \forall c \in CM \qquad \text{(PMDEF)}$$

$r \in R$: a set of regions
$c \in C$: a set of commodities (also referred to as c' and C')
$c \in CM(\subset C)$: a set of imported commodities

[4]The equations in the "international trade block" are involved only if you chose a global model.

PWM_c : world import price (c.i.f) of commodity c
$PM_{r,c}$: composite commodity price (including import tax and transaction costs)
$dis\_imp_{r,c}$: price difference of the import commodity c
$tm_{r,c}$: import tariff rate
EXR_r : exchange rate country r

The import price is the price paid by domestic users for imported commodities (exclusive of the sales tax). Equation (PMDEF) states that it is a transformation of the world price of these imports, considering the exchange rate and import tariffs plus transaction costs (the cost of trade inputs needed to move the commodity from the border to the demander) per unit of the import. The domain of the equation is the set of imported commodities (a subset of the commodity set). The model includes one equation like (PMDEF) for every imported commodity.

Note that the notational principles make it possible to distinguish between variables (upper-case Latin letters) and parameters (lower-case Latin letters). This means that the exchange rate and the domestic-import price are flexible, while the tariff rate and the world import price are fixed. The fixedness of the world import price stems from the "small-country" assumption. That is, for all its imports, the assumed share of world trade for the modeled country is so small that it faces an infinitely elastic supply curve at the prevailing world price.

Export price:

$$PE_{r,c} = PWE_c \cdot dis\_exp_{r,c} \cdot (1 - te_{r,c}) \cdot \overline{EXR_r}, \qquad \forall r \in R, c \in CE \qquad \text{(PEDEF)}$$

$c \in CE(\subset C)$: a set of exported commodities (with domestic production)
$PE_{r,c}$: export price of commodity c
PWE_c : f.o.b. export world price
$dis\_exp_{r,c}$: price difference of the export commodity c
$te_{r,c}$: export tax rate

The export price is the price received by domestic producers when they sell their output in export markets. This equation is similar in structure to the import price definition. The main difference is that the tax reduces the price received by the domestic producers of exports (instead of adding to the price paid by domestic demanders of imports). The domain of the equation is the set of exported commodities, all of which are produced domestically.

Demand price of domestic nontraded goods:

$$PDD_{r,c} = PDS_{r,c}, \qquad \forall r \in R, c \in CD \qquad \text{(PDDDEF)}$$

$c \in CD(\subset C)$: a set of commodities with domestic sales of domestic output
$PDD_{r,c}$: demand price for commodity produced and sold domestically
$PDS_{r,c}$: supply price for commodity produced and sold domestically

Equation (PDDDEF) defines the demand prices as the supply price.

Absorption:

$$PQ_{r,c} \cdot QQ_{r,c} = PDD_{r,c} \cdot QD_{r,c} + PM_{r,c} \cdot QM_{r,c}, \quad \forall r \in R, c \in (CD \cup CM) \tag{PQDDEF}$$

$PQ_{r\ c,}$: composite commodity price excluding sales tax
$QQ_{r,c}$: quantity of goods supplied to domestic market (composite supply)
$QD_{r,c}$: quantity sold domestically of domestic output
$QM_{r,c}$: quantity of imports of commodity

Absorption is total domestic spending on a commodity at domestic demander prices. Equation (PQDDEF) defines it exclusive of the sales tax. Absorption is expressed as the sum of spending on domestic output and imports at the demand prices, *PDD* and *PM*. The prices *PDD* and *PM* include the cost of trade inputs but exclude the commodity sales tax.

The equation as a whole applies to all commodities that are imported and/or have domestic sales of domestic output (the union of the sets *CD* and *CM*). It does not apply to commodities for which the entire output volume is exported. Each of the two terms on the right-hand side applies only to its relevant set (*CD* and *CM*, respectively). In the *GAMS* code, *PM* and *QM* are fixed at zero for commodities that are not elements in the set *CM*; similarly *PDD* and *QD* are fixed at zero for commodities that are not elements in the set *CD*. This approach is followed throughout: all variables that should be excluded from the model are fixed at zero.

Commodity market monetary balance:

$$PQ_{r,c} \cdot QQ_{r,c} = PQD_{r,c} \cdot \left(\begin{array}{l} \sum\limits_{a \in A} dfpq_{r,c,a} \cdot QINT_{r,c,a} + \sum\limits_{h \in H} pfdq_{r,c,h} \cdot QH_{r,c,h} \\ + pfdq_{r,c,"gov"} \cdot QG_{r,c} + pfdq_{r,c,"S-I"} \cdot QINV_{r,c} \end{array} \right), \quad \forall r \in R, c \in CX \tag{PQDEF}$$

$PQD_{r,c}$: composite commodity price excluding sales tax
$dfpq_{r,c,i}$: price differences of commodity price among inputs sectors
$QINT_{r,c,a}$: quantity of commodity *c* as intermediate input to activity *a*
$QH_{r,c,h}$: quantity of consumption of marketed commodity *c* for household *h*
$QG_{r,c}$: government consumption demand for commodity
$QINV_{r,c}$: quantity of fixed investment demand for commodity

Marketed output with stock change:

$$QX2_{r,c} = QX_{r,c} + stch_{r,c}, \quad \forall r \in R, c \in CX \tag{QX2DEF}$$

$QX_{r,c}$: aggregate marketed quantity of domestic output of commodity
$QX2_{r,c}$: aggregate marketed quantity of domestic output of commodity including stock change
$stch_{r,c}$: stock change of commodity c (positive)

Marketed output value with stock change:

$$PX2_{r,c} \cdot QX2_{r,c} = PX_{r,c} \cdot QX_{r,c}, \qquad \forall r \in R, c \in CX \qquad \text{(PXDEF)}$$

$PX_{r,c}$: aggregate producer price for commodity
$PX2_{r,c}$: aggregate producer price for commodity including stock change effects

The above two equations describe the relationship between output and stock changes and those of market value.

Marketed output value:

$$PX2_{r,c} \cdot QX2_{r,c} = PDS_{r,c} \cdot QD_{r,c} + PE_{r,c} \cdot QE_{r,c}, \qquad \forall r \in R, c \in CX \quad \text{(PX2DEF)}$$

$c \in CX(\subset C)$: a set of commodities with domestic output
$QE_{r,c}$: quantity of exports

For each domestically produced commodity, the marketed output value at producer prices is stated as the sum of the values of domestic sales and exports. Domestic sales and exports are valued at the prices received by the suppliers, *PDS* and *PE*, both of which have been adjusted downwards to account for the cost of trade inputs.

The domain limitation to domestically produced commodities (the elements in the set *CX*) has to be stated explicitly given that the model includes a category of imported commodities without domestic production. The domestic part applies only to elements in *CD*, whereas the export part applies only to elements in *CE*. *PX* and *QX* are referred to as "aggregate" values since they may apply to an aggregation of output from different domestic producers of the same commodity. By dividing through by *QX2*, this equation could be rewritten as an explicit definition of *PX*.

Activity price:

$$PA_{r,a} \cdot QA_{r,a} \cdot \left(1 + RQUOQA_{r,a}\right) = \sum_{c \in C} PXAC_{r,a,c} \cdot QXAC_{r,a,c}, \qquad \forall r \in R, a \in A \qquad \text{(PADEF)}$$

$a \in A$: a set of activities
$PA_{r,a}$: activity price (gross revenue per activity unit)
$QA_{r,a}$: quantity (level) of activity
$PXAC_{r,a,c}$: producer price of commodity c for activity a
$QXAC_{r,a,c}$: marketed output quantity of commodity c from activity a
$RQUOQA_{r,a}$: shadow subsidies of the fixed activity level

The gross revenue per activity unit, the activity price, is the return from selling the output or outputs of the activity, defined as yields per activity unit multiplied by activity-specific commodity prices, summed over all commodities. This allows for the fact that activities may produce multiple commodities.

If the activity level is fixed, the complementary subsidies are generated as *RQUOQA*.

Aggregate nonenergy intermediate input price:

$$QINTA_{r,a} \cdot PINTA_{r,a} = \sum_{c \in CNEN} dfpq_{r,c,a} \cdot PQD_{r,c} \cdot QINT_{r,c,a} \cdot (1 + tqd_{r,c,a}) \quad \forall r \in R, a \in A, c \in CNEN$$

(PINTADEF)

$c \in CNEN$: a set of nonenergy commodities
$PINTA_{r,a}$: aggregate intermediate input price for activity a
$QINTA_{r,a}$: quantity of aggregate intermediate input
$tq_{r,c,ac}$: rate of sales tax (as share of composite price inclusive of sales tax), suffix ac including activity a and institution i

The activity-specific aggregate intermediate input price shows the cost of disaggregated intermediate inputs per unit of aggregate intermediate input. It depends on composite commodity prices and intermediate input coefficients, which show the quantity of input commodity c per unit of aggregate intermediate input (not per unit of output).

Activity revenue and costs (nonenergy transformation sector):

$$\begin{aligned} PA_{r,a} \cdot (1 - ta_{r,a}) \cdot QA_{r,a} = {} & PVAE_{r,a} \cdot QVAE_{r,a} + PINTA_{r,a} \cdot QINTA_{r,a} \\ & + PRES_{r,a} \cdot QRES_{r,a} + GHGCA\_NENE_{r,a} \\ & + VRENCAP_{r,a} \cdot QA_{r,a} + \sum_{emcm \in EMCM} QRED_{r,\mathrm{emcm},a}, \\ & \forall r \in R, a \in \mathrm{ACES} \end{aligned}$$

(PVADEF)

Activity revenue and costs (energy transformation sector):

$$\begin{aligned} PA_{r,a} \cdot (1 - ta_{r,a}) \cdot QA_{r,a} = {} & PVA_{r,a} \cdot QVA_{r,a} + PINTA_{r,a} \cdot QINTA_{r,a} + PENE_{r,a} \cdot QENE_{r,a} \\ & + PRES_{r,a} \cdot QRES_{r,a} + GHGCA\_NENE_{r,a} \\ & + VRENCAP_{r,a} \cdot QA_{r,a} \sum_{emcm \in EMCM} QRED_{r,\mathrm{emcm},a} \\ & \forall r \in R, a \in ALEO \end{aligned}$$

(PVADEF)

$a \in ACES(\subset A)$: a set for nonenergy transformation
$a \in ALEO(\subset A)$: a set for energy transformation
$ta_{r,a}$: tax rate for activity
$PVAE_{r,a}$: price of (aggregate) energy and value-added bundle (nonenergy transformation sector)

$QVAE_{r,a}$: quantity of (aggregate) energy and value-added bundle (nonenergy transformation sector)
$PVA_{r,a}$: price of (aggregate) value added
$QVA_{r,a}$: quantity of (aggregate) value added
$PENE_{r,a}$: price of (aggregate) energy input
$QENE_{r,a}$: quantity of (aggregate) energy input
$GHGCA\_NENE_{r,a}$: GHG emission cost-related biomass burning and CCS negative emissions of activity *a* in region *r*
$VRENCAP_{r,a}$: Rent of electricity capacity activity *a* in region *r*
$PRES_{r,a}$: price of resource input
$QRES_{r,a}$: quantity of resource input
$QRED_{r,emcm,a}$: input of counter emission reduction countermeasures of activity *a* and measure *emcm*
$emcm \in EMCM$: a set of emission reduction countermeasures (CCS)

Activity cost is different between the energy transformation sector and nonenergy transformation sector as shown in the previous chapter. The difference is energy and value-added treatment. For each activity, total revenue net of taxes is fully exhausted by payments for value added and intermediate inputs. Given the above definitions of *PA*, *PVA, PENE, PINTA*, *PENE*, and *PRES*, Equation (PVADEF) implicitly defines the value-added price, *PVA*.

If we had GHG emission constraints, each activity is levied on to its GHG emissions. The GHG emission cost related with biomass burning is represented as *GHGCA_NENE*. GHG cost related to energy consumption is included in energy cost.

Moreover, sometimes, activity level is constrained by political decisions; for example, nuclear power plant construction is not determined only by economic rationality. In such cases, a rent is absorbed by the activity as *VRENCAP*.

The emission reduction countermeasures for CCS technology cost are *QRED*.

Resource input price:

$$pres\_base_{r,a} = PRES_{r,a}, \qquad \forall r \in R, a \in A \qquad \text{(PRESDEF)}$$

Resource inputs price is assumed equal as the activity price.

$pres\_base_{r,a}$: resource price (normally 1)

Consumer price index:

$$CPI_r = \sum_{c \in C} \left(PQD_{r,c} \cdot dfpq_{r,c,"hurb"} \cdot \left(1 + tqd_{r,c,"hurb"}\right) + PGHG_r \cdot \sum_{g \in G} gwp_{r,g} \cdot efffc_{r,c,"hurb",g} \right) \cdot cwts_{r,c} \quad \forall r \in R, \forall g \in G \qquad \text{(CPIDEF)}$$

$cwts_{r,c}$: weight of commodity c in the consumer price index
CPI_r : consumer price index (exogenous variable)

Producer price index for nontraded market output:

$$DPI_r = \sum_{c \in C} PDS_{r,c} \cdot dwts_{r,c} \qquad \forall r \in R \qquad \text{(DPIDEF)}$$

$dwts_{r,c}$: weight of commodity c in the producer price index
DPI_r : producer price index for domestically marketed output

Equations (CPIDEF) and (DPIDEF) define the consumer price index and the producer price index for domestically marketed output. The *CPI* is fixed and functions as the numéraire in the basic model version; alternatively, the *DPI* may be fixed. A numéraire is required since the model is homogeneous of degree zero in prices. A doubling of the value of the numéraire would double all prices but leave all real quantities unchanged. All simulated price and income changes should be interpreted as changes vis-à-vis the numéraire price index.

Export price index:

$$EPI_r = \sum_{c \in C} PE_{r,c} \cdot ewts_{r,c} \qquad \forall r \in R \qquad \text{(EPIDEF)}$$

$ewts_{r,c}$: weight of commodity c in the export price index
EPI_r : export price index

Import price index:

$$MPI_r = \sum_{c \in C} PM_{r,c} \cdot mwts_{r,c} \qquad \forall r \in R \qquad \text{(MPIDEF)}$$

$mwts_{r,c}$: weight of commodity c in the import price index
MPI_r : import price index

Governmental consumption price index:

$$GPI_r = \sum_{c \in C} PQD_{r,c} \cdot dfpq_{r,c,\text{"gov"}} \cdot \left(1 + tqd_{r,c,\text{"gov"}}\right) \cdot gwts_{r,c} \qquad \forall r \in R$$

(GPIDEF)

$gwts_{r,c}$: weight of commodity c in the government price index
GPI_r : government price index

Capital formation price index:

$$IPI_r = \sum_{c \in C} PQD_{r,c} \cdot dfpq_{r,c,"S-I"} \cdot \left(1 + tqd_{r,c,"S-I"}\right) \cdot iwts_{r,c} \qquad \forall r \in R \quad \text{(IPIDEF)}$$

$iwts_{r,c}$: weight of commodity c in the capital formation price index
IPI_r : capital formation price index

Export, import, governmental consumption, and capital formation price indices are defined as above.

12.4.2 Production Block

The production and trade block covers four categories: domestic production and input use; the allocation of domestic output to home consumption, the domestic market, and exports; the aggregation of supply to the domestic market (from imports and domestic output sold domestically); and the definition of the demand for trade inputs that is generated by the distribution process.

Production is carried out by activities that are assumed to maximize profits subject to their technology, taking prices (for their outputs, intermediate inputs, and factors) as given. In other words, it acts in a perfectly competitive setting. The CGE model includes the first-order conditions for profit maximization by producers.

Leontief technology: demand for aggregate value added (energy transformation sector)

$$QVA_{r,a} = iva_{r,a} \cdot QA_{r,a}, \qquad \forall r \in R, a \in ALEO \qquad \text{(LEOAGGVA)}$$

Leontief technology: demand for aggregate energy input (energy transformation sector)

$$QENE_{r,a} = iena_{r,a} \cdot QA_{r,a}, \qquad \forall r \in R, a \in ALEO \qquad \text{(LEOAGGENE)}$$

Energy and value-added bundle (nonenergy transformation sector)

$$QVAE_{r,a} = ivae_{r,a} \cdot QA_{r,a}, \qquad \forall r \in R, a \in ACES \qquad \text{(LEOAGGVAE)}$$

Leontief technology: demand for aggregate nonenergy intermediate input

$$QINTA_{r,a} = inta_{r,a} \cdot QA_{r,a}, \qquad \forall r \in R, a \in A \qquad \text{(LEOAGGINT)}$$

Leontief technology: demand for resource input

$$QRES_{r,a} = ires_{r,a} \cdot QA_{r,a}, \qquad \forall r \in R, a \in A \qquad \text{(LEOAGGRES)}$$

$iva_{r,a}$: quantity of value added per activity unit
$iena_{r,a}$: quantity of aggregate energy input per activity unit
$ivae_{r,a}$: quantity of value-added energy composite per activity unit
$inta_{r,a}$: quantity of aggregate nonenergy intermediate input per activity unit
$ires_{r,a}$: quantity of aggregate resource input per activity unit

As for the energy transformation sectors, Leontief functions at the top of the conventional inputs are equations (LEOAGGVA), (LEOAGGENE), (LEOAGGINT), and (LEOAGGRES), where the demands for value added, the aggregate intermediate inputs, the aggregate energy input, and resource inputs are defined as Leontief functions of the activity level. For the nonenergy transformation sectors, the equations (LEOAGGVA) and (LEOAGGENE) are replaced by equation (LEOAGGVAE).

Energy and value-added composite:

$$QVAE_{r,a} = \alpha_{r,a}^{vae} \cdot \left(\delta_{r,a}^{vae} \cdot QVA_{r,a}^{-\rho_{r,a}^{vae}} + \left(1 - \delta_{r,a}^{vae}\right) \cdot QENE_{r,a}^{-\rho_{r,a}^{vae}}\right)^{-\frac{1}{\rho_{r,a}^{vae}}}, \quad \forall r \in R, a \in ACES$$

(CESVAENE)

Energy and value-added input CES technology: energy – value-added input ratio

$$QVA_{r,a} = QENE_{r,a} \cdot \left(\frac{\delta_{r,a}^{vae}}{1 - \delta_{r,a}^{vae}} \cdot \frac{PENE_{r,a}}{PVA_{r,a}}\right)^{\frac{1}{1+\rho_{r,a}^{vae}}}, \qquad \forall r \in R, a \in ACES$$

(CESVAENEFOC)

$\alpha_{r,a}^{vae}$: efficiency parameter in the CES energy and value-added function
$\delta_{r,a}^{vae}$: CES energy and value-added function share parameter in activity a
$\rho_{r,a}^{vae}$: CES energy and value-added function exponent.

Energy and value-added composite balance:

$$QVAE_{r,a} \cdot PVAE_{r,a} = QENE_{r,a} \cdot PENE_{r,a} + QVA_{r,a} \cdot PVA_{r,a}, \qquad \forall r \in R, a \in ACES$$

(PVEDEF)

Nonenergy transformation sectors determine their energy and value-added inputs as shown previously. CES function is applied to these inputs. If there is an industry using no energy, the value added is defined as below:

Energy and value-added composite (nonenergy use sector):

$$QVAE_{r,a} = QVA_{r,a}, \qquad \forall r \in R, a \in ACES \qquad \text{(CESVAENE2)}$$

Value-added and factor demands: non-power supply activities

$$QVA_{r,a} = \alpha_{r,a}^{va} \cdot \left(\sum_{f \in F} \delta_{r,a}^{va} \cdot \left(fcmult_{r,f,a} \cdot QF_{r,f,a} \right)^{-\rho_{r,a}^{va}} \right)^{-\frac{1}{\rho_{r,a}^{va}}}, \qquad \forall r \in R, a \in A$$

(CESVAPRD)

Value-added and factor demands: power supply activities

$$PVA_{r,a} \cdot (1 - tva_{r,a}) \cdot QVA_{r,a} = \sum_{f \in F} WF_{r,f} \cdot WFDIST_{r,f,a}, \qquad \forall r \in R, a \in A$$

(CESVAFOC2)

Factor demand: non-power supply activities

$$WF_{r,f} \cdot WFDIST_{r,f,a} = PVA_{r,a} \cdot (1 - tva_{r,a}) \cdot QVA_{r,a} \cdot \left(\sum_{f \in F'} \delta_{r,a}^{va} \cdot \left(fcmult_{r,f,a} \cdot QF_{f\ r,a} \right)^{-\rho_{r,a}^{va}} \right)^{-1} \cdot \delta_{r,a}^{va} \cdot fcmult_{r,f,a}{}^{-\rho_{r,a}^{va}} \cdot QF_{r,f,a}{}^{-\rho_{r,a}^{va}-1},$$
$$\forall r \in R, a \in A, f \in F$$

(CESVAFOC)

Factor demand: power supply activities

$$QF_{r,f,a} = ivfa_{r,f,a} \cdot QVA_{r,a}, \qquad \forall r \in R, a \in A, f \in F \qquad \text{(FCTLEO)}$$

Factor cost: power supply activities

$$\sum_{f} WF_{r,f} \cdot WFDIST_{r,f,a} \cdot QF_{r,f,a} = PVA_{r,a} \cdot (1 - tva_{r,a}) \cdot QVA_{r,a}, \qquad \forall r \in R, a \in A$$

(FCTLEO2)

$f \in F(=F')$: a set of factors
$tva_{r,a}$: rate of value-added tax for activity a

$\alpha^{va}_{r,a}$: efficiency parameter in the CES value-added function
$\delta^{va}_{r,a}$: CES value-added function share parameter for factor f in activity a
$\rho^{va}_{r,a}$: CES value-added function exponent
$ivfa_{r,f,a}$: input coefficient of factors for Leontief inputs
$fcmult_{r,f,a}$: productivity shifter of factor f in activity a
$QF_{r,f,a}$: quantity demanded of factor f from activity a
$WF_{r,f}$: average price of factor
$WFDIST_{r,f,a}$: factor price distortion factor for factor f in activity a

Equation (CESVAPRD) states that, for each activity, the quantity of value added is a CES function of disaggregated factor quantities. According to equation (CESVAFOC), activities demand factors at the point where the marginal cost of each factor (defined on the left-hand side as the activity-specific factor price) is equal to the marginal revenue product (net of intermediate input costs) of the factor. In the GAMS code, the domain of equation (CESVAFOC) is limited to the factor-activity combinations that appear in the base-year SAM. Similar domain restrictions apply to other equations that are defined over mappings between multiple indices. The exponent ρ^{va} is a transformation of the elasticity of factor substitution: the higher this elasticity, the smaller the value of ρ^{va} and the larger the optimal change in the ratios between different factor quantities in response to changes in relative factor prices. The fact that the average factor price is an endogenous variable while the activity-specific "wage-distortion" factor is exogenous reflects the treatment of factor markets.

Regarding with power sectors, we assume the Leontief type of factor input function, so they are slightly different equations as (FCTLEO) and (FCTLEO2).

Capital aggregation: perfect substitution

$$QF_{r,"ccap",a} = QF_{r,"ncap",a} + QF_{r,"cap",a} \cdot COPR_{r,a}, \qquad \forall r \in R, a \in A \tag{CAPAGG5\_PS}$$

Capital aggregation: new industry

$$QF_{r,"ccap",a} = QF_{r,"ncap",a}, \qquad \forall r \in R, a \in A \tag{CAPAGG4}$$

Capital aggregation: base year

$$QF_{r,"ccap",a} = QF_{r,"cap",a}, \qquad \forall r \in R, a \in A \tag{CAPAGG1}$$

Capital cost balance:

$$\begin{aligned} QF_{r,f,a} \cdot WF_{r,f} \cdot WFDIST_{r,f,a} = {} & QF_{r,"ncap",a} \cdot WF_{r,"ncap"} \cdot WFDIST_{r,"ncap",a} \\ & + QF_{r,"cap",a} \cdot COPR_{r,a} \cdot WF_{r,"cap"} \cdot WFDIST_{r,"cap",a}, \\ & \forall r \in R, a \in A, f \in FCCAP \end{aligned} \tag{CAPAGG2}$$

Capital rate of return for new and old:

$$WF_{r,"ncap"} \cdot WFDIST_{r,"ncap",a} \geq WF_{r,"cap"}$$
$$WFDIST_{r,"cap",a} \perp QF_{r,\}ncap\},a} \geq 0, \qquad \forall r \in R, a \in A \qquad \text{(CAPAGG3\_PS)}$$

Capital operation ratio:

$$COPR_{r,a} = \left(\frac{WF_{r,"cap"} \cdot WFDIST_{r,"cap",a}}{WF_{r,"ncap"} \cdot WFDIST_{r,"ncap",a}} \right)^{\varsigma_{r,a}}, \qquad \forall r \in R, a \in A \quad \text{(CAPOPR)}$$

$f \in FCAP(=F)$: a set of capital (new and old; "ncap" and "cap")
$COPR_{r,a}$: operation ratio
$\varsigma_{r,a}$: a parameter for operation ratio

Capital vintage is taken into account and the old and new capitals are aggregated. The first equation represents quantity of operated capital "ccap" is sum of new and old capital quantity. The old capital has operation rate COPR. COPR works only if the rate of return of a sector is less than country average rate of return. In other words the sector does not require any new capital. This is described in the second equation. The third equation defines operation rate which is a function of a ratio of old and new capital's rate of return.

Disaggregated intermediate input demand:

$$QINT_{r,c,a} \cdot \frac{faceff_{r,c,a,"old"} \cdot QF_{r,"CAP",a} + faceff_{r,c,a,"new"} \cdot QF_{r,"NCAP",a}}{QF_{r,"CAP",a} + QF_{r,"NCAP",a}}$$
$$= ica_{r,c,a} \cdot QINTA_{r,a}, \qquad \forall r \in R, a \in A, c \in CNEN$$

(INTDEM)

$ica_{r,c,a}$: quantity of c per unit of aggregate intermediate input a
$faceff_{r,c,a,*}$: old and new capital efficiency parameter (only used for power sectors)

Emission reduction countermeasures (CCS is a good example) are installable when the emission is constrained. The emission reduction inputs is *QRED* and their cost ($/ton$CO_2$eq) is η.

CCS service intermediate input demand:

$$QINT_{r,c,a} = \sum_{emcm \in CCSMAP} QRED_{r,emcm,a}, \qquad \forall r \in R, a \in A, c \in C\_CCS$$

(CCSINPUT)

$emcm \in EMCM$: a subset of emission reduction countermeasures
C_CCS : a set of CCS service

Reduction measures (for nonenergy-related GHG emissions):

$$QRED_{r,emcm,a} = \xi^{\max}_{\text{emcm},a} \cdot \eta_{emcm,a} \cdot \sum_{g \in G} gwp_g \cdot EMALI_{r,a,g}, \quad \forall r \in R, a \in A, emcm \in EMCM0 \qquad \text{(GHG\_ABATEMENT)}$$

$emcm \in EMCM0(\subset EMCM)$: a subset of emission reduction countermeasures which are for nonenergy-related emissions

Reduction measures (for energy-related GHG emissions):

$$QRED_{r,emcm,a} = \xi^{\max}_{emcm,a} \cdot \eta_{emcm,a} \cdot \sum_{c \in ENE} \sum_{g \in G} gwp_g \cdot QINT_{r,c,a} \cdot enur_{r,c,a} \cdot efffc_{r,c,a,g}, \quad \forall r \in R, a \in A, emcm \in EMCM1 \qquad \text{(GHG\_ABATEMENT2)}$$

$emcm \in EMCM1(\subset EMCM)$: a subset of emission reduction countermeasures which are for energy-related emissions

Reduction measures (for biomass power plant GHG absorption):

$$QRED_{r,emcm,a} = \xi^{\max}_{emcm,a} \cdot \eta_{emcm,a} \cdot \sum_{g \in G} gwp_g \cdot EMBII_{r,a,g}, \quad \forall r \in R, a \in A, emcm \in EMCM2 \qquad \text{(GHG\_ABATEMENT3)}$$

$emcm \in EMCM2(\subset EMCM)$: a subset of emission reduction countermeasures which are for biomass power plant absorption

CCS installation rate is determined by the following equation:

CCS-equipped share:

$$TSCT_{r,a,tc} = \alpha^{tech}_{a,tc} \cdot PGHG_r^{\beta_{r,a,tc}} \cdot \left(\sum_{tc \in TC} \alpha^{\text{tech}}_{a,tc} \cdot PGHG_r^{\beta^{\text{tech}}_{r,a,tc}} \right)^{-1}, \forall r \in R, a \in A, tc \in TC \qquad \text{(TECHTOTSHARECCS)}$$

$tc \in TC$: a set of technology

$TSCT_{r,a,tc}$: technology tc's share in sector a and region r

$\alpha^{tech}_{a,tc}$: share parameter for technology Logit selection function

$\beta^{tech}_{r,a,tc}$: exponent of technology Logit selection

Energy input technology share:

$$QINT_{r,c,a} = \frac{QENE_{r,a} \cdot \beta^{inden}_{r,c,a} \cdot \left\{ PQD_{r,c} \cdot (1 + tqd_{r,c,a}) + PGHG_r \cdot \sum_{g \in G} gwp_g \cdot enur_{r,c,a} \cdot efffc_{r,c,a,g} \right\}^{el^{inden}_{r,c,a}}}{\sum_{cp \in ENE} \beta^{inden}_{r,cp,a} \cdot \left\{ PQD_{r,cp} \cdot (1 + tqd_{r,cp,a}) + PGHG_r \cdot \sum_{g \in G} gwp_g \cdot enur_{r,cp,a} \cdot efffc_{r,cp,a,g} \right\}^{el^{inden}_{r,cp,a}}},$$
$$\forall r \in R, c \in ENE, a \in A$$

(ENELOGIT)

Energy input costs:

$$PENE_{r,a} \cdot QENE_{r,a} = \sum_{c \in C\_ENE} \left\{ PQD_{r,c,a} \cdot (1 + tqd_{r,c,a}) + PGHG_r \cdot \sum_{g \in G} gwp_g \cdot enur_{r,c,a} \cdot efffc_{r,c,a,g} \right\} \cdot QINT_{r,c,a},$$
$$\forall r \in R, a \in A$$

(PENEDEF)

$PGHG_r$	: GHG emission price in region r (US$/t$CO_2$)
$a \in ACES\_ENE(\subset A)$	: a set of activities with a CES function at energy nest
$\beta^{inden}_{r,c,a}$	: share parameter of Logit function for industrial activity energy source selection
$el^{inden}_{r,c,a}$	: price elasticity parameter of Logit function for industrial activity energy source selection
$efffc_{r,c,ac,g}$	: emission factors for emissions fossil fuel combustion by sector ac consuming of goods c
$enur_{r,c,ac}$	: energy-used ratio (1-nonenergy-use ratio)
gwp_g	: global warming potential of gas g

Activities excluding energy transformation sector, which is specified by a set of *ACES_ENE*, have CES energy input technology. Transport sector is an exception for this energy specification and it will be explained later. The energy input is disaggregated into each energy source. The energy source share is determined by Logit function as equation (ENELOGIT). This equation has two parameters: one determines base share and the other represents price elasticity. Price includes tax and carbon emission tax which is formulated as multiplying energy-used ratio, emission coefficient, and GWP (global warming potential) associated with carbon tax rate. These parameters are explained later in detail.

Energy consumption of energy transformation sector:

$$QINT_{r,c,a} = QENE_{r,a} \cdot iene_{r,c,a}, \qquad \forall r \in R, a \in \mathrm{ALEO\_}ENE, c \in ENE \qquad \text{(LEOENEELY)}$$

$a \in ALEO\_ENE (\subset A)$: a set of activities with a CES function at energy nest
$iene_{r,c,a}$: energy commodity consumption ratio

Transport sector has different production structures. Transport sector provides two types of transport services: passenger and freight. The transport service is provided by 11 modes.

The freight transport mode demand is caused by industrial activity and household consumption formulated as follows:

Freight transport generated by industrial activity:

$$QDTRS_{r,tr,a} = QINT_{r,"COM\_TRS",a} \cdot trscvf_{r,tr,a}, \qquad \forall r \in R, a \in A, tr \in TR\_FRT \qquad \text{(INTTRNS)}$$

Freight transport generated by household consumption:

$$QDTRS_{r,tr,h} = QH_{r,"COM\_TRS",h} \cdot trscvf_{r,tr,h}, \qquad \forall r \in R, h \in H, tr \in TR\_FRT \qquad \text{(HOUTRNS)}$$

Total freight transport:

$$QDTRST_{r,tr} = \sum_{ac \in AC} QDTRS_{r,tr,ac}, \qquad \forall r \in R, tr \in TR\_FRT \qquad \text{(TRSFRTEQ)}$$

$QDTRS_{r,tr,ac}$: freight transport service demand by sector *ac* and modes
$QDTRST_{r,tr}$: total transport service demand by modes
$trscvf_{r,tr,ac}$: transport service demand by modes
$tr \in TR$: a set of transport mode

Passenger transport excluding household own car driving is formulated as follows:

Passenger transport (excluding household passenger car):

$$QDTRST_{r,tr} = QA_{r,"trs"} \cdot trscef_{r,tr}, \qquad \forall r \in R, tr \in TR\_PSS \qquad \text{(TRSPSSEQ)}$$

$trscef_{r,tr}$: passenger transport demand coefficient

Energy consumption of transport sector is derived from the transport volume and energy price. Its energy source is determined by Logit function.

Transport energy demand:

$$TRS\_ENE_{r,tr} = QDTRST_{r,tr} \cdot trseneeffi_{r,tr} \cdot \left(\frac{PENE\_TR_{r,tr}}{pene\_tr\_base_{r,tr}} \right)^{el^{trspr}_{r,tr}},$$

$$\forall r \in R, a \in A, tr \in TR \qquad \text{(TRSENEEQ)}$$

$TRS\_ENE_{r,tr}$: transport energy demand by modes
$el^{trspr}_{r,tr}$: transport energy demand price elasticity
$trseneeffi_{r,tr}$: transport energy efficiency
$PENE\_TR_{r,tr}$: energy price of transport mode *tr*
$pene\_tr\_base_{r,tr}$: energy price of transport mode *tr* in base year

Transport energy price by mode:

$$PENE\_TR_{r,tr} = \frac{\sum_{c \in C} TRS\_ENE\_FL_{r,tr,c} \cdot \left\{ PQD_{r,c} \cdot \left(1 + tqd_{r,c,"TRS"}\right) + PGHG_r \cdot \sum_{g \in G} gwp_g \cdot enur_{r,c,"trs"} \cdot efffc_{r,c,"trs",g} \right\}}{\sum_c TRS\_ENE\_FL_{r,tr,c}},$$

$$\forall r \in R, a \in A, tr \in TR \qquad \text{(PENE\_TR\_DEF)}$$

Transport energy source:

$$TRS\_ENE\_FL_{r,c,tr} = TRS\_ENE_{r,tr} \cdot \frac{\beta^{trsen}_{r,c,tr} \cdot \left\{ PQD_{r,c} \cdot \left(1 + tqd_{r,c,"TRS"}\right) + PGHG_r \cdot \sum_{g \in G} gwp_g \cdot enur_{r,c,"trs"} \cdot efffc_{r,c,"trs",g} \right\}^{el^{trsen}_{r,c,tr}}}{\sum_{cp \in ENE} \beta^{trsen}_{r,cp,tr} \cdot \left\{ PQD_{r,cp} \cdot \left(1 + tqd_{r,cp,"TRS"}\right) + PGHG_r \cdot \sum_{g \in G} gwp_g \cdot enur_{r,cp,"trs"} \cdot efffc_{r,cp,"trs",g} \right\}^{el^{trsen}_{r,cp,tr}}},$$

$$\forall r \in R, c \in ENE, tr \in TR \qquad \text{(TRSENEFLEQ)}$$

$TRS\_ENE\_FL_{r,c,tr}$: transport energy demand by modes and energy sources
$\beta^{trsen}_{r,c,tr}$: share parameter of Logit function for transport mode energy source selection
$el^{trsen}_{r,c,a}$: price elasticity parameter of Logit function for transport mode energy source selection

Reclassify the energy demand of transport activity to intermediate inputs and total energy consumption.

Transport total energy consumption:

$$QENE_{r,a} = \sum_{tr\in TR} TRS\_ENE_{r,tr}, \qquad \forall r \in R, a \in A\_TRS \qquad \text{(CESVAENE\_TRS)}$$

Transport total energy consumption by fuel:

$$QINT_{r,c,a} = \sum_{tr\in TR} TRS\_ENE\_FL_{r,c,tr}, \qquad \forall r \in R, c \in ENE, a \in A\_TRS$$

(QENEFUEL_TRS)

Commodity production and allocation:

$$QXAC_{r,a,c} = \theta_{r,a,c} \cdot QA_{r,a}, \qquad \forall r \in R, a \in A, c \in CX \qquad \text{(COMPRDFN)}$$

$_{r,a,c}$: yield of output c per unit of activity *a*

On the right-hand side, production quantities, disaggregated by activity, are defined as yields times activity levels. On the left-hand side, these quantities are allocated to market sales. Note that this equation permits (1) any commodity to be produced by one or more activities and (2) any activity to produce one or more commodities.

Commodity production and allocation (nonenergy commodities):

$$QX_{r,c} = \alpha^{ac}_{r,c} \cdot \left(\sum_{a\in A} \delta^{ac}_{r,a,c} \cdot QXAC_{r,a,c}^{\;-\rho^{ac}_{r,c}} \right)^{-\frac{1}{\rho^{ac}_{r,c}}}, \qquad \forall r \in R, c \in (CX - ENE)$$

(OUTAGGFN)

$\alpha^{ac}_{r,c}$: shift parameter for domestic commodity aggregation function
$\delta^{ac}_{r,a,c}$: shift parameter for domestic commodity aggregation function
$\rho^{ac}_{r,c}$: domestic commodity aggregation function exponent
$c \in ENE$: a set of energy commodities (COM_COA, COM_OIL, COM_GAS, COM_P_P, COM_COP, COM_ELY, COM_GDT)

First-order condition for output aggregation function (nonenergy commodities):

$$PXAC_{r,a,c} = PX_{r,c} \cdot QX_{r,c} \cdot \left(\sum_{ap\in A} \delta^{ac}_{r,ap,c} \cdot QXAC_{r,ap,c}^{\;-\rho^{ac}_{r,ap}} \right)^{-1} \cdot \delta^{ac}_{r,a,c} \cdot QXAC_{r,a,c}^{\;-\rho^{ac}_{r,a}-1},$$

$$\forall r \in R, a \in A, c \in (CX - ENE) \qquad \text{(OUTAGGFOC)}$$

Aggregate marketed production of any commodity excluding energy commodity is defined as a CES aggregate of the marketed output levels of the different activities producing the commodity (equation (OUTAGGFN)). The optimal quantity of the commodity from each activity source is inversely related to the activity-specific price (equation (OUTAGGFOC)). *QX* appears as the output, sold at the price, *PX*, and produced with the inputs, *QXAC*, that are purchased at the prices, *PXAC*.

More specifically, the choice between commodities from different sources is cast as an optimization problem. Equations (OUTAGGFN) and (OUTAGGFOC) are the first-order conditions for maximizing profits from selling the aggregate output, *QX*, at the price, *PX*, subject to the aggregation function and the disaggregated commodity prices, *PXAC*. A decline in the price, *PXAC*, of one activity relative to others would shift demand in its favor without totally eliminating demand for other, higher-priced sources. The degree of substitutability between different producers depends on the value of ρ_a^{ac} which is a transformation of the elasticity of substitution.

It should be noted that, for the case where there is a single producer of a given commodity, the value of the share parameter, δ_a^{ac}, would be unity and, as a result, $QXAC = QX$ and $PXAC = PX$, irrespective of the values for the elasticity and the exponent. Or if we assume perfect substitution among the production sectors, equation (OUTAGGFN) is substituted by (OUTAGG_PERFECT_SUB2).

Single sector commodity source:

$$PXAC_{r,a,c} = PX_{r,c}, \qquad \forall r \in R, a \in A, c \in (CX - ENE) \qquad \text{(OUTAGGFOC2)}$$

Perfect substitution commodity sources:

$$\sum_{a \in A} QXAC_{r,a,c} = QX_{r,c}, \qquad \forall r \in R, a \in A \qquad \text{(OUTAGG\_PERFECT\_SUB2)}$$

Share of commodity production and allocation (energy commodities):

$$SHAC_{r,a,c} = \frac{\psi_{r,a,c}^{ac} PXAC_{r,a,c}{}^{\eta_{r,a,c}^{ac}}}{\sum_{ap \in A} \psi_{r,ap,c}^{ac} PXAC_{r,ap,c}{}^{\eta_{r,ap,c}^{ac}}}, \qquad \forall r \in R, c \in (CX \cap ENE)$$

$$\text{(PRD\_MATERIAL\_BAL\_SH)}$$

$SHAC_{r,a,c}$: share of the commodity *c* produced by activity *a*
$\psi_{r,a,c}^{ac}$: share parameter of the commodity *c* produced by activity *a*
$\eta_{r,a,c}^{ac}$: elasticity of domestic commodity aggregation

Commodity production and allocation (energy commodities):

$$QXAC_{r,a,c} = QX_{r,c} \cdot SHAC_{r,a,c}, \qquad \forall r \in R, a \in A, c \in (CX \cap ENE)$$

(PRD_MATERIAL_BAL)

Aggregate marketed production of the energy commodity is defined as a share function. The share is determined by Logit function. The volume is calculated by multiplying the share and the total produced commodity volume.

Balance of the output and commodity aggregate (energy commodities):

$$\sum_{a \in A} QXAC_{r,a,c} \cdot PXAC_{r,a,c} = QX_{r,c} PX_{r,c}, \qquad \forall r \in R, c \in (CX \cap ENE)$$

(PRDSUM_MATERIAL_BAL)

Output transformation (CET) function (nonenergy commodities):

$$QX2_{r,c} = \alpha^t_{r,c} \cdot \left(\delta^t_{r,c} \cdot DELTA\_AD2_{r,c} \cdot QE_{r,c}{}^{\rho^t_{r,c}} + (1 - \delta^t_{r,c} \cdot DELTA\_AD2_{r,c}) \cdot QD_{r,c}{}^{\rho^t_{r,c}} \right)^{\frac{1}{\rho^t_{r,c}}},$$
$$\forall r \subset R, c \subset (CE \cap CD \quad ENE)$$

(CET)

Export constraint:

$$qe\_up_{r,c} \geq QE_{r,c} \perp DELTA\_AD2_{r,c} \geq 1, \qquad \forall r \in R, c \in C\{qe\_up_{r,c} \geq 0\}$$

(CET_B2)

$\alpha^t_{r,c}$	: a CET function shift parameter
$\delta^t_{r,c}$	: a CET function share parameter
$\rho^t_{r,c}$	: a CET function exponent
$DELTA\_AD2_{r,c}$	: adjustment variable for export constraint
$qe\_up_{r,c}$	: export constraint

Equations (CET) and (CET_B2) address the allocation of marketed domestic output, defined in equation (OUTAGGFN), to two alternative destinations: domestic sales and exports for nonenergy commodities. Equation (CET) reflects the assumption of imperfect transformability between these two destinations. The CET function, which applies to commodities that are both exported and sold domestically, is identical to a CES function except for negative elasticities of substitution. The elasticity of transformation between the two destinations is a transformation ofρ^t, for which the lower limit is one. The values are restricted to assure that the isoquant corresponding to the output transformation function is concave to the origin.

If export quantity constraints are put, adjustment variable *DELTA_AD2* would work.

Export-domestic supply ratio (nonenergy commodities):

$$\frac{QE_{r,c}}{QD_{r,c}} = \left(\frac{PE_{r,c}}{PDS_{r,c}} \cdot \frac{1 - \delta^t_{r,c} \cdot DELTA\_AD2_{r,c}}{\delta^t_{r,c} \cdot DELTA\_AD2_{r,c}}\right)^{\frac{1}{\rho^t_{r,c}-1}}, \qquad \forall r \in R, c \in (CE \cap CD - ENE)$$

(ESUPPLY)

Equation (ESUPPLY) defines the optimal mix between exports and domestic sales. Equations (PX2DEF), (CET), and (ESUPPLY) constitute the first-order conditions for maximization of producer revenues given the two prices and subject to the CET function and a fixed quantity of domestic output. Note that equation (ESUPPLY) assures that an increase in the export-domestic price ratio generates an increase in the export-domestic supply ratio (i.e., a shift toward the destination that offers the higher return).

Output transformation for domestically sold outputs without exports and for exports without domestic sales:

$$QX2_{r,c} = QD_{r,c} + QE_{r,c}, \quad \forall r \in R, c \in (CE \cap CEN) \cup (CD \cap CDN) \quad \text{(CET2,CET3)}$$

$c \in CEN(\subset C)$: non-exported commodities (complement of CE)
$c \in CDN(\subset C)$: commodities without domestic market sales of domestic output (complement of CD)

This equation replaces the CET function for domestically produced commodities that do not have both exports and domestic sales. It allocates the entire output volume to one of these two destinations.

Share of the domestically sold and export (energy commodities):

$$SHQE_{r,c} = \frac{\psi^t_{r,c} PE_{r,c}{}^{\eta^t_{r,c}}}{\psi^t_{r,c} PE_{r,c}{}^{\eta^t_{r,c}} + \psi 2^t_{r,c} PDS_{r,c}{}^{\eta^t_{r,c}}}, \qquad \forall r \in R, c \in (CE \cap CD \cap ENE)$$

(EXP_MATERIAL_BAL_SH)

$SHQE_{r,c}$: share of domestically sold and export commodity c
$\psi^t_{r,c}$: scale parameter for export share of commodity c
$\psi 2^t_{r,c}$: scale parameter for domestically produced share of commodity c
$\eta^t_{r,c}$: elasticity of domestic produced commodity aggregation

Exported energy commodities:

$$QE_{r,c} = QX2_{r,c} \cdot SHQE_{r,c}, \qquad \forall r \in R, c \in (CE \cap CD \cap ENE)$$

(EXP_MATERIAL_BAL)

Domestically sold energy commodities:

$$QD_{r,c} = QX2_{r,c} \cdot (1 - SHQE_{r,c}), \qquad \forall r \in R, c \in (CE \cap CD \cap ENE)$$

(DOME_MATERIAL_BAL)

The allocation of the domestically produced energy commodities is determined with the price change comparing with the previous year. The share of the export to produced commodities is described as the above equation. The volume is calculated by multiplying the share and the total produced commodity volume.

Composite supply (Armington) function (nonenergy commodities):

$$QQ_{r,c} = \alpha^q_{r,c} \cdot \left(\delta^q_{r,c} \cdot QM_{r,c}{}^{-\rho^q_{r,c}} + \left(1 - \delta^q_{r,c}\right) \cdot QD_{r,c}{}^{-\rho^q_{r,c}}\right)^{-\frac{1}{\rho^q_{r,c}}},$$
$$\forall r \in R, c \in (CM \cap CD - ENE)$$

(ARMINGTON)

$\alpha^q_{r,c}$: an Armington function shift parameter
$\delta^q_{r,c}$: an Armington function share parameter
$\rho^q_{r,c}$: an Armington function exponent

Imperfect substitutability between imports and domestic output sold domestically is captured by a CES aggregation function in which the composite commodity that is supplied domestically is "produced" by domestic and imported commodities entering this function as "inputs." When the domain of this function is limited to commodities that are both imported and produced domestically, it is often called an "Armington" function, named after the originator of the idea of using a CES function for this purpose. The elasticity of substitution between commodities from these two sources is a transformation of for which the lower limit is minus one.

Import-domestic demand ratio (nonenergy commodities):

$$\frac{QM_{r,c}}{QD_{r,c}} = \left(\frac{PDD_{r,c}}{PM_{r,c}} \cdot \frac{\delta^q_{r,c}}{1 - \delta^q_{r,c}}\right)^{\frac{1}{\rho^q_{r,c}+1}}, \qquad \forall r \in R, c \in (CM \cap CD - ENE)$$

(COSTMIN)

Equation (COSTMIN) defines the optimal mix between imports and domestic output. Its domain is thus limited to imports with domestic production. Note that the equation assures that an increase in the domestic-import price ratio generates an increase in the import-domestic demand ratio (that is, a shift away from the source that becomes more expensive). Together, equations (PQDDEF), (ARMINGTON), and (COSTMIN) constitute the first-order conditions for cost-minimization given the two prices and subject to the Armington function and a fixed quantity of the composite commodity.

Composite supply for nonimported outputs and nonproduced imports:

$$QQ_{r,c} = QD_{r,c} + QM_{r,c}, \qquad \forall r \in R, c \in (CD \cap CMN) \cup (CM \cap CDN)$$

(ARMINGTON2, ARMINGTON3)

$c \in CMN(\subset C)$: a set of nonimported commodities

The Armington function is replaced by equation (ARMINGTON2, ARMINGTON3) for the union of commodities that have either imports or domestic sales of domestic output but not both. For any commodity in this category, it imposes equality between "composite supply" and one of the variables on the right-hand side.

Share of the domestically sold and imported (energy commodities):

$$SHQM_{r,c} = \frac{\psi^m_{r,c} PM_{r,c}{}^{\eta^m_{r,c}}}{\psi^m_{r,c} PM_{r,c}{}^{\eta^m_{r,c}} + \psi 2^m_{r,c} PDD_{r,c}{}^{\eta^m_{r,c}}}, \qquad \forall r \in R, c \in (CM \cap CD \cap ENE)$$

(IMP_MATERIAL_BAL_SH)

$SHQM$: share of domestically sold and imported commodity c
$\psi^m_{r,c}$: scale parameter for import share of commodity c
$\psi 2^m_{r,c}$: scale parameter for domestically produced goods share of commodity c
$\eta^m_{r,c}$: elasticity of domestic consumption commodity aggregation

Imported energy commodities:

$$QM_{r,c} = QQ_{r,c} \cdot SHQM_{r,c}, \qquad \forall r \in R, c \in (CM \cap CD \cap ENE)$$

(IMP_MATERIAL_BAL)

Domestically sold energy commodities:

$$QD_{r,c} = QQ_{r,c} \cdot (1 - SHQM_{r,c}), \qquad \forall r \in R, c \in (CM \cap CD \cap ENE)$$

(DOMI_MATERIAL_BAL)

The aggregation of imported and domestically produced energy commodities is determined with the price change compared to the previous year being the same as the export share determination.

12.4.3 Land Block

The land productivity is exogenously determined if you switch on the land disaggregation (which is specified as LANDAGG in program).

Land input of activity:

$$QF_{r,f,a} = landeff_{r,f,a} \cdot QA_{r,a}, \qquad \forall r \in R, f \in FLND, a \in A \qquad \text{(LANDINPUT)}$$

$landeff_{r,f,a}$: land productivity coefficient

The land cost is equal to the summation of all AEZ cost.

AEZ aggregation for activity a:

$$QF_{r,f,a} \cdot WF_{r,f} \cdot WFDIST_{r,f,a} = \sum_{fl \in FL} PLAND_{r,fl,a} \cdot QLAND_{r,fl,a}, \quad \forall r \in R, f \in FLND, a \in A \qquad \text{(LANDAGGAEZ)}$$

$PLAND_{r,fl,a}$: land price of activity a and AEZ fl
$QLAND_{r,fl,a}$: land quantity of activity a and AEZ fl

The agricultural activities determine the share of AEZ based on Logit function and the land price.

AEZ selection for activity a:

$$QLAND_{r,fl,a} = QF_{r,''lnd'',a} \cdot \frac{\delta^{ls}_{r,fl,a} \cdot PLAND_{r,fl,a}{}^{\rho^{ls}_{r,a}}}{\sum_{flp \in FL} \delta^{ls}_{r,flp,a} \cdot PLAND_{r,flp,a}{}^{\rho^{ls}_{r,a}}}, \qquad \forall r \in R, fl \in FL, a \in A \qquad \text{(LANDAEZSEL)}$$

$\delta^{ls}_{r,fl,a}$: share parameter of AEZ input
$\rho^{ls}_{r,a}$: exponent parameter of AEZ input function

Crop field-aggregated price is as follows:

Crop land price:

$$QCROP_{r,fl} \cdot PCROP_{r,fl} = \sum_{a \in AAGR} PLAND_{r,fl,a} \cdot QLAND_{r,fl,a}, \qquad \forall r \in R, fl \in FL \qquad \text{(CROPLANDPRICE)}$$

$PCROP_{r,fl}$: crop field AEZ-aggregated land price
$QCROP_{r,fl}$: crop field AEZ-aggregated land use

Crop land selection is determined by the Logit as specified as follows:

AEZ land crop selection:

$$QLAND_{r,fl,a} = QCROP_{r,fl} \cdot \frac{\delta^{ltc}_{r,fl,a} \cdot PLAND_{r,fl,a}{}^{\rho^{ltc}_{r,fl}}}{\sum\limits_{ap \in AAGR} \delta^{ltc}_{r,fl,ap} \cdot PLAND_{r,fl,ap}{}^{\rho^{ltc}_{r,fl}}}, \quad \forall r \in R, fl \in FL, a \in AAGR \tag{CROPSEL}$$

$\delta^{ltc}_{r,fl}$: share parameter of crop AEZ input
$\rho^{ltc}_{r,a}$: exponent parameter crop input

Grazing field-aggregated price is as follows:

Grazing land price:

$$QGRZ_{r,fl} \cdot PGRZ_{r,fl} = \sum_{a \in ALIV} PLAND_{r,fl,a} \cdot QLAND_{r,fl,a}, \qquad \forall r \in R, fl \in FL \tag{GRAZPRICE}$$

$PGRZ_{r,fl}$: grazing AEZ-aggregated land price
$QGRZ_{r,fl}$: grazing AEZ-aggregated land use

Grazing land selection is determined by the Logit as specified as follows:

AEZ land grazing selection:

$$QLAND_{r,fl,a} = QGRZ_{r,fl} \cdot \frac{\delta^{ltc}_{r,fl,a} \cdot PLAND_{r,fl,a}{}^{\rho^{ltc}_{r,fl}}}{\sum\limits_{ap \in ALIV} \delta^{ltc}_{r,fl,ap} \cdot PLAND_{r,fl,ap}{}^{\rho^{ltc}_{r,fl}}}, \quad \forall r \in R, fl \in FL, a \in ALIV \tag{GRAZSEL}$$

$\delta^{ltc}_{r,fl}$: share parameter of grazing AEZ input
$\rho^{ltc}_{r,fl}$: exponent parameter grazing input

Crop and grazing land-use decision is made by the following equations:

Crop and grazing selection for AEZ (crop):

$$QCROP_{r,fl} = QAGRT_{r,fl} \cdot \frac{\delta^{ltt}_{r,fl,"CROP"} \cdot PCROP_{r,fl}{}^{\rho^{ltt}_{r,fl}}}{\delta^{ltt}_{r,fl,"CROP"} \cdot PCROP_{r,fl}{}^{\rho^{ltt}_{r,fl}} + \delta^{ltt}_{r,fl,"GRZ"} \cdot PGRZ_{r,fl}{}^{\rho^{ltt}_{r,fl}}}, \quad \forall r \in R, fl \in FL \tag{CROPGRAZSEL}$$

$QAGRT_{r,fl}$: agriculture AEZ-aggregated land use
$\delta^{ltt}_{r,fl}$: share parameter of crop and grazing input
$\rho^{ltt}_{r,fl}$: exponent parameter crop and grazing input

Crop and grazing selection for AEZ (grazing):

$$QGRZ_{r,fl} = QAGRT_{r,fl} \cdot \frac{\delta^{ltt}_{r,fl,"GRZ"} \cdot PGRZ_{r,fl}{}^{\rho^{ltt}_{r,fl}}}{\delta^{ltt}_{r,fl,"CROP"} \cdot PCROP_{r,fl}{}^{\rho^{ltt}_{r,fl}} + \delta^{ltt}_{r,fl,"GRZ"} \cdot PGRZ_{r,fl}{}^{\rho^{ltt}_{r,fl}}},$$
$$\forall r \in R, fl \in FL$$

(CROPGRAZSEL2)

Aggregation of agriculture (in case missing either crop or graze):

$$QAGRT_{r,fl} = QCROP_{r,fl} + QGRZ_{r,fl}, \qquad \forall r \in R, fl \in FL \quad \text{(AGRAGGNON)}$$

Aggregated agriculture land price is derived from the next equation.

Aggregated agricultural price:

$$QAGR_{r,fl} \cdot PAGRT_{r,fl} = QCROP_{r,fl} \cdot PCROP_{r,fl} + QGRZ_{r,fl} \cdot PGRZ_{r,fl}, \quad \forall r \in R, fl \in FL$$

(AGRLANDPRICE)

$PAGRT_{r,fl}$: agriculture AEZ-aggregated land price

Forestry or agricultural use is determined by following equations:

Agriculture and forestry selection for AEZ (agriculture):

$$QAGRT_{r,fl} = QLANT_{r,fl} \cdot \frac{\delta^{ltf}_{r,fl,"AGR"} \cdot PAGRT_{r,fl}{}^{\rho^{ltf}_{r,fl}}}{\delta^{ltf}_{r,fl,"AGR"} \cdot PAGRT_{r,fl}{}^{\rho^{ltf}_{r,fl}} + \delta^{ltf}_{r,fl,"FRS"} \cdot PLAND_{r,fl,"FRS"}{}^{\rho^{ltf}_{r,fl}}},$$
$$\forall r \in R, fl \in FL$$

(AGRFRSSEL)

$QLANT_{r,fl}$: total AEZ-aggregated land use
$\delta^{ltf}_{r,fl}$: share parameter of agriculture and forestry input
$\rho^{ltf}_{r,fl}$: exponent parameter agriculture and forestry input

Agriculture and forestry selection for AEZ (forestry):

$$QLAND_{r,fl,a} = QLANDT_{r,fl} \cdot \frac{\delta^{ltf}_{r,fl,"AGR"} \cdot PLAND_{r,fl,"FRS"}{}^{\rho^{ltf}_{r,fl}}}{\delta^{ltf}_{r,fl,"AGR"} \cdot PAGRT_{r,fl}{}^{\rho^{ltf}_{r,fl}} + \delta^{ltf}_{r,fl,"FRS"} \cdot PLAND_{r,fl,"FRS"}{}^{\rho^{ltf}_{r,fl}}},$$
$$\forall r \in R, fl \in FL, A \in AFRS$$

(AGRFRSSEL2)

Aggregation of agriculture (in case missing agriculture):

$$QLANDT_{r,fl} = QLAND_{r,fl,"FRS"}, \qquad \forall r \in R, fl \in FL \qquad \text{(AGGLAND\_NON)}$$

Aggregation of agriculture (in case missing forestry):

$$QLANDT_{r,fl} = QAGRT_{r,fl}, \qquad \forall r \in R, fl \in FL \qquad \text{(AGGLAND\_NON2)}$$

Total land-use price is expressed by the following:

Aggregated total land price:

$$\begin{aligned} QLANDT_{r,fl} \cdot PLANDT_{r,fl} = {} & QAGRT_{r,fl} \cdot PAGRT_{r,fl} \\ & + QLAND_{r,fl,"FRS"} \cdot PLAND_{r,fl,"FRS"}, \quad \forall r \in R, fl \in FL \end{aligned} \qquad \text{(AGGLANDPRICE)}$$

$PLANDT_{r,fl}$: total AEZ-aggregated land price

Total land use is determined by the following equation:

Total land:

$$QLANDT_{r,fl} = qlandtotara_{r,fl} \cdot \frac{\delta^{ara}_{r,fl,"USE"} \cdot PLANDT_{r,fl}^{\rho^{ara}_{r,fl}}}{\delta^{ara}_{r,fl,"USE"} \cdot PLANDT_{r,fl}^{\rho^{ara}_{r,fl}} + \delta^{ara}_{r,fl,"NUS"} \cdot plandt\_ini_{r,fl}^{\rho^{ara}_{r,fl}}},$$
$$\forall r \in R, fl \in FL$$

(TOTLAND)

$qlandtotara_{r,fl}$: total AEZ arable land
$plandt\_ini_{r,fl}$: total unused land price
$\delta^{ara}_{r,fl,*}$: share parameter of arable land input
$\rho^{ara}_{r,fl}$: exponent parameter of arable land input

Land use change emissions by AEZ:

$$LUCHEM_{r,fl} = LUCHEM\_P_{r,fl} + LUCHEM\_N_{r,fl}, \qquad \forall r \in R, fl \in FL$$

(EQ_LUCHEM)

$LUCHEM_{r,fl}$: land-use change emissions in AEZ fl region r
$LUCHEM\_P_{r,fl}$: land-use change emissions (positive) in AEZ fl region r
$LUCHEM\_N_{r,fl}$: land-use change emissions (negative) in AEZ fl region r

Positive land-use change emissions:

$$LUCHEM\_P_{r,fl} = \max\left[0, \left(qfrspre_{r,fl} - QPRMLANDT_{r,fl,"frs"}\right) \cdot ecoefluc_{r,fl,"frs"}\right], \quad \forall r \in R, fl \in FL$$

(EQ_LUCHEM_P)

$QPRMLANDT_{r,fl}$: primary land area AEZ *fl*, region *r*

$qfrspre_{r,fl,sprl}$: previous year's primary land area AEZ *fl*, region *r*, land classification *sprl*

$ecoefluc_{fl,sprl}$: land-use change emission coefficient AEZ *fl*, region *r*, land classification *sprl*

Negative land-use change emissions:

$$LUCHEM\_N_{r,fl} = -\sum_{y \in Yold} prluch_{y,r,fl} \cdot cf\_v2t \cdot cdst\_wd \cdot 11/3 \cdot \left[\exp\left[\alpha_{fl}^{bms} - \frac{30}{\{(Pyear - baseyear + 1) - Year_y\}}\right] - \exp\left[\alpha_{fl}^{bms} - \frac{30}{\{(Pyear - \text{baseyear}) - Year_y\}}\right]\right],$$

$$\forall r \in R, fl \in FL$$

(EQ_LUCHEM_N)

Pyear : the calculation year

Baseyear : base year

$Year_y$: year order (ex; base year is 1)

$prluch_{y,r,fl}$: prior year's land-use change

cf_v2t : conversion factor from volume to ton

$cdst\_wd$: conversion factor for carbon density

α^{bms}_{fl} : a parameter of biomass stock

12.4.4 Institution Block

Factor income:

$$YF_{r,f} = \sum_{a \in A} WF_{r,f} \cdot WFDIST_{r,f,a} \cdot QF_{r,f,a} + transfr\_f_{r,f,"ROW"} \cdot EXR_r, \quad \forall r \in R, f \in F$$

(YFDEF)

$YF_{r,f}$: income of factor *f*

$transfr\_f_{r,f}$: factor transfer from abroad

Institutional factor incomes:

$$YIF_{r,i,f} = shif_{r,i,f} \cdot \left(\left(1 - tf_{r,f} \right) \cdot YF_{r,f} - transfr_{r,"ROW",f} \cdot EXR_r \right), \quad \forall r \in R, i \in INSD, f \in F \qquad \text{(YIFDEF)}$$

$i \in INS$	: a set of institutions (domestic and rest of the world)
$i \in INSD(\subset INS)$	: a set of institutions (domestic and rest of the world)
$YIF_{r,i,f}$	: income to domestic institution i from factor f
$shif_{r,i,f}$	: share of domestic institution i in income of factor f
$tf_{r,f}$	: direct tax rate for factor f
$transfr_{r,f}$	: factor transfer to abroad

Equation (YFDEF) defines the total income of each factor. In equation (YIFDEF), this income is splitted among domestic institutions in fixed shares after payment of direct factor taxes and transfers to the rest of the world. The latter are fixed in foreign currency and transformed into domestic currency by multiplying by the exchange rate. This equation makes reference to the set of domestic institutions (households, enterprises, and the government), a subset of the set of institutions, which also includes the rest of world.

Income of nongovernmental domestic institution:

$$\begin{aligned} YI_{r,i} = & \sum_{f \in F} YIF_{r,i,f} + TRII\_Resource_{r,i} + shincome_{r,i} \cdot GHGTCOST_r \\ & + VRENCAPTOT_{r,i} - (PGHG\_G + PGHG\_IMP\_QUO_r - PGHG\_EXP\_QUO_r) \cdot \\ & \quad GHG\_IMP_r \times EXR_r \cdot shincome_{r,i} + shres_{r,i} \cdot \sum_{a \in A} PRES_{r,a} \cdot QRES_{r,a} \\ & + shincome_{r,i} \cdot \sum_{a \in A} QENE_{r,a} \cdot PENE_{r,a} \cdot \left(\frac{1}{1 - ADEEI_{r,a}} - 1 \right) \\ & + shincome_{r,i} \cdot \sum_{a \in A} RQUOQA_{r,a} \cdot QA_{r,a} \cdot PA_{r,a}, \\ & \forall r \in R, i \in INSDNG \end{aligned} \qquad \text{(YIDEF)}$$

$i \in INSDNG(\subset INSD)$	: a set of domestic nongovernment institutions
$YI_{r,I}$	: income of institution i (in the set INSDNG)
$TRII\_Resource_{r,I}$	: transfers to institution i
$GHGTCOST_r$	: GHG emission cost
$VRENCAPTOT_r$	: rent related to electricity capacity
$shincome_{r,I}$	:total income share of GHG emission cost for institution i

$shres_{r,I}$	: resource income share of institution i
GHG_IMP_r	: GHG emission credit import (net)
$PGHG\_G_r$	: global GHG emission price
$PGHG\_IMP\_QUO_r$	: GHG emission price generated by import quota
$PGHG\_EXP\_QUO_r$	: GHG emission price generated by export quota

Domestic nongovernment institutions form a subset of the set of domestic institutions. The total income of any domestic nongovernment institution is the sum of factor incomes (defined in equation (YIDEF), transfers from other domestic nongovernment institutions, the balance of payment of GHG emission trading, CCS installation cost, rent generated by the quota of the activity level, and electricity generation capacity rent. This rent total is defined below (only for enterprise):

Total rent of electricity capacity:

$$VRENCAPTOT_{r,''ent''} = \sum_{a \in A} VRENCAP_{r,a} \quad \forall r \in R \qquad \text{(VRENCAPIN)}$$

Household consumption expenditures:

$$\begin{aligned} EH_{r,h} = \; & (1 - shii\_use_{r,h})(1 - MPS_{r,h}) \cdot (1 - TINS_{r,h}) \cdot YI_{r,h} \\ & - PGHG_r \cdot \sum_{g \in G} gwp_{r,g} \cdot (EMALH_{r,h,g} + EMBIH_{r,h,g}) \\ & \forall r \in R, h \in H \end{aligned} \qquad \text{(EHDEF)}$$

$h \in H(\subset INSDNG)$	: a set of households
$EH_{r,h}$	: household consumption expenditures
$shii\_use_{r,I}$	: share of net income of i
$MPS_{r,I}$	: marginal propensity to save for domestic nongovernment institution (exogenous variable)
$TINS_{r,I}$	: direct tax rate for institution i

Among the domestic nongovernment institutions, only households demand commodities. Household expenditure is divided into two sources: one is energy and the other is nonenergy commodities. Nonenergy commodity expenditure is determined by LES or alternatively by AIDADS. The energy expenditure has two classifications. One is car fuel use expenditure and the other is other energy service demands such as space heating, space cooling, cooking, and so on.

The passenger car made by household is formulated by the income level and its elasticity.

Passenger car by household:

$$QCARU_{r,h} = pcaru_{r,h} \cdot \left(\frac{EH_{r,h}/CPI_r}{EH\_base_{r,h}/CPI\_base_r} \right)^{pasch_{r,h}}, \qquad \forall r \in R, h \in H$$

(HCARPAS)

Energy consumption caused by passenger car in household[5]**:**

$$QCARUENET_{r,h} = QCARU_{r,h} \cdot careneeff_{r,h}, \qquad \forall r \in R, h \in H$$

(QCARENETDEF)

Energy price for passenger car in household:

$$PENE\_H\_CAR_{r,h} \cdot QCARENET_{r,h} = \sum_{c \in C} QCARUENE_{r,h,c} \cdot \left(PQD_{r,c} \cdot dfpq_{r,c,h} \cdot (1 + tqd_{r,c,h}) + \sum_{g \in G} PGHG_r \cdot gwp_g \cdot efffc_{r,c,h,g} \right), \quad \forall r \in R, h \in H$$

(CARENEPRCDEF)

Energy fuel consumption caused by passenger car in household:

$$QCARUENE_{r,h,c} = QCARUENET_{r,h} \frac{\beta^{carh}_{r,c,h} \cdot \left\{ PQD_{r,c} \cdot (1 + tqd_{r,c,h}) + PGHG_r \cdot \sum_{g \in G} gwp_g \cdot enur_{r,c,h} \cdot efffc_{r,c,h,g} \right\}^{el^{carh}_{r,c,h}}}{\sum_{cp} \beta^{carh}_{r,cp,h} \cdot \left\{ PQD_{r,cp} \cdot (1 + tqd_{r,cp,h}) + +PGHG_r \cdot \sum_{g \in G} gwp_g \cdot enur_{r,cp,h} \cdot efffc_{r,cp,h,g} \right\}^{el^{carh}_{r,cp,h}}}$$

$$\forall r \in R, h \in H, c \in ENE$$

(QCARENEDEF)

$QCARU_{r,h}$	: household passenger car service demand
$pcaru_{r,h}$	: household passenger car service demand in base year
$pasch_{r,h}$	: income elasticity of passenger car service demand
$QCARUENET_{r,h}$	: household passenger car energy use
$careneeff_{r,h}$	: household passenger car energy coefficient
$\beta^{carh}_{r,c,h}$	: share parameter of Logit function for transport mode energy source selection
$el^{carh}_{r,c,h}$	: price elasticity parameter of Logit function for transport mode energy source selection
$EH\_base_{r,h}$	: household expenditure of base year
CPI_base_r	: base year's CPI

[5]This equation is currently not used.

Household consumption spending (LES):

$$PQH_{r,ch} \cdot QCH_{r,ch,h} = PQH_{r,ch} \cdot \gamma^{m}_{r,ch,h} + \beta^{m}_{r,ch,h} \cdot \left(EH_{r,h} - \sum_{c' \in C} PQH_{r,ch'} \cdot \gamma^{m}_{r,c'h,h} \right), \quad \forall r \in R, \quad c \in C, h \in H$$

(HMDEM2)

$\gamma^{m}_{r,ch,h}$: subsistence consumption of marketed commodity c for household h
$\beta^{m}_{r,ch,h}$: marginal share of consumption spending on marketed commodity c for household h

If the end-use device selection is considered, the above equation is slightly changed. The concept of the following equation is energy device selection that is made outside of the utility function.

$$PQH_{r,ch} \cdot QCH_{r,ch,h} = PQH_{r,ch} \cdot \gamma^{m}_{r,ch,h} + \beta^{m}_{r,ch,h} \cdot \left(EH_{r,h} - \sum_{c' \in C} PQH_{r,ch'} \cdot \gamma^{m}_{r,c'h,h} - \sum_{c' \in C\_ENE} \left(PQD_{r,c'} \cdot dfpq_{r,c',h} \cdot (1 + tqd_{r,c',h}) + \sum_{g \in G} PGHG_{r} \cdot gwp_{g} \cdot efffc_{r,c',h,g} \right) \cdot QH_{r,c',h} \right)$$
$$\forall r \in R, c \in C, h \in H$$

(HMDEM2)

Household goods price:

$$QCH_{r,ch,h} \cdot PQH_{r,ch,h} = \sum_{c \subset wch_{r,ch,c,h}} (QH_{r,c,h} - QCARENE_{r,c,h}) \cdot \left(PQD_{r,c} \cdot dfpq_{r,c,h} \cdot (1 + tqd_{r,c,h}) + \sum_{g \in G} PGHG_{r} \cdot gwp_{g} \cdot efffc_{r,c,h,g} \right), \quad \forall r \in R, ch \in CH, h \in H$$

(HPMAP)

Household goods consumption mapping:

$$QH_{r,c,h} = \sum_{ch \in wch_{r,ch,c,h}} QCH_{r,ch,h}, \quad \forall r \in R, c \in C, h \in H \qquad \text{(HMMAP)}$$

$PQH_{r,ch,h}$: price of household category commodity ch
$QCH_{r,ch,h}$: household consumption of household commodity category ch
$ch \in CH$: household commodity category ch

It is assumed that each houschold maximizes a "Stone-Geary" utility function subject to a consumption expenditure constraint. The resulting first-order conditions, equations (HMDEM2), are referred to as LES (linear expenditure system) functions since spending on individual commodities is a linear function of total consumption spending, *EH*. Note that the energy commodity price includes GHG emission price and additional energy efficiency improvement is also considered.

One thing that should be noted here is household commodity category *ch*. This is the same as c except for energy commodity. Energy commodity is aggregated as one "energy." Therefore, this function determines total energy consumption rather than that of each fuel type. Fuel share is determined by Logit function.

The energy source selection is made by Logit function as follows:

Household energy source consumption:

$$QH_{r,h,c} = QCARENE_{r,h,c} + QCH_{r,\text{"com\_ene"},h} \frac{\beta^{enec}_{r,c,h} \cdot \left\{ PQD_{r,c} \cdot (1 + tqd_{r,c,h}) + PGHG_r \cdot \sum_{g \in G} gwp_g \cdot enur_{r,c,h} \cdot efffc_{r,c,h,g} \right\}^{el^{enec}_{r,c,h}}}{\sum_{cp} \beta^{enec}_{r,cp,h} \cdot \left\{ PQD_{r,cp} \cdot (1 + tqd_{r,cp,h}) + +PGHG_r \cdot \sum_{g \in G} gwp_g \cdot enur_{r,cp,h} \cdot efffc_{r,cp,h,g} \right\}^{el^{enec}_{r,cp,h}}},$$
$$\forall r \in R, h \in H, c \in ENE$$

(HMDEM_ENE_FL)

As mentioned, the consumption function can be switched from LES to AIDADS in order to better perform the change of household expenditure behavior once their income increases. The AIDADS function is described in equations AIDADS_UTI, AIDADS_MCS, and AIDADS_CNS for utility, budget share parameter, and the consumption demand, respectively.

Household utility (AIDADS):

$$Vu_{r,ch} = \sum_{ch} V\mu_{r,ch,h} \cdot \ln \left(\frac{QCH_{r,ch,c}}{poph_{r,h}} - \theta h_{r,ch,h} \right) - (\ln A + 1), \quad \forall r \in R, h \in H$$

(AIDADS_UTI)

Household marginal share (AIDADS):

$$V\mu_{r,ch,h} = \frac{\left[\alpha h_{r,ch,h} + \beta h_{r,ch,h} \cdot e^{uh_{r,h}} \right]}{\left[1 + e^{uh_{r,h}} \right]}, \quad \forall r \in R, ch \in CH, h \in H$$

(AIDADS_MCS)

Household consumption spending (AIDADS):

$$
\begin{aligned}
QCH_{r,ch,h} \cdot PQH_{r,ch,h} = \; & poph_{r,h} \cdot \\
& \left[\begin{array}{l} \theta h_{r,ch,h} \cdot PQH_{r,ch,h} + V\mu_{r,ch,h} \cdot \\ \left(\begin{array}{l} EH_h - \sum_{c' \in C\_ENE} \left(PQD_{r,c'} \cdot dfpq_{r,c',h} \cdot (1 + tqd_{r,c',h}) + \sum_{g \in G} PGHG_r \cdot gwp_g \cdot efffc_{r,c',h,g} \right) \cdot QCARENE_{r,c',h} \\ - \sum_{chp \in CH} PQH_{r,chp,h} \cdot \theta h_{r,chp,h} \end{array} \right) \end{array} \right] \\
& \forall r \in R, ch \in CH, h \in H
\end{aligned}
\qquad \text{(AIDADS\_CNS)}
$$

$Vu_{r,h}$: utility of household h defined by AIDADS

$V\mu_{r,ch,h}$: marginal share of consumption spending on household commodity category ch for household h

$\theta h_{r,ch,h}$: subsistence consumption of household commodity category ch for household h

$\alpha h_{r,ch,h}$: a parameter for AIDADS

$\beta h_{r,ch,h}$: a parameter for AIDADS

$poph_{r,h}$: population of household h

Investment demand:

$$QINV_{r,c} = IADJ_r \cdot \overline{qinv_{r,c}}, \quad \forall r \in R, c \in C \qquad \text{(INVDEM)}$$

$IADJ_r$: investment adjustment factor (exogenous variable)

$\overline{qinv_{r,c}}$: base-year quantity of fixed investment demand

Fixed investment demand is defined as the base-year quantity multiplied by an adjustment factor. For the basic model version, the adjustment factor is exogenous, in effect also making the investment quantity exogenous. Inventory investment is also included in the model, but is treated as an exogenous demand.

Government consumption demand:

$$QG_{r,c} = \overline{GADJ_r} \cdot \overline{qg_{r,c}}, \quad \forall r \in R, c \in C \qquad \text{(GOVDEM)}$$

$\overline{GADJ_r}$: government consumption adjustment factor (exogenous variable)

$\overline{qg_{r,c}}$: government consumption adjustment factor (exogenous variable)

Similarly, government consumption demand, in which the main component tends to be the services provided by the government labor force, is also defined as the base-year quantity multiplied by an adjustment factor. This factor is also exogenous and, hence, the quantity of government consumption is fixed.

Government revenue:

$$
\begin{aligned}
YG_r = & \sum_{i \in INSDNG} TINS_{r,i} \cdot YI_{r,i} + \sum_{f \in F} tf_{r,f} \cdot YF_{r,f} \\
& + \sum_{a \in A} ta_{r,a} \cdot PA_{r,a} \cdot QA_{r,a} + \sum_{a \in A} tva_{r,a} \cdot PVA_{r,a} \cdot QVA_{r,a} \\
& + \sum_{c \in CM} tm_{r,c} \cdot PWM_c \cdot dis\_imp_{r,c} \cdot QM_{r,c} \cdot \overline{EXR}_r \\
& + \sum_{c \in CE} te_{r,c} \cdot PWE_c \cdot dis\_\exp_{r,c} \cdot QE_{r,c} \cdot \overline{EXR}_r \\
& + \sum_{c \in C} \sum_{a \in A} tqd_{r,c,a} \cdot dfpq_{r,c,a} \cdot PQD_{r,c} \cdot QINT_{r,c,a} \\
& + \sum_{c \in C} \sum_{h \in H} tqd_{r,c,h} \cdot dfpq_{r,c,h} \cdot PQD_{r,c} \cdot QH_{r,c,h} \\
& + \sum_{c \in C} tqd_{r,c,''gov''} \cdot dfpq_{r,c,''gov''} \cdot PQD_{r,c} \cdot QG_{r,c} \\
& + \sum_{c \in C} tqd_{r,c,''S-I''} \cdot dfpq_{r,c,''S-I''} \cdot PQD_{r,c} \cdot QINV_{r,c} \\
& + \sum_{f \in F} YIF_{r,''gov'',f} + TRII\_Resource_{r,''gov''} - TRII\_Use_{r,''gov''} \\
& + GHGTCOST_r \cdot shincome_{r,''gov''} + GHG\_IMP_r \cdot EXR_r \cdot \\
& (PGHG\_IMP\_QUO_r - PGHG\_EXP\_QUO_r) + shres_{r,''gov''} \cdot \sum_{a \in A} PRES_{r,a} \cdot QRES_{r,a} \quad \forall r \in R
\end{aligned}
$$

(YGDEF)

YG_r : government revenue

Total government revenue is the sum of revenues from taxes, factors, and transfers from the rest of the world. Emission trade cost is paid by government.

Government expenditure:

$$EG_r = \sum_{c \in C} PQD_{r,c} \cdot dfpq_{r,c,''gov''} \cdot \left(1 + tqd_{r,c,''gov''}\right) \cdot QG_{r,c} \quad \forall r \in R \qquad \text{(EGDEF)}$$

EG_r : government expenditures
$TRII\_Use_{r,i}$: transfers from institution i

Total government spending is the sum of government spending on consumption and transfers.

Transfer use:

$$TRII\_Use_{r,i} = shii\_use_{r,i} \cdot (1 - MPS_{r,i}) \cdot (1 - TINS_{r,i}) \cdot YI_{r,i}, \quad \forall r \in R, i \in INSDNG$$

(TRIIDEF2_H, TRIIDEF2_E)

Transfers between domestic nongovernment institutions are paid as fixed shares of the total institutional incomes net of direct taxes and savings. The values of *MPS* and *TINS* are defined in separate equations.

Government transfer use:

$$TRII\_Use_{r,"gov"} = \overline{trnsfr\_CRT_{r,"gov"}} \cdot CPI_r \qquad \forall r \in R \qquad \text{(TRIIDEF2\_G)}$$

$\overline{trnsfr\_CRT_{r,"gov"}}$: governmental transfer in base year

Governmental transfer is the production of CPI and that of base year.

Transfer resource:

$$TRII\_Resource_{r,i} = shii\_resource_{r,i} \cdot \left(crt\_in_r \cdot \overline{EXR_r} - crt\_out_r \cdot \overline{EXR_r} + \sum_{i'} TRII\_Use_{r,i'} \right), \quad \forall r \in R, i \in INSD \qquad \text{(TRIIDEF3)}$$

$shii\_resource_{r,I}$: a ratio of transfer to institution i of total transfer in a country
crt_in_r : transfer from rest of the world
crt_out_r : transfer to rest of the world

12.4.5 International Trade Block

Imported commodity:

$$QWM_{r,c} = QM_{r,c}, \quad \forall r \in R, c \in CM \qquad \text{(QWMDEF)}$$

$QWM_{r,c}$: quantity of imports of commodity

Internationally traded commodity equals to each country's imports.

Exported commodity:

$$QWE_{r,c} = QE_{r,c}, \quad \forall r \in R, c \in CE \qquad \text{(QWEDEF)}$$

$QWE_{r,c}$: quantity of exports of commodity

Quantity of domestic export is the sum of commodity export and international trade service (it is only for transport sector).

World trade nominal balance:

$$\sum_{r \in R} (1 - tw_{r,c}) \cdot PWM_c \cdot QWM_{r,c} \cdot dis\_imp_{r,c} = \sum_{r \in R} PWE_c \cdot QWE_{r,c} \cdot dis\_exp_{r,c} - \sum_{c \in C\_TRS} PTRS_c \cdot QTRS_c, \quad \forall c \in (CM \cap CE) \qquad \text{(WMEQ)}$$

$tw_{r,c}$: international trade cost ratio
$PTRS_c$: price of international trade service
$QTRS_c$: quantity of international trade service

World nominal trade balance is shown in equation (WMEQ).

World trade volume balance:

$$\sum_{r \in R} QWM_{r,c} = \left(1 - \lambda_c^w\right) \cdot \sum_{r \in R} \left(QWE_{r,c} - QTRS_c\right), \quad \forall c \in (CM \cap CE) \quad \text{(WMQB)}$$

λ_c^w : depreciation rate of traded commodity.

Total imported commodity in the world is equal to that of exported excluding depreciation.

Transport service demand:

$$QTRS_c = tsh_c \cdot \sum_r QWE_{r,c}, \qquad \forall c \in C\_TRS \qquad \text{(WTTR)}$$

CIF and FOB relationship:

$$PTRS_c \cdot QTRS_c = \sum_{r'} tw_{r',c} \cdot PWM_c \cdot QWM_{r',c} \cdot dis\_imp_{r',c}, \quad \forall c \in C\_TRS \qquad \text{(PTRSDEF)}$$

tsh_c : share of international trade service to world total international trade service
$c \in C\_TRS$: a set of transport service

The transport service is assumed to be proportional to the world total export of transport. The nominal international trade service provided by a country r is assumed to produce a constant share ratio and world total international trade service.

12.4.6 Biomass Consumption as Fuel Combustion

The biomass is still the main energy source in some developing countries and they are a main factor of air pollutants emission. The biomass consumption is calculated as below:

Biomass consumption (household):

$$TBH_{r,h} = poph_{r,h} \cdot bioc_{r,h} \cdot biod_{r,h} \quad \forall r \in R, h \in H \qquad \text{(TRDBIO\_CONS\_HOU)}$$

Biomass consumption (industry):

$$TBI_{r,a} = QA_{r,a} \cdot bioc_{r,a} \cdot biod_{r,a} \quad \forall r \in R, a \in A \qquad \text{(TRDBIO\_CONS\_IND)}$$

$TBH_{r,h}$: biomass consumption by household h
$TBI_{r,a}$: biomass consumption by activity a
$bioc_{r,ac}$: biomass consumption coefficient to the activity level of sector ac
$biod_{r,ac}$: decreasing rate of biomass consumption of sector ac

The biomass consumption is defined as activity level multiplied by two coefficients. One is the base-year-calibrated coefficient *bioc* and the other is decreasing rate of its use *biod*. Equation (TRDBIO_CONS_HOU) represents household biomass consumption and the activity level is defined as number of population. The industrial biomass consumption is shown in equations (TRDBIO_CONS_IND).

12.4.7 Air Pollutants and GHG Emissions

As is mentioned already in section 0, the air pollutants and GHG emissions are classified into two groups (related to activity level and fuel combustion). We are going to formulate these two groups separately.

Emissions related to activity level (industrial activity):

$$\begin{aligned} EMALI_{r,a,g,emsc} &= QA_{r,a} \cdot efacl_{r,a,g,emsc} \cdot \left(1 - NERED_{r,a,g,emsc}\right) \\ &- \sum_{emcm \in EMCM} \left({QRED_{r,emcm,a}} / {\eta_{emcm,a}} \right), \quad \forall r \in R, a \in A, g \in G, emsc \in EMSC \end{aligned}$$

$$\text{(EMI\_FROM\_IND\_PRO, EMI\_FROM\_IND\_PRO2)}$$

$g \in G$: a set of emission gases
$emsc \in EMSC$: a set of emission sources
$EMALI_{r,a,g,emsc}$: emissions nonenergy-related emission by industrial activity a, energy source *emsc*
$efacl_{r,ac,g,emsc}$: emission factors for emissions related to activity level by sector ac, energy source *emsc*
$NERED_{r,a,g,emsc}$: emission reduction caused by the GHG emission price, energy source *emsc*

Additional emission reductions related to activity level (industrial activity):

$$NERED_{r,a,g,emsc} = 1 - (PGHG_r + 1)^{-\sigma^{ghg}_{r,a,g,emsc}}, \ \forall r \in R, a \in A, g \in G, emsc \in EMSC$$

(EFFIMPR_PRICE_CONS)

$\sigma^{ghg}_{r,a,g,emsc}$: elasticity of the additional emission reductions of nonenergy-related emissions

Emissions related to activity level (household):

$$EMALH_{r,h,g,emsc} = poph_{r,h} \cdot efacl_{r,h,g,emsc} \quad \forall r \in R, h \in H, g \in G, emsc \in EMSC$$

(EMI_FROM_HOU_TOT)

$EMALH_{r,h,g.emsc}$: emissions related to activity level by household *h*, emission source *emsc*

The emissions related to activity level such as CO_2 emission from cement production is calculated by multiplying the activity level by the emission factor *efacl*. However, nonenergy-related GHG emission related to activity levels such as CH_4 emission from rice fields and CO_2 emission from the cement industry is defined at the top nest of the production function.

If an industrial sector can install CCS technology, it reduces the emission. In addition, there is additional emission reduction in mitigation case as *NERED*.

Emissions related to fossil fuel combustion (industrial activity):

$$EMFFI_{r,c,a,g} = QINT_{r,c,a} \cdot enur_{r,c,a} \cdot efffc_{r,c,a,g} \quad \forall r \in R, c \in ENE, a \in A, g \in G$$

(EMI_FROM_FF_IND)

$EMFFI_{r,c,a,g}$: emissions related to fossil fuel combustion emitted by industrial activity *a* consuming of goods *c*

Emissions related to fossil fuel combustion (household):

$$EMFFH_{r,c,h,g} = QH_{r,c,h} \cdot efffc_{r,c,h,g} \quad \forall r \in R, c \in ENE, h \in H, g \in G$$

(EMI_FROM_FF_HOU)

$EMFFH_{r,c,h,g}$: emissions related to fossil fuel combustion emitted by household *h* consumption of goods *c*

The CO_2 emitted by industrial activity is calculated in equation (EMI_FROM_FF_IND). The energy commodity inputs are multiplied by emission factor *efffc* and "energy-used ratio" *enur* represents the ratio of combustion to intermediate input of the energy commodity. Only the chemical industry is the only sector assumed to use fossil fuel as nonenergy.

The emission factor *efffc* of CO_2 is multiplied by three kinds of parameters which are "flag of fossil fuel transformation," "carbon fraction," and "CO_2 emission coefficient."

The emission factor *efffc* of non-CO_2 is calibrated by EDGAR 4.1 emission inventory data[6] and Bond et al. (2007) with EMEP/EEA air pollutant emission inventory guidebook 2009 (EEA 2009) and IPCC guideline (IPCC 2006).

Emissions related to biomass combustion (industrial activity):

$$EMBII_{r,a,g} = TBI_{r,a} \cdot efbio_{r,a,g} \quad \forall r \in R, a \in A, g \in G \qquad \text{(TRDBIO\_EMI\_IND)}$$

$EMBII_{r,a,g}$: emissions related to biomass combustion emitted by industrial activity a

$efbio_{r,ac,g}$: emission factors for emissions fossil fuel combustion by sector ac

Emissions related to biomass combustion (household activity):

$$EMBIH_{r,h,g} = TBH_{r,h} \cdot efbio_{r,h,g} \quad \forall r \in R, h \in H, g \in G \qquad \text{(TRDBIO\_EMI\_HOU)}$$

$EMBIH_{r,h,g}$: emissions related to biomass combustion emitted by household h

The emission from biomass combustion is estimated to multiply biomass consumption (*EMBII* and *EMBIH*) by emission factor *efbio*.

Nonenergy-related gas emissions:

$$EMNEG_{r,g} = ecfneng_{r,g} \cdot \left(\frac{GDP_r}{gdp\_base_r}\right)^{ecf\_gdpelas_{r,g}} \cdot PGHG_r^{-\sigma ghg\_hfc_{r,g}} \qquad \text{(EMNEGDEF)}$$

$EMNEG_{r,g}$: nonenergy-related emissions C2F6, SF6, CF4, and HFCs

$ecfneng_{r,g}$: emission coefficient of nonenergy-related emissions

GDP_r : GDP of region r

gdp_base_r : base-year GDP of region r

$ecf\_gdpela_{r,r}$: elasticity of Fgas to GDP increase in region r, gas g

$\sigma ghg\_hfc_{r,g}$: parameter for a MAC curve of international transport CO_2 reduction

[6]Download from http://edgar.jrc.ec.europa.eu/index.php on 4 Aug. 2010.

GHG emission total in a region:

$$GHGT_r = \sum_{g \in G} gwp_g \cdot \left\{ \begin{array}{l} \left(\sum_{h \in H} \sum_{c \in C} QH_{r,c,h} \cdot enur_{r,c,h} \cdot efffc_{r,c,h,g} + \sum_{a \in A} \sum_{c \in C} QINT_{r,c,a} \cdot enur_{r,c,a} \cdot efffc_{r,c,a,g} \right) \\ + \sum_{emsc \in EMSC} \left(\sum_{h \in H} EMALH_{r,h,g,emsc} + \sum_{a \in A} EMALI_{r,a,g,emsc} \right) \\ + \left(\sum_{h \in H} EMBIH_{r,h,g} + \sum_{a \in A} EMBII_{r,a,g} \right) \\ + EMNEG_{r,g} \\ + \sum_{fl \in FL} LUCHEM_{r,fl} \end{array} \right\} \quad \forall r \in R$$

(GHG_FROM_TOT)

$GHGT_r$: GHG emission from region r (CO_2 equivalent)

Region total GHG emission (CO2 equivalent) is the summation of industry and household emissions weighted by gwp_g (CO_2=1, CH_4=25, N_2O=298, C2F6, SF6, CF4). Currently, GHG coverage is Kyoto gases.

GHG emission includes emission trading:

$$GHGT\_CT_r = GHGT_r - GHGT\_IMP_r \cdot \quad \forall r \in R \qquad \text{(GHG\_TOT\_CT)}$$

$GHGT\_CT_r$: GHG emission from region r (CO_2 equivalent) includes emission permit import.

GHG emission permits can be imported from foreign countries. In reality, the amount of emission trading is constrained to a certain level which can be treated as an import or export quota. In addition, these import and export quota make the emission price higher or lower considering global and domestic emission prices as below.

GHG emission importing trading upper limit:

$$\overline{ghgt\_imp\_cap_r} - GHGT\_IMP_r \geq 0 \quad \perp PGHG\_IMP\_QUO_r \geq 0 \quad \forall r \in R$$

(GHGT_IMPDEF)

GHG emission exporting trading upper limit:

$$GHGT\_IMP_r - \overline{ghgt\_exp\_cap_r} \geq 0 \quad \perp PGHG\_EXP\_QUO_r \geq 0 \quad \forall r \in R$$

(GHGT_EXPDEF)

GHG emission price and international price:

$$PGHG_r = EXR_r \cdot (PGHG\_G + PGHG\_IMP\_QUO_r - PGHG\_EXP\_QUO_r) \quad \forall r \in R$$

(GHGT_IMPDEF2)

$\overline{ghgt\_impcap_r}$: GHG emission trading (import) limit
$\overline{ghgt\_expcap_r}$: GHG emission trading (export) limit

If emission trade is equal to the limit, the domestic GHG emission price (*PGHG*) will be different from global price (*PGHG_G*).

GHG emission constraint:

$$\overline{ghgc_r} - GHGT\_CT_r \geq 0 \quad \perp PGHG_r \geq 0 \quad \forall r \in R$$

(GHG_CONSTRAINT_REGION)

$\overline{ghgc_r}$: GHG emission constraint

If the regional total GHG emission is constrained by $\overline{ghgc_r}$, the GHG emission price is complementary variable of this equation.

GHG emission cost of nonenergy (industry):

$$GHGCA\_NENE_{r,a} = PGHG_r \cdot \sum_{g \in G} gwp_g \cdot \left(EMBII_{r,a,g} + \sum_{emsc \in EMSC} EMALI_{r,a,g,emsc} \right) \quad \forall r \in R, a \in A$$

(GHG_NENE_COST_A)

$GHGCA\_NENE_{r,a}$: GHG emission cost related to nonenergy consumption

GHG total cost:

$$GHGTCOST_r = \sum_{a \in A} \left(QINT_{r,c,a} \cdot enur_{r,c,a} \cdot efffc_{r,c,a,g} \right) + GHGCA\_NENE_{r,c,a} + \sum_{h \in H} PGHG_r \cdot \sum_{g \in G} gwp_g \cdot \left(\sum_{c \in C} QH_{r,c,h} \cdot efffc_{r,c,h,g} + EMALH_{r,h,g} + EMBIH_{r,h,g} \right) \quad \forall r \in R$$

(GHGCOSTDEF)

GHG cost consists of those of activity and household.

When the GHG constrained on only global total emissions, the following equations are introduced instead of the above.

Global GHG emission constraint:

$$\overline{ghgtot\_c} - \sum_{r \in R} GHGT\_CT_r \geq 0 \quad \perp PGHG\_G \geq 0$$

(GHG_CONSTRAINT_TOTAL)

$\overline{ghgtot\_c}$: global GHG emission constraint
$PGHG\_G$: GHG emission price corresponding to the global emission constraint

Global GHG emission constraint price:

$$PGHG\_G = PGHG_r, \quad \forall r \in R \qquad \text{(GHG\_CONSTRAINT\_PRICE)}$$

Global GHG emission trading total:

$$\sum_{r \in R} GHG\_IMP_r = 0 \qquad \text{(GHG\_WORLD\_TR)}$$

Net GHG emission trading should be zero globally.

GHG emission related to the international aviation and marine bankers are calculated by following equation:

GHG emission related to the international transport:

$$EMFFINT_{tr,g} = \sum_{c \in C} INTTRSENE_{tr,c} \cdot efint\_trs_{tr,c,g} \qquad \text{(GAS\_FROM\_FF\_INT)}$$

$EMFFINT_{tr,g}$: gas g emissions from international transport
$efint\_trs_{tr,c,g}$: international transport emissions coefficients

The energy use in international transport is as follows:

International transport energy demand:

$$INTTRSENE_{tr,c} = inttrenecoef_{tr,c} \cdot \sum_{r} QDTRST_{r,tr} \cdot PGHG_r^{-\sigma ghg\_int}$$

(INTTRSENEDEF)

$Inttrenecoeft_{r,c}$: international transport energy coefficient
σghg_int : parameter for a MAC curve of international transport CO2 reduction

GHG emission related to the international marine and aviation bankers are calculated by following equation.

12.4.8 System Constraint Block

Factor markets:

$$\sum_{a\in A} QF_{r,f,a} = \overline{QFS}_{r,f}, \quad \forall r \in R, f \in F \qquad \text{(FACEQUIL)}$$

$\overline{QFS}_{r,f}$: quantity supplied of factor (exogenous variable)

Equation (FACEQUIL) imposes equality between the total quantity demanded and the total quantity supplied for each factor other than capital. In the basic model version, all demand variables are flexible, while the supply variable is fixed. The factor wage, *WF*, is the equilibrating variable that assures that this equation is satisfied: an increase in *WF* raises the wage paid by each activity, *WF WFDIST*, which is inversely related to the quantities of factor demand, *QF*. All factors other than old capital are mobile between the demand activities.

Composite commodity markets:

$$QQ_{r,c} - \left(QX2_{r,c} + QM_{r,c}\right)\cdot loss_{r,c} + stch2_{r,c} = \sum_{a\in A} QINT_{r,c,a} + \sum_{h\in H} QH_{r,c,h} + QG_{r,c} + QINV_{r,c}, \quad \forall r \in R, c \in C \qquad \text{(COMEQUIL)}$$

$loss_{r,c}$: distribution loss rate
$stch2_{r,c}$: stock change of commodity *c* (negative)

Equation (COMEQUIL) imposes equality between quantities supplied (from equations (ARMINGTON), (COSTMIN) and (ARMINGTON2, ARMINGTON3)) and demanded of the composite commodity. The demand side includes endogenous terms (from equations (INTDEM)), (HMMAP), (INVDEM) and (GOVDEM)) and a new exogenous term for stock change. Among the endogenous terms, *QG* and *QINV* are fixed in the basic model version (compare with equations (INVDEM) and (GOVDEM)). The composite commodity supply, *QQ*, drives demands for domestic marketed output, *QD*, and imports, *QM*. The market-clearing variables are the quantities of import supply, for the import side, and the two interrelated domestic prices, *PDD* and *PDS*, for domestic market output.

Current-account balance for the rest of the world, in foreign currency:

$$\sum_{c \in CM} PWM_c \cdot dis\_imp_{r,c} \cdot QM_{r,c} + transfr\_crt\_out_r$$
$$+ GHG\_IMP_r \cdot PGHG_r + \sum_{f \in F} transfr_{r,"ROW",f} = \sum_{c \in CE} PWE_c \cdot dis\_exp_{r,c} \cdot QE_{r,c}$$
$$+ transfr\_crt\_in_r + \overline{FSAV_r} + \sum_{f \in F} transfr\_f_{r,f,"ROW"}$$
$$\forall r \in R, f \in F$$

(CURACCBAL)

$\overline{FSAV_r}$: foreign savings (FCU) (exogenous variable)
$transfr\_crt\_out_r$: current transfer to the rest of the world
$transfr\_crt\_in_r$: current transfer from the rest of the world

The current-account balance, which is expressed in foreign currency, imposes equality between the country's spending and its earning of foreign exchange. For the basic model version, foreign savings is fixed; the (real) exchange rate (*EXR*) serves the role of equilibrating variable to the current-account balance. The fact that all items except imports and exports are fixed means that the trade deficit also is fixed. Alternatively, the exchange rate may be fixed and foreign savings unfixed. In this case, the trade deficit is free to vary.

Government balance:

$$YG_r = EG_r + GSAV_r \qquad \forall r \in R \qquad \text{(GOVBAL)}$$

$GSAV_r$: government savings

The government balance imposes equality between current government revenue and the sum of current government expenditures (not including government investment) and savings. Savings may be negative. The alternative mechanisms for maintaining this balance are associated with equation (TINSDEF).

Direct tax rate:

$$TINS_{r,i} = \overline{tins_{r,i}} \cdot \left(1 + \overline{TINSADJ_r} \cdot tins01_{r,i}\right) + \overline{DTINS_{r,i}} \cdot tins01_{r,i},$$
$$\forall r \in R, i \in INSDNG$$

(TINSDEF)

$\overline{tins_{r,i}}$: rate of direct tax on domestic institutions *i*
$\overline{TINSADJ_r}$: direct tax scaling factor (= 0 for base, exogenous variable)
$tins01_{r,i}$: 0.1 parameter with 1 for institutions with potentially flexed direct tax rates
$\overline{DTINS_{r,i}}$: change in domestic institution tax share (= 0 for base, exogenous variable)

Equation (TINSDEF) defines the direct tax rates of domestic nongovernment institutions. For the basic model version, all variables on the right-hand side are fixed, in effect fixing the values for the direct tax rate variable for all institutions. In this setting, government savings is the endogenous variable that clears the government balance.

In the GAMS implementation of the standard model, two alternative closure rules are coded for the government balance. For both alternatives, government savings is fixed. In the first case, *DTINS* is the flexible variable that clears the government balance by scaling the base-year tax rates of each tax-paying institution. In this setting, the rates will change by a uniform number of (percentage) points for all institutions with a value of 1 for the parameter *tins*01 (that is, for all institutions with potentially flexed direct tax rates). Hence, the initial tax rate has no impact on the rate change. In the second case, *TINSADJ* becomes a variable, while *DTINS* is fixed. For this closure, the changes in *TINS* are relatively large for institutions with relatively large base-year rates (if they have a value of 1 for *tins*01).

Institutional savings rates:

$$MPS_{r,i} = \overline{mps_{r,i}} \cdot \left(1 + \overline{MPSADJ_r} \cdot mps01_{r,i}\right) + DMPS_r \cdot mps01_{r,i}, \quad \forall r \in R, i \in INSDNG \qquad \text{(MPSDEF)}$$

$\overline{mps_{r,i}}$: base savings rate for domestic institution i

$\overline{MPSADJ_r}$: savings rate scaling factor (= 0 for base)

$mps01_{r,i}$: 0–1 parameter with 1 for institutions with potentially flexed direct tax rates

$DMPS_r$: 0–1 parameter with 1 for institutions with potentially flexed direct tax rates

Equation (MPSDEF) defines the savings rates of domestic nongovernment institutions. Its structure is the same as that of equation (TINSDEF). Whether one or none of the variables *MPSADJ* and *DMPS* is flexible depends on the closure rule for the savings-investment balance. For the basic model version, *DMPS* is flexible, permitting *MPS* to be adjusted by a uniform rate for selected (one or more) nongovernment institutions.

Savings-investment balance:

$$\sum_{i \in INSDNG} MPS_{r,i} \cdot (1 - TINS_{r,i}) \cdot YI_{r,i} + GSAV_r + \overline{FSAV_r} \cdot \overline{EXR_r} = \sum_{c \in C} PQ_{r,c} \cdot dfpq_{r,c,"S-I"} \cdot \left(1 + tq_{r,c,"S-I"}\right) \cdot QINV_{r,i} + WALRAS_r \quad \forall r \in R \qquad \text{(SAVINVBAL)}$$

Equation (SAVINVBAL) states that total savings and total investment have to be equal. Total savings is the sum of savings from domestic nongovernment institutions, the government, and the rest of the world, with the last item converted into domestic currency. Total investment is the sum of the values of fixed investment (gross fixed capital formation) and stock changes.

In the basic model version, the flexible variable, *DMPS*, performs the task of clearing this balance (compare with equation (MPSDEF)). None of the other items in the savings-investment balance is free to vary to assure that the balance holds. Given that the balancing role is performed by the savings side, this closure represents a case of "investment-driven" savings. In the GAMS code, additional savings-investment closures have also been programed.

Global investment balance:

$$\sum_{r \in R} FSAV_r = 0 \qquad \text{(GLOBALINVBAL)}$$

Global investment should be balanced.

Up to this point, the matrix of this model as stated is not square; the number of equations exceeds the number of variables by one. However, the model satisfies Walras' law: one equation is functionally dependent on the others and can be dropped. The savings-investment balance or the current-account balance is commonly eliminated. After eliminating one equation, the model becomes square and, in the absence of errors in formulation, a unique solution typically exists. Instead of dropping one equation, it is also possible to add one variable to the macroeconomic balance equations. The solution value of this variable should be zero. If not, one or more equations are not satisfied and a general equilibrium solution has not been found. This is the approach followed in the GAMS version of the model. A variable called *WALRAS* is added to the savings-investment balance. No equation is dropped.

After this adjustment, the model presented is complete and self-contained. In the basic model version, three more equations (and three new variables that appear in them) are added. The reason for including these is that they permit the formulation of balanced savings-investment closures 4 and 5. We will return to the closure issue later, after presenting the new equations and their notation.

12.4.9 Activity Constraint Block

Climate policy analysis requires us to simulate power sectors activities in detail; however, their activity levels are decided not only by economic rationality but also political decision. Therefore, the power share or activity level should be constrained exogenously.

Activity constraint (upper boundary):

$$renew\_up_{r,a} - QA_{r,a} \cdot \theta_{r,a,"COM\_ELY"} \geq 0 \quad VRENCAP_{r,a} \geq 0,$$
$$\forall r \in R, a \in A = \{renew\_up_{r,a} > 0\}$$

(RENEWABLE_CAP)

$renew\_up_{r,a}$: capacity of the activity level a (for power sector energy)

Alternatively, we could have the constraints for power generation share.

$$sh\_ely\_up_{r,a} \cdot \sum_{ap \in A} \left(\theta_{r,ap,"ely"} \cdot QA_{r,ap}\right) - \theta_{r,a,"ely"} \cdot QA_{r,a} \geq 0,$$
$$VRENCAP_{r,a} \geq 0, \qquad \forall r \in R, a \in A = \{sh\_ely\_up_{r,a} > 0\}$$

(POWER_SHARE_CAP)

$sh\_ely\_up_{r,a}$: power generation share of activity a

Activity constraint (QUOTA for aggregated region and activity):

$$quotaqa_{ragg,aagg} = \sum_{r \in Map\_Ragg(r,ragg)} \sum_{a \in Map\_aagg(a,aagg)} \sum_{c \in C} QA_{r,a} \cdot \theta_{r,a,c} \geq 0$$
$$\perp RQUOQA\_agg_{ragg,aagg} \geq 0, \forall ragg \in Ragg,$$
$$aagg \in Aagg$$

(QUOQA_agg)

Activity constraint (QUOTA shadow price):

$$RQUOQA_{r,a} = \sum_{ragg \in Map\_Ragg(r,ragg)} \sum_{aagg \in Map\_aagg(a,aagg)} RQUOQA\_agg_{ragg,aagg} \geq 0,$$
$$\forall r \in R, a \in A$$

(QUOQA)

$aagg \in Aagg$: a set of aggregated activity
$ragg \in Ragg$: a set of aggregated regions
$quotaqa_{ragg,aagg}$: quota of aggregated region $ragg$ and aggregated activity $aagg$

In some cases, activities output can be constrained by a quota.

12.4.10 End-Use Energy Device Module

This model has an option to determine final energy demand based on detailed energy device specification. Basically the devices are selected by Logit function with each device's cost. The cost consists of investment cost and operation cost, and the former cost is annualized by using discount rate and lifetime of individual devices. The latter is energy, carbon emission cost and other operation cost like labor wages. The above device selection is assumed to be made only for new investment devices. The current year's old device stock is passed from previous year. The list of the technology energy service and their units are provided in Chap. 13, Table 13.3.

End-use device stock

End-use device stock is defined as the sum of stock remained from previous year and new investment. The stock of previous year is depleted by inverse of the lifetime and discounted by rate of not working devices:

$$END\_STK_{r,ac,l} = \left(\left(1 - 1/\tau_l \right) \cdot stka\_pre_{r,ac,l} \cdot \left(1 - \sum_{i \in LI(l,i)} END\_QEOR_{r,ac,i} \right) + END\_QR_{r,ac,l} \right) \quad \text{(EQ\_END\_STK)}$$

$END\_STK_{r,ac,l}$	: end-use device stock
τ_l	: lifetime of device l
$END\_QEOR_{r,ac,i}$	: end-use device (not working)
$stka\_pre_{r,ac,l}$	: previous year's stock of combination of device l in sector ac
$END\_QR_{r,ac,l}$	: end-use device new investment
$i \in I$	: energy service i
$l \in L$	: end-use device l
$LI(l,i)$	: mapping device l and energy service i

End-use device stock operation

End-use device stock operation is derived from working stock and operation ratio:

$$\left(1 + \lambda^d_{r,ac,l}\right) \cdot END\_STK_{r,ac,l} = END\_QXD_{r,ac,l} \quad \text{(EQ\_END\_OPE)}$$

$\lambda^d_{r,ac,l}$	: operating rate of device l in sector ac
$END\_QXD_{r,ac,l}$	: end-use device stock operation

End-use device service supply

Service demand is equal to the product of end-use device stock operation and supply output per operation unit considering service efficiency:

$$\left(1 + \varphi_{r,ac,i}\right) \cdot \sum_{l \in LI(l,i)} \left(ad_{r,ac,l,i} \cdot END\_QXD_{r,ac,l} \right) = QSD_{r,ac,i} \quad \text{(EQ\_END\_SRD)}$$

$\varphi_{r,ac,i}$: a measure of service efficiency of service type i in sector ac
$ad_{r,ac,l,i}$: supply output of service i per operating unit of device l in sector ac (same as specific service output)
$QSD_{r,ac,i}$: service demand

Industry service demand

The industrial service demand is derived based on base year's service demand and the change of quantity of activity compared with base year. *indserincel* is the elasticity determining the effects from the change of quantity of activity to service demand:

$$QSD_{r,a,i} = sd\_base_{r,a,i} \cdot \left(\frac{QA_{r,a}}{QA\_base_{r,a}}\right)^{indserincel_{r,a}} \quad \forall a \notin A\_TRS \qquad \text{(EQ\_END\_SRD\_IND)}$$

$sd\_base_{r,a,i}$: base service demand coefficient of industry
$indserincel_{r,a}$: output elasticity of energy service demand of industry

Residential service demand

The residential service demand is derived based on base year's service demand and the changes of household income and consumer price index compared with base year. Similar to *indeserincel* in industrial service demand, *resserincel* is the elasticity parameter.

$$QSD_{r,h,i} = sd\_base_{r,h,i} \cdot \left(\frac{EH_{r,h}}{EH\_base_{r,h}} \cdot \frac{CPI\_base_r}{CPI_r}\right)^{resserincel_{r,h}} \qquad \text{(EQ\_END\_SRD\_RSD)}$$

$sd\_base_{r,h,i}$: base service demand coefficient of household
$resserincel_{r,h}$: income elasticity of energy service demand of household

Transport service demand

The transport service demand is derived based on base year's service demand:

$$QSD_{r,a,i} = sd\_base_{r,a,i} \cdot \frac{\sum_{tr \in MTR(tr,i)} QDTRST_{r,tr}}{\sum_{tr \in MTR(tr,i)} QDTRST\_base_{r,tr}} \quad \forall a \in A\_TRS$$

$$\text{(EQ\_END\_SRD\_TRS)}$$

End-use device new investment selection

End-use device as new investment is selected based on device's price by Logit function. The device selection is made for each energy service demand.

$$\sum_{i \in LI(l,i)} \left(\left(1 + \varphi_{r,ac,i}\right) \cdot ad_{r,ac,l,i} \cdot \left(1 + \lambda^{d}_{r,ac,l}\right) \cdot END\_QR_{r,ac,l} \right)$$

$$= \sum_{i \in LI(l,i)} \left(\frac{\mu_{r,ac,l} \cdot (END\_PCOST_{r,ac,l})^{\eta^{end}_{r,ac,l}}}{\sum_{ll \in LI(ll,i)} \left(\mu_{r,ac,ll} \cdot (END\_PCOST_{r,ac,ll})^{\eta^{end}_{r,ac,ll}} \right)} \cdot END\_QRT_{r,ac,i} \right)$$

(EQ_END_LOGIT)

$END\_PCOST_{r,ac,l}$	: end-use device price
$\eta^{end}_{r,ac,l}$	: exponent of Logit function
$END\_QRT_{r,ac,i}$	: end-use total new investment for service

End-use device price

End-use device price is the sum of annualized investment cost, operation cost, and enegy cost. The energy cost is calculated as the product of energy use by device and energy price including carbon tax. Note that emissions from biomass fuel are not considered as GHG.

$$END\_PCOST_{r,ac,l} = invc_{r,ac,l} + \Bigg(oped_{r,ac,l} + \sum_{K\_END} \Bigg(ied_{r,ac,l,K\_END} \cdot \Bigg(END\_PED_{r,K\_END,ac} + \Bigg(1 - \sum_{i \in LI(l,i)} BFRATIO_{r,ac,i} \Bigg) \cdot PGHG_{r} \cdot \sum_{g \in G} \left(gwp_{r,g} \cdot emcoef\_end_{r,ac,K\_END,g} \right) \Bigg) \Bigg) \Bigg) \cdot \left(1 + \lambda^{d}_{r,ac,l}\right)$$

(EQ_END_PRICE)

$invc_{r,ac,l}$	: annualized investment cost per unit of combination of device l in sector ac and region r
$oped_{r,ac,l}$	: operating cost except for energy per unit of combination of device l in sector ac and region r
ied_{r,ac,l,K_END}	: energy use of energy kind k per operating unit (or specific energy input) of device l in sector ac region r
$END\_PED_{r,K\_END,ac}$	: end-use energy price
$BFRATIO_{r,ac,i}$	: end-use biofuel ratio

$emcoef\_end_{r,ac,K\_END,g}$: emission coefficient for end-use device
$k \in K\_END$: energy kind for endues device

End-use total new investment for service

The new investment for each energy service should be greater than 0. Once it reaches zero, the old stock would not work and ENDQEOR has value. This mechanism is formulated with MCP.

$$END\_QRT_{r,ac,i} \geq 0 \perp END\_QEOR_{r,ac,i} \geq 0 \qquad \text{(EQ\_END\_OPR)}$$

End-use energy price:

$$END\_PED_{r,K\_END,ac} = end\_ped\_base_{r,K\_END,ac} \cdot \sum_{c \in ENEENDMAP(c,K\_END)} \left(\frac{PQD_{r,c}}{PQD\_base_{r,c}} \right) \qquad \text{(EQ\_END\_ENEPRI)}$$

$end\_ped\_base_{r,K\_END,ac}$: base energy price of end-use stock
$c \subset ENEENDMAP(c, K\_END)$: ...

Energy use of service i in end-use energy:

$$END\_ENE_{r,ac,i,K\_END} = \sum_{l \in LI(l,i)} ied_{r,ac,l,K\_END} \cdot END\_QXD_{r,ac,l} \qquad \text{(EQ\_END\_ENERGY)}$$

END_ENE_{r,ac,i,K_END} : energy use of service i in end-use energy

Energy use of service i in CGE energy

The energy kinds used for end-use device are more aggregated than CGE's original commodity classification. For example, the coal in end-use device consists of crude coal and coal producst (COM_COA and COM_COP). Thus, the energy consumption specified by end-use devices should be disaggregated. In this model, that disaggregation is made by Logit function as below.

$$QEND\_ENE_{r,c,ac,i} = \sum_{\substack{KAGG\_END \in MK\_AGGC \\ (KAGG\_END,c)}} \left(\sum_{\substack{K\_END \in MK\_AGGEND \\ (KAGG\_END,K\_END)}} (END\_ENE_{r,ac,i,K\_END}) \cdot \frac{\mu^{ea}_{r,c,ac,i} \cdot \left(dfpq_{r,c,ac} \cdot PQD_{r,c} \cdot (1 + tqd_{r,c,ac}) + PGHG_r \cdot \sum_{g \in G} \left(gwp_{r,g} \cdot enur_{r,c,ac} \cdot efffc_{r,c,ac,g} \right) \right)^{\eta^{ea}_{r,c,ac}}}{\sum_{\substack{cp \in MKAGGC \\ (KAGG\_END,CP)}} \left(\mu^{ea}_{r,cp,ac,i} \cdot \left(dfpq_{r,cp,ac} \cdot PQD_{r,cp} \cdot (1 + tqd_{r,cp,ac}) + PGHG_r \cdot \sum_{g \in G} \left(gwp_{r,g} \cdot enur_{r,cp,ac} \cdot efffc_{r,cp,ac,g} \right) \right)^{\eta^{ea}_{r,cp,ac}} \right)} \right)$$

(EQ_END_ENERGY_MAP)

$QEND\_ENE_{r,c,ac,i}$: energy use of service i in CGE energy

Industry end-use energy mapping

The energy consumption specified by end-use devices is mapped to intermediate inputs and total energy consumption of each industrial and transport sector:

$$QINT_{r,c,a} = \sum_{i} QEND\_ENE_{r,c,a,i}$$ (EQ_END_ENERGY_INDMAP)

Industry end-use energy total:

$$QENE_{r,a} = \sum_{c \in C\_ENE(c)} QINT_{r,c,a}$$ (EQ_END_ENERGY_INDTOT)

Transport end-use energy mapping:

$$TRS\_ENE\_FL_{r,tr,c} = \sum_{i \in MTR\_I(tr,i)} QEND\_ENE_{r,c,\text{"TRS"},i} \quad \forall tr \in TR$$

(EQ_END_ENERGY_TRSMAP)

Transport end-use energy total:

$$TRS\_ENE_{r,tr} = \sum_{c} TRS\_ENE\_FL_{r,tr,c}$$ (EQ_END_ENERGY_TRSTOT)

Residential end-use energy mapping:

$$QH_{r,c,h} - QCARENE_{r,h,c} = \sum_{i} QEND\_ENE_{r,c,h,i}$$

(EQ_END_ENERGY_RSDMAP)

Residential car end-use energy mapping:

The energy consumption for passenger car owned by household is allocated from transport to household consumption.

$$QCARENE_{r,h,c} = \sum_{i \in MTR\_I("PC",i)} QEND\_ENE_{r,c,"TRS",i}$$

(EQ_END_ENERGY_CARMAP)

$trspc\_enecoef_{r,h,c}$: passenger transport industry and residential efficiency difference

Residential car end-use energy total:

$$QCARENET_{r,h} = \sum_{c} QCARENE_{r,h,c} \qquad \text{(EQ\_END\_ENERGY\_CARTOT)}$$

End-use biofuel ratio

The biofuel ratio is defined as below.

$$BFRATIO_{r,ac,i} \cdot \sum_{c \in MKaggc("OIL",c)} QEND\_ENE_{r,c,ac,i} = QEND\_ENE_{r,"COM\_BIO",ac,i}$$

(EQ_END_BIORATIO)

Mathematical Statement

Sets

$a \in A$: a set of activities
$a \in ACES(\subset A)$: a set for nonenergy transformation
$a \in ACES\_ENE(\subset A)$: a set of activities with a CES function at energy nest
$a \in ALEO(\subset A)$: a set for energy transformation
$a \in ALEO\_ENE(\subset A)$: a set of activities with a CES function at energy nest
$aagg \in Aagg$: a set of aggregated activity

$c \in C$	: a set of commodities (also referred to as *c'* and *C'*)
C_CCS	: a set of CCS service
$c \in CD(\subset C)$	: a set of commodities with domestic sales of domestic output
$c \in CDN(\subset C)$	: commodities without domestic market sales of domestic output (complement of CD)
$c \in CE(\subset C)$	: a set of exported commodities (with domestic production)
$c \in CEN(\subset C)$	: non-exported commodities (complement of CE)
$c \in CM(\subset C)$	: a set of imported commodities
$c \in CMN(\subset C)$	: a set of nonimported commodities
$c \in CNEN$	: a set of nonenergy commodities
$c \in C\_TRS$	: a set of transport service
$c \in CX(\subset C)$	: a set of commodities with domestic output
$c \in ENE$	: a set of energy commodities (COM_COA, COM_OIL, COM_GAS, COM_P_P, COM_COP, COM_ELY, COM_GDT)
$c \in ENEENDMAP(c, K\_END)$	: Mapping for enduse energy kinds and commodity
$ch \in CH$	: household commodity category *ch*
$emcm \in EMCM$	: a subset of emission reduction countermeasures
$emcm \in EMCM0(\subset EMCM)$	: a subset of emission reduction countermeasures which are for nonenergy-related emissions
$emcm \in EMCM1(\subset EMCM)$	: a subset of emission reduction countermeasures which are for energy-related emissions
$emcm \in EMCM2(\subset EMCM)$	: a subset of emission reduction countermeasures which are for biomass power plant absorption
$emcm \in EMCM$	: a set of emission reduction countermeasures (CCS)
$f \in F(=F')$	: a set of factors
$f \in FCAP(=F)$	: a set of capital (new and old, "ncap" and "cap")
$f \in F_{lab}$	: labor (skilled and unskilled labor)
$g \in G$	: a set of emission gases
$h \in H(\subset INSDNG)$	: a set of households
$i \in INS$	: a set of institutions (domestic and rest of the world)
$i \in INSD(\subset INS)$	: a set of institutions (domestic and rest of the world)
$i \in INSDNG(\subset INSD)$	: a set of domestic nongovernment institutions
$emsc \in EMSC$	: a set of emission sources
$i \in I$	: energy service *i*
$k \in K\_END$	: energy kind for end-use device
$l \in L$	: end-use device *l*
$LI(l, i)$	: mapping device *l* and energy service *i*

$r \in R$: a set of regions
$ragg \in Ragg$: a set of aggregated regions
$t \in T$: a set of time series
$tc \in TC$: a set of technology
$tr \in TR$: a set of transport mode

Parameters

Latin Letters

$ad_{r,ac,l,i}$: supply output of service i per operating unit of device l in sector ac (same as specific service output)
$aeei^{t}_{r,c,ac}$: annual AEEI rate of energy commodity c, sector ac, time t, and region r
$Baseyear$: base year
$bioc_{r,ac}$: biomass consumption coefficient to the activity level of sector ac
$biod_{r,ac}$: decreasing rate of biomass consumption of sector ac
$careneeff_{r,h}$: household passenger car energy coefficient
$cdst\_wd$: conversion factor for carbon density
cf_v2t : conversion factor from volume to ton
CPI_base_{r} : base year's CPI
crt_in_{r} : transfer from rest of the world
crt_out_{r} : transfer to rest of the world
$cwts_{r,c}$: weight of commodity c in the consumer price index
dep^{t}_{r} : capital depreciation rate in time t and region r
$dfpq_{r,c,I}$: price differences of commodity price among inputs sectors
$dis\_exp_{r,c}$: price difference of the export commodity c
$dis\_imp_{r,c}$: price difference of the import commodity c
$dwts_{r,c}$: weight of commodity c in the producer price index
$ecfneng_{r,g}$: emission coefficient of nonenergy-related emissions
$ecf\_gdpela_{r,r}$: elasticity of Fgas to GDP increase in region r, gas g.
$ecoefluc_{fl,sprl}$: land-use change emission coefficient AEZ fl, region r, land classification $sprl$
$efacl_{r,ac,g,emsc}$: emission factors for emissions related to activity level by sector ac, energy source $emsc$,
$efbio_{r,ac,g}$: emission factors for emissions fossil fuel combustion by sector ac
$efffc_{r,c,ac,g}$: emission factors for emissions fossil fuel combustion by sector ac consuming of goods c
$efint\_trs_{tr,c,g}$: international transport emissions coefficients
$EH\_base_{r,h}$: household expenditure of base year
$el^{carh}_{r,c,h}$: price elasticity parameter of Logit function for transport mode energy source selection

$el_{r,c,a}^{inden}$: price elasticity parameter of Logit function for industrial activity energy source selection
$el_{r,tr}^{pssincome}$: passenger transport income elasticity
$el_{r,c,a}^{trsen}$: price elasticity parameter of Logit function for transport mode energy source selection
$el_{r,tr}^{trspr}$: transport energy demand price elasticity
$emcoef\_end_{r,ac,K\_END,g}$: emission coefficient for end-use device
$end\_ped\_base_{r,K\_END,ac}$: base energy price of end-use stock
$enur_{r,c,ac}$: energy-used ratio (1-nonenergy-use ratio)
$ewts_{r,c}$: weight of commodity c in the export price index
$faceff_{r,c,a,*}$: old and new capital efficiency parameter (only used for power sectors)
$facgr_{r,f}^{t}$: expected factor input growth rate (calculated by 12.5.1 and 12.5.2)
$fcmult_{r,f,a}$: productivity shifter of factor f in activity a
$gdpgr_{r}^{t}*$: expected GDP growth target (annual growth rate)
gdp_base_{r} : base-year GDP of region r.
$\overline{ghgc_{r}}$: GHG emission constraint
$\overline{ghgt\_expcap_{r}}$: GHG emission trading (export) limit
$\overline{ghgt\_impcap_{r}}$: GHG emission trading (import) limit
gwp_{g} : global warming potential of gas g
$gwts_{r,c}$: weight of commodity c in the government price index
$ica_{r,c,a}$: quantity of c per unit of aggregate intermediate input a
ied_{r,ac,l,K_END} : energy use of energy kind k per operating unit (or specific energy input) of device l in sector ac region r
$iena_{r,a}$: quantity of aggregate energy input per activity unit
$iene_{r,c,a}$: energy commodity consumption ratio
$indserincel_{r,a}$: output elasticity of energy service demand of industry
$inta_{r,a}$: quantity of aggregate nonenergy intermediate input per activity unit
$inttrenecoeft_{r,c}$: international transport energy coefficient
$invc_{r,ac,l}$: annualized investment cost per unit of combination of device l in sector ac and region r
$ires_{r,a}$: quantity of aggregate resource input per activity unit
$iva_{r,a}$: quantity of value added per activity unit
$ivae_{r,a}$: quantity of value-added energy composite per activity unit
$ivfa_{r,f,a}$: input coefficient of factors for Leontief inputs
$iwts_{r,c}$: weight of commodity c in the capital formation price index
$labor\_stock_{r}^{t}$: labor stock in time t and region r

$labgr_r^t$: labor stock (annual) growth rate in time t and region r
$landeff_{r,f,a}$: land productivity coefficient
$loss_{r,c}$: distribution loss rate
$mps01_{r,i}$: 0–1 parameter with 1 for institutions with potentially flexed direct tax rates
$mwts_{r,c}$: weight of commodity c in the import price index
$oped_{r,ac,l}$: operating cost except for energy per unit of combination of device l in sector ac and region r
$pasch_{r,h}$: income elasticity of passenger car service demand
$pcaru_{r,h}$: household passenger car service demand in base year
$pene\_tr\_base_{r,tr}$: energy price of transport mode tr in base year
$plandt\_ini_{,\, r,fl}$: total unused land price
$poph_{r,h}$: population of household h
$pres\_base_{r,a}$: resource price (normally 1)
$prluch_{y,r,fl}$: prior year's land-use change
$Pyear$: the calculation year
$qe\_up_{r,c}$: export constraint
$qfrspre_{r,fl,sprl}$: previous year's primary land area AEZ fl, region r, land classification $sprl$
$qlandtotara_{r,ft}$: total AEZ arable land
$quotaqa_{ragg,aagg}$: quota of aggregated region $ragg$ and aggregated activity $aagg$
$renew\_up_{r,a}$: capacity of the activity level a (for power sector energy)
$resserincel_{r,h}$: income elasticity of energy service demand of household
$sd\_base_{r,a,i}$: base service demand coefficient of industry
$sd\_base_{r,h,i}$: base service demand coefficient of household
$sh\_ely\_up_{r,a}$: power generation share of activity a
$shii\_resource_{r,I}$: a ratio of transfer to institution i of total transfer in a country
$shii\_use_{r,I}$: share of net income of i
$shif_{r,i,f}$: share of domestic institution i in income of factor f
$shincome_{r,I}$: total income share of GHG emission cost for institution i
$shres_{r,I}$: resource income share of institution i
$stch_{r,c}$: stock change of commodity c (positive)
$stch2_{r,c}$: stock change of commodity c (negative)
$stka\_pre_{r,ac,l}$: previous year's stock of combination of device l in sector ac
$ta_{r,a}$: tax rate for activity
$te_{r,c}$: export tax rate
$tf_{r,f}$: direct tax rate for factor f
$tins01_{r,i}$: 0.1 parameter with 1 for institutions with potentially flexed direct tax rates

$tm_{r,c}$	: import tariff rate
$tq_{r,c,ac}$	: rate of sales tax (as share of composite price inclusive of sales tax). Suffix *ac* includes activity *a* and institution *i*
$transfr\_crt\_in_r$	: current transfer from rest of the world
$transfr\_crt\_out_r$	: current transfer to rest of the world
$transfr_{r,f}$	: factor transfer to abroad
$transfr\_f_{r,f}$	: factor transfer from abroad
$trscvf_{r,tr,ac}$	: transport service demand by modes
$trseneeffi_{r,tr}$	: transport energy efficiency
$trspc\_enecoef_{r,h,c}$	: passenger transport industry and residential efficiency difference
$trscef_{r,tr}$	: passenger transport demand coefficient
tsh_c	: share of international trade service to world total international trade service
$tva_{r,a}$	: rate of value-added tax for activity *a*
$tw_{r,c}$	: international trade cost ratio
$Year_y$	: year order (ex; base year is 1)

Greek Letters

$\alpha^{ac}_{r,c}$	: shift parameter for domestic commodity aggregation function
$\alpha^{q}_{r,c}$	: an Armington function shift parameter
$\alpha^{t}_{r,c}$	: a CET function shift parameter
$\alpha^{tech}_{a,tc}$	: share parameter for technology Logit selection function
$\alpha^{va}_{r,a}$	: efficiency parameter in the CES value-added function
$\alpha^{va}_{r,a}*$	: adjusted efficiency parameter in the CES value-added function
$\alpha^{vae}_{r,a}$	: efficiency parameter in the CES energy and value-added function
$\alpha^{bms}{}_{fl}$	: a parameter of biomass stock
$\beta^{carh}_{r,c,h}$	: share parameter of Logit function for household car energy source selection
$\beta^{inden}_{r,c,a}$	: share parameter of Logit function for industrial activity energy source selection
$\beta^{m}_{r,ch,h}$	: marginal share of consumption spending on marketed commodity ch for household *h*
$\beta^{tech}_{r,a,tc}$	: exponent of technology Logit selection
$\beta^{trsen}_{r,c,tr}$	: share parameter of Logit function for transport mode energy source selection
$\lambda^{d}_{r,ac,l}$	: operating rate of device *l* in sector *ac*
λ^{w}_{c}	: depreciation rate of traded commodity *c*
$\rho^{ac}_{r,c}$	: domestic commodity aggregation function exponent
$\rho^{ara}_{r,fl}$	: exponent parameter of arable land input

$\rho^{ltc}_{r,a}$: exponent parameter crop input
$\rho^{ltc}_{r,fl}$: exponent parameter grazing input
$\rho^{ltf}_{r,fl}$: exponent parameter agriculture and forestry input
$\rho^{q}_{r,c}$: an Armington function exponent
$\rho^{t}_{r,c}$: a CET function exponent
$\rho^{ls}_{r,a}$: exponent parameter of AEZ input function
$\rho^{ltt}_{r,fl}$: exponent parameter crop and grazing input
$\rho^{va}_{r,a}$: CES value-added function exponent
$\rho^{vae}_{r,a}$: CES energy and value-added function exponent
$\delta^{ac}_{r,a,c}$: shift parameter for domestic commodity aggregation function
$\delta^{ara}_{r,fl,*}$: share parameter of arable land input
$\delta^{ltc}_{r,fl}$: share parameter of crop AEZ input
$\delta^{ltc}_{r,fl}$: share parameter of grazing AEZ input
$\delta^{ls}_{r,fl,a}$: share parameter of AEZ input
$\delta^{ltf}_{r,fl}$: share parameter of agriculture and forestry input
$\delta^{ltt}_{r,fl}$: share parameter of crop and grazing input
$\delta^{q}_{r,c}$: an Armington function share parameter
$\delta^{t}_{r,c}$: a CET function share parameter
$\delta^{va}_{r,a}$: CES value-added function share parameter for factor f in activity a
$\delta^{vae}_{r,a}$: CES energy and value-added function share parameter in activity a
$\theta_{r,a,c}$: yield of output c per unit of activity *a*
$\psi^{ac}_{r,a,c}$: share parameter
$\psi^{m}_{r,c}$: scale parameter for import share of commodity *c*
$\psi 2^{m}_{r,c}$: scale parameter for domestically produced goods share of commodity *c*
$\psi^{t}_{r,c}$: scale parameter for export share of commodity *c*
$\psi 2^{t}_{r,c}$: scale parameter for domestically produced share of commodity *c*
$\eta^{ac}_{r,a,c}$: elasticity of domestic commodity aggregation
$\eta^{end}_{r,ac,l}$: exponent of Logit function
$\eta^{m}_{r,c}$: elasticity of domestic consumption commodity aggregation
$\eta^{t}_{r,c}$: elasticity of domestic produced commodity aggregation
$\sigma^{ghg}_{r,a,g,emsc}$: elasticity of the additional emission reductions of nonenergy-related emissions
σghg_int : parameter for a MAC curve of international transport CO2 reduction
$\sigma ghg\_hfc_{r,g}$: parameter for a MAC curve of international transport CO2 reduction
$\varsigma_{r,a}$: a parameter for operation ratio
$\theta h_{r,ch,h}$: subsistence consumption of household commodity category *ch* for household *h*

$\alpha h_{r,ch,h}$: a parameter for AIDADS
$\beta h_{r,ch,h}$: a parameter for AIDADS
τ_l : lifetime of device l
$\varphi_{r,ac,i}$: a measure of service efficiency of service type i in sector ac region r

Exogenous Variables

$\overline{DTINS_{r,i}}$: change in domestic institution tax share (= 0 for base; exogenous variable)
$\overline{FSAV_r}$: foreign savings (FCU) (exogenous variable)
$\overline{GADJ_r}$: government consumption adjustment factor (exogenous variable)
$\overline{ghgc_r}$: GHG emission constraint
$\overline{ghgt\_impcap_r}$: GHG emission trading (import) limit
$\overline{ghgt\_\mathrm{exp}cap_r}$: GHG emission trading (export) limit
$\overline{ghgtot\_c}$: global GHG emission constraint
$\overline{mps_{r,i}}$: base savings rate for domestic institution i
$\overline{MPSADJ_r}$: savings rate scaling factor (= 0 for base)
$\overline{QFS_{r,f}}$: quantity supplied of factor (exogenous variable)
$\overline{qg_{r,c}}$: government consumption adjustment factor (exogenous variable)
$\overline{qinv_{r,c}}$: base-year quantity of fixed investment demand
$\overline{tins_{r,i}}$: rate of direct tax on domestic institutions i
$\overline{TINSADJ_r}$: direct tax scaling factor (= 0 for base; exogenous variable)
$\overline{trnsfr\_CRT_{r,"gov"}}$: governmental transfer in base year

Endogenous Variables

$BFRATIO_{r,ac,i}$: end-use biofuel ratio
$COPR_{r,a}$: operation ratio
CPI_r : consumer price index (exogenous variable)
$DELTA\_AD2_{r,c}$: adjustment variable for export constraint
$DMPS_r$: 0–1 parameter with 1 for institutions with potentially flexed direct tax rates
DPI_r : producer price index for domestically marketed output
EG_r : government expenditures
$EH_{r,h}$: household consumption expenditures
$EH\_base_{r,h}$: household expenditure of base year
$EMALI_{r,a,g,emsc}$: emissions nonenergy-related emission by industrial activity a, energy source $emsc$
$EMALH_{r,h,g.emsc}$: emissions related to activity level by household h, emission source $emsc$

$EMBIH_{r,h,g}$	: emissions related to biomass combustion emitted by household h
$EMBII_{r,a,g}$	: emissions related to biomass combustion emitted by industrial activity a
$EMFFH_{r,c,h,g}$	: emissions related to fossil fuel combustion emitted by household h consumption of goods c
$EMFFI_{r,c,a,g}$	: emissions related to fossil fuel combustion emitted by industrial activity a consuming of goods c
$EMFFINT_{tr,g}$	: gas g emissions from international transport
$EMNEG_{r,g}$	: nonenergy-related emissions C2F6, SF6, CF4, and HFCs
END_ENE_{r,ac,i,K_END}	: energy use of service i in end-use energy
$END\_PCOST_{r,ac,l}$	: end-use device price
$END\_PED_{r,K\_END,ac}$	: end-use energy price
$END\_QEOR_{r,ac,i}$	: end-use device (not working)
$END\_QR_{r,ac,l}$	: end-use device new investment
$END\_QRT_{r,ac,i}$	: end-use total new investment for service
$END\_QXD_{r,ac,l}$	: end-use device stock operation
$END\_STK_{r,ac,l}$	: end-use device stock
EPI_r	: export price index
EXR_r	: exchange rate country r
GDP_r	: GDP of region r
$GHGCA\_NENE_{r,a}$	: GHG emission cost related biomass burning and CCS negative emissions of activity a in region r
$GHGCA\_NENE_{r,a}$	: GHG emission cost related to nonenergy consumption
$GHGTCOST_r$	: GHG emission cost
$GHGT\_CT_r$	: GHG emission from region r (CO_2 equivalent) includes emission permit import
$GHGT_r$	: GHG emission from region r (CO_2 equivalent)
GHG_IMP_r	: GHG emission credit import (net)
GPI_r	: government price index
$GSAV_r$	: government savings
$IADJ_r$	: investment adjustment factor (exogenous variable)
IPI_r	: capital formation price index
MPI_r	: import price index
$MPS_{r,I}$	: marginal propensity to save for domestic nongovernment institution (exogenous variable)
$NERED_{r,a,g,emsc}$	: emission reduction caused by the GHG emission price, energy source *emsc*
$PAGRT_{r,fl}$	: agriculture AEZ-aggregated land price
$PA_{r,a}$	: activity price (gross revenue per activity unit)
$PCROP_{r,fl}$	: crop field AEZ-aggregated land price
$PDD_{r,c}$	: demand price for commodity produced and sold domestically
$PDS_{r,c}$	: supply price for commodity produced and sold domestically

$PENE_{r,a}$	: price of (aggregate) energy input
$PE\_base_{r,c}$	: base year's PE
$PE_{r,c}$	: export price of commodity *c*
$PENE^{bau}_{r,a}$	: energy price (BaUcase)
$PENE\_TR_{r,tr}$	: energy price of transport mode *tr*
$PGHG\_EXP\_QUO_r$	: GHG emission price generated by export quota
$PGHG\_G$	: GHG emission price corresponding to the global emission constraint
$PGHG\_G_r$	: global GHG emission price
$PGHG\_IMP\_QUO_r$	: GHG emission price generated by import quota
$PGHG_r$	: GHG emission price in region *r* (US$/t$CO_2$)
$PGRZ_{r,fl}$	: grazing AEZ-aggregated land price
$PINTA_{r,a}$	: aggregate intermediate input price for activity *a*
$PLANDT_{r,fl}$	: total AEZ-aggregated land price
$PLAND_{r,fl,a}$	: land price of activity *a* and AEZ *fl*
$PM\_base_{r,c}$	: base year's PM
$PM_{r,c}$	: composite commodity price (including import tax and transaction costs)
$PQD\_base_{r,c}$	: base year's PQD
$PQD_{r,c}$	: composite commodity price excluding sales tax
$PQH_{r,ch,h}$	: price of household category commodity *ch*
$PQ_{r,c}$	: composite commodity price excluding sales tax
$PRES_{r,a}$	: price of resource input
$PTRS_c$	: price of international trade service
$PVAE_{r,a}$	: price of (aggregate) energy and value-added bundle (nonenergy transformation sector)
$PVA_{r,a}$	: price of (aggregate) value added
PWE_c	: f.o.b. export world price
PWM_c	: world import price (c.i.f) of commodity c
$PX2_{r,c}$	: aggregate producer price for commodity including stock change effects
$PXAC_{r,a,c}$	: producer price of commodity *c* for activity *a*
$PX_{r,c}$	: aggregate producer price for commodity
$QAGRT_{r,fl}$	: agriculture AEZ-aggregated land use
$QA_{r,a}$	: quantity (level) of activity
$QCARU_{r,h}$	: household passenger car service demand
$QCARUENET_{r,h}$	: household passenger car energy use
$QCH_{r,ch,h}$	: household consumption of household commodity category *ch*
$QCROP_{r,fl}$	: crop field AEZ-aggregated land use
$QDTRST_{r,tr}$	: total transport service demand by modes
$QDTRS_{r,tr,ac}$	: freight transport service demand by sector *ac* and modes
$QD_{r,c}$	: quantity sold domestically of domestic output
$QENE_{r,a}$	: quantity of (aggregate) energy input

$QEND\_ENE_{r,c,ac,i}$	: energy use of service *i* in CGE energy
$QE_{r,c}$	: quantity of exports
$QF_{r,f,a}$	: quantity demanded of factor f from activity *a*
$QGRZ_{r,fl}$	: grazing AEZ-aggregated land use
$QG_{r,c}$	: government consumption demand for commodity
$QH_{r,c,h}$	: quantity of consumption of marketed commodity *c* for household *h*
$QINTA_{r,a}$	: quantity of aggregate intermediate input
$QINT_{r,c,a}$	: quantity of commodity *c* as intermediate input to activity *a*
$QINV_{r,c}$	: quantity of fixed investment demand for commodity
$QLAND_{r,fl,a}$	: land quantity of activity *a* and AEZ *fl*
$QLANT_{r,fl}$	: total AEZ-aggregated land use
$QM_{r,c}$	: quantity of imports of commodity
$QPRMLANDT_{r,fl}$	: primary land area AEZ *fl*, region *r*
$QQ_{r,c}$	: quantity of goods supplied to domestic market (composite supply)
$QRED_{r,emcm,a}$	: input of counter emission reduction countermeasures of activity *a* and measure *emcm*
$QRES_{r,a}$	: quantity of resource input
$QSD_{r,ac,i}$	: service demand
$QTRS_{c}$	: quantity of international trade service
$QVAE_{r,a}$	: quantity of (aggregate) energy and value-added bundle (nonenergy transformation sector)
$QVA_{r,a}$	: quantity of (aggregate) value added
$QWE_{r,c}$	: quantity of exports of commodity
$QWM_{r,c}$	: quantity of imports of commodity
$QX2_{r,c}$	: aggregate marketed quantity of domestic output of commodity including stock change
$QXAC_{r,a,c}$	: marketed output quantity of commodity *c* from activity *a*
$QX_{r,c}$	: aggregate marketed quantity of domestic output of commodity
$RQUOQA_{r,a}$	: shadow subsidies of the fixed activity level
$SHAC_{r,a,c}$	: share of the commodity *c* produced by activity *a*
$SHQE_{r,c}$	: share of domestically sold and export commodity *c*
$SHQM$	: share of domestically sold and imported commodity *c*
$TBH_{r,h}$	: biomass consumption by household *h*
$TBI_{r,a}$	: biomass consumption by activity *a*
$TINS_{r,I}$	: direct tax rate for institution *i*
$TRII\_Resource_{r,I}$	: transfers to institution *i*
$TRII\_Use_{r,I}$	: transfers from institution *i*
$TRS\_ENE\_FL_{r,c,tr}$	: transport energy demand by modes and energy sources
$TRS\_ENE_{r,tr}$	: transport energy demand by modes
$TSCT_{r,a,tc}$	: technology *tc*'s share in sector a and region r

$VRENCAP_{r,a}$	: rent of electricity capacity activity a in region r
$VRENCAPTOT_r$	: rent related to electricity capacity
$Vu_{r,h}$	: utility of household h defined by AIDADS
$V\mu_{r,ch,h}$	: marginal share of consumption spending on household commodity category ch for household h
$WFDIST_{r,f,a}$	: factor price distortion factor for factor f in activity a
$WF_{r,f}$	: average price of factor
$YF_{r,f}$	: income of factor f
YG_r	: government revenue
$YI_{r,I}$	: income of institution i (in the set INSDNG)
$YIF_{r,i,f}$	: income to domestic institution i from factor f

Equation

Import price:

$$PM_{r,c} = PWM_c \cdot dis\_imp_{r,c} \cdot (1 + tm_{r,c}) \cdot EXR_r, \qquad \forall c \in CM$$

Export price:

$$PE_{r,c} = PWE_c \cdot dis\_\exp_{r,c} \cdot (1 - te_{r,c}) \cdot \overline{EXR_r}, \qquad \forall r \in R, c \in CE$$

Demand price of domestic nontraded goods:

$$PDD_{r,c} = PDS_{r,c}, \qquad \forall r \in R, c \in CD$$

Absorption:

$$PQ_{r,c} \cdot QQ_{r,c} = PDD_{r,c} \cdot QD_{r,c} + PM_{r,c} \cdot QM_{r,c}, \qquad \forall r \in R, c \in (CD \cup CM)$$

Commodity market monetary balance:

$$PQ_{r,c} \cdot QQ_{r,c} = PQD_{r,c} \cdot \begin{pmatrix} \sum_{a \in A} dfpq_{r,c,a} \cdot QINT_{r,c,a} + \sum_{h \in H} pfdq_{r,c,h} \cdot QH_{r,c,h} \\ + pfdq_{r,c,"gov"} \cdot QG_{r,c} + pfdq_{r,c,"S-I"} \cdot QINV_{r,c} \end{pmatrix},$$
$$\forall r \in R, c \in CX$$

Marketed output with stock change:

$$QX2_{r,c} = QX_{r,c} + stch_{r,c}, \qquad \forall r \in R, c \in CX$$

Marketed output value with stock change:

$$PX2_{r,c} \cdot QX2_{r,c} = PX_{r,c} \cdot QX_{r,c}, \qquad \forall r \in R, c \in CX$$

Marketed output value:

$$PX2_{r,c} \cdot QX2_{r,c} = PDS_{r,c} \cdot QD_{r,c} + PE_{r,c} \cdot QE_{r,c}, \qquad \forall r \in R, c \in CX$$

Activity price:

$$PA_{r,a} \cdot QA_{r,a} \cdot \left(1 + RQUOQA_{r,a}\right) = \sum_{c \in C} PXAC_{r,a,c} \cdot QXAC_{r,a,c}, \qquad \forall r \in R, a \in A$$

Aggregate Nonenergy intermediate input price:

$$QINTA_{r,a} \cdot PINTA_{r,a} = \sum_{c \in CNEN} dfpq_{r,c,a} \cdot PQD_{r,c} \cdot QINT_{r,c,a} \cdot \left(1 + tqd_{r,c,a}\right)$$
$$\forall r \in R, a \in A, c \in CNEN$$

Activity revenue and costs (nonenergy transformation sector):

$$\begin{aligned} PA_{r,a} \cdot (1 - ta_{r,a}) \cdot QA_{r,a} = {} & PVAE_{r,a} \cdot QVAE_{r,a} + PINTA_{r,a} \cdot QINTA_{r,a} \\ & + PRES_{r,a} \cdot QRES_{r,a} + GHGCA\_NENE_{r,a} \\ & + VRENCAP_{r,a} \cdot QA_{r,a} + \sum_{emcm \in EMCM} QRED_{r,emcm,a}, \\ & \forall r \in R, a \in ACES \end{aligned}$$

Activity revenue and costs (energy transformation sector):

$$\begin{aligned} PA_{r,a} \cdot (1 - ta_{r,a}) \cdot QA_{r,a} = {} & PVA_{r,a} \cdot QVA_{r,a} + PINTA_{r,a} \cdot QINTA_{r,a} \\ & + PENE_{r,a} \cdot QENE_{r,a} + PRES_{r,a} \cdot QRES_{r,a} \\ & + GHGCA\_NENE_{r,a} \\ & + VRENCAP_{r,a} \cdot QA_{r,a} \sum_{emcm \in EMCM} QRED_{r,emcm,a}, \\ & \forall r \in R, a \in ALEO \end{aligned}$$

Resource input price:

$$pres\_base_{r,a} = PRES_{r,a}, \qquad \forall r \in R, a \in A$$

Consumer price index:

$$CPI_r = \sum_{c\in C}\left(PQD_{r,c}\cdot dfpq_{r,c,"hurb"}\cdot\left(1+tqd_{r,c,"hurb"}\right)+PGHG_r\cdot\sum_{g\in G} gwp_{r,g}\cdot efffc_{r,c,"hurb",g}\right)\cdot cwts_{r,c}$$
$$\forall r\in R, \forall g\in G$$

Producer price index for nontraded market output:

$$DPI_r = \sum_{c\in C} PDS_{r,c}\cdot dwts_{r,c} \qquad \forall r\in R$$

Export price index:

$$EPI_r = \sum_{c\in C} PE_{r,c}\cdot ewts_{r,c} \qquad \forall r\in R$$

Import price index:

$$MPI_r = \sum_{c\in C} PM_{r,c}\cdot mwts_{r,c} \qquad \forall r\in R$$

Governmental consumption price index:

$$GPI_r = \sum_{c\in C} PQD_{r,c}\cdot dfpq_{r,c,"gov"}\cdot\left(1+tqd_{r,c,"gov"}\right)\cdot gwts_{r,c} \qquad \forall r\in R$$

Capital formation price index:

$$IPI_r = \sum_{c\in C} PQD_{r,c}\cdot dfpq_{r,c,"S-I"}\cdot\left(1+tqd_{r,c,"S-I"}\right)\cdot iwts_{r,c} \qquad \forall r\in R$$

Leontief technology: demand for aggregate value added (energy transformation sector)

$$QVA_{r,a} = iva_{r,a}\cdot QA_{r,a}, \qquad \forall r\in R, a\in ALEO$$

Leontief technology: demand for aggregate energy input (energy transformation sector)

$$QENE_{r,a} = iena_{r,a}\cdot QA_{r,a}, \qquad \forall r\in R, a\in ALEO$$

Energy and value-added bundle (nonenergy transformation sector):

$$QVAE_{r,a} = ivae_{r,a} \cdot QA_{r,a}, \qquad \forall r \in R, a \in ACES$$

Leontief technology: demand for aggregate nonenergy intermediate input

$$QINTA_{r,a} = inta_{r,a} \cdot QA_{r,a}, \qquad \forall r \in R, a \in A$$

Leontief technology: demand for resource input:

$$QRES_{r,a} = ires_{r,a} \cdot QA_{r,a}, \qquad \forall r \in R, a \in A$$

Energy and value-added composite:

$$QVAE_{r,a} = \alpha^{vae}_{r,a} \cdot \left(\delta^{vae}_{r,a} \cdot QVA_{r,a}{}^{-\rho^{vae}_{r,a}} + \left(1 - \delta^{vae}_{r,a}\right) \cdot QENE_{r,a}{}^{-\rho^{vae}_{r,a}}\right)^{-\frac{1}{\rho^{vae}_{r,a}}}, \qquad \forall r \in R, a \in ACES$$

Energy and value-added input CES technology: energy – value-added input ratio

$$QVA_{r,a} = QENE_{r,a} \cdot \left(\frac{\delta^{vae}_{r,a}}{1 - \delta^{vae}_{r,a}} \cdot \frac{PENE_{r,a}}{PVA_{r,a}}\right)^{\frac{1}{1+\rho^{vae}_{r,a}}}, \qquad \forall r \in R, a \in ACES$$

Energy and value-added composite balance:

$$QVAE_{r,a} \cdot PVAE_{r,a} = QENE_{r,a} \cdot PENE_{r,a} + QVA_{r,a} \cdot PVA_{r,a}, \qquad \forall r \in R, a \in ACES$$

Energy and value-added composite (nonenergy use sector):

$$QVAE_{r,a} = QVA_{r,a}, \qquad \forall r \in R, a \in ACES$$

Value-added and factor demands: non-power supply activities

$$QVA_{r,a} = \alpha^{va}_{r,a} \cdot \left(\sum_{f \in F} \delta^{va}_{r,a} \cdot \left(fcmult_{r,f,a} \cdot QF_{r,f,a}\right)^{-\rho^{va}_{r,a}}\right)^{-\frac{1}{\rho^{va}_{r,a}}}, \qquad \forall r \in R, a \in A$$

Value-added and factor demands: power supply activities

$$PVA_{r,a} \cdot (1 - tva_{r,a}) \cdot QVA_{r,a} = \sum_{f \in F} WF_{r,f} \cdot WFDIST_{r,f,a}, \qquad \forall r \in R, a \in A$$

Factor demand: non-power supply activities

$$WF_{r,f} \cdot WFDIST_{r,f,a} = PVA_{r,a} \cdot (1 - tva_{r,a}) \cdot QVA_{r,a} \cdot \left(\sum_{f \in F'} \delta_{r,a}^{va} \cdot \left(fcmult_{r,f,a} \cdot QF_{f\ r,a} \right)^{-\rho_{r,a}^{va}} \right)^{-1} \cdot \delta_{r,a}^{va} \cdot fcmult_{r,f,a}{}^{-\rho_{r,a}^{va}} \cdot QF_{r,f,a}{}^{-\rho_{r,a}^{va}-1},$$
$$\forall r \in R, a \in A, f \in F$$

Factor demand: power supply activities

$$QF_{r,f,a} = ivfa_{r,f,a} \cdot QVA_{r,a}, \quad \forall r \in R, a \in A, f \in F$$

Factor cost: power supply activities

$$\sum_{f} WF_{r,f} \cdot WFDIST_{r,f,a} \cdot QF_{r,f,a} = PVA_{r,a} \cdot (1 - tva_{r,a}) \cdot QVA_{r,a}, \quad \forall r \in R, a \in A$$

Capital aggregation: perfect substitution

$$QF_{r,''ccap'',a} = QF_{r,''ncap'',a} + QF_{r,''cap'',a} \cdot COPR_{r,a}, \quad \forall r \in R, a \in A$$

Capital aggregation: new industry

$$QF_{r,''ccap'',a} = QF_{r,''ncap'',a}, \quad \forall r \in R, a \in A$$

Capital aggregation: base year

$$QF_{r,''ccap'',a} = QF_{r,''cap'',a}, \quad \forall r \in R, a \in A$$

Capital cost balance:

$$\begin{aligned} QF_{r,f,a} \cdot WF_{r,f} \cdot WFDIST_{r,f,a} = {} & QF_{r,''ncap'',a} \cdot WF_{r,''ncap''} \cdot WFDIST_{r,''ncap'',a} \\ & + QF_{r,''cap'',a} \cdot COPR_{r,a} \cdot WF_{r,''cap''} \cdot WFDIST_{r,''cap'',a}, \\ & \forall r \in R, a \in A, f \in FCCAP \end{aligned}$$

Capital rate of return for new and old:

$$WF_{r,''ncap''} \cdot WFDIST_{r,''ncap'',a} \geq WF_{r,''cap''} \cdot WFDIST_{r,''cap'',a} \perp QF_{r,\}ncap\},a} \geq 0,$$
$$\forall r \in R, a \in A$$

Capital operation ratio

$$COPR_{r,a} = \left(\frac{WF_{r,"cap"} \cdot WFDIST_{r,"cap",a}}{WF_{r,"ncap"} \cdot WFDIST_{r,"ncap",a}} \right)^{\varsigma_{r,a}}, \quad \forall r \in R, a \in A$$

Disaggregated intermediate input demand:

$$QINT_{r,c,a} \cdot \frac{faceff_{r,c,a,"old"} \cdot QF_{r,"CAP",a} + faceff_{r,c,a,"new"} \cdot QF_{r,"NCAP",a}}{QF_{r,"CAP",a} + QF_{r,"NCAP",a}}$$
$$= ica_{r,c,a} \cdot QINTA_{r,a}, \qquad \forall r \in R, a \in A, c \in CNEN$$

CCS service intermediate input demand:

$$QINT_{r,c,a} = \sum_{emcm \in CCSMAP} QRED_{r,emcm,a}, \quad \forall r \in R, a \in A, c \in C\_CCS$$

Reduction measures (for nonenergy-related GHG emissions):

$$QRED_{r,emcm,a} = \xi^{\max}_{emcm,a} \cdot \eta_{emcm,a} \cdot \sum_{g \in G} gwp_g \cdot EMALI_{r,a,g},$$
$$\forall r \in R, a \in A, emcm \in EMCM0$$

Reduction measures (for energy-related GHG emissions):

$$QRED_{r,emcm,a} = \xi^{\max}_{emcm,a} \cdot \eta_{emcm,a} \cdot \sum_{c \in ENE} \sum_{g \in G} gwp_g \cdot QINT_{r,c,a} \cdot enur_{r,c,a} \cdot efffc_{r,c,a,g},$$
$$\forall r \in R, a \in A, emcm \in EMCM1$$

Reduction measures (for biomass power plant GHG absorption):

$$QRED_{r,emcm,a} = \xi^{\max}_{emcm,a} \cdot \eta_{emcm,a} \cdot \sum_{g \in G} gwp_g \cdot EMBII_{r,a,g},$$
$$\forall r \in R, a \in A, emcm \in EMCM2$$

CCS-equipped share:

$$TSCT_{r,a,tc} = \alpha^{tech}_{a,tc} \cdot PGHG_r^{\beta_{r,a,tc}} \cdot \left(\sum_{tc \in TC} \alpha^{tech}_{a,tc} \cdot PGHG_r^{\beta^{tech}_{r,a,tc}} \right)^{-1}, \quad \forall r \in R, a \in A, tc \in TC$$

Energy input technology share:

$$QINT_{r,c,a} = QENE_{r,a} \cdot \frac{\beta^{inden}_{r,c,a} \cdot \left\{ PQD_{r,c} \cdot (1 + tqd_{r,c,a}) + PGHG_r \cdot \sum_{g \in G} gwp_g \cdot enur_{r,c,a} \cdot efffc_{r,c,a,g} \right\}^{el^{inden}_{r,c,a}}}{\sum_{cp \in ENE} \beta^{inden}_{r,cp,a} \cdot \left\{ PQD_{r,cp} \cdot (1 + tqd_{r,cp,a}) + PGHG_r \cdot \sum_{g \in G} gwp_g \cdot enur_{r,cp,a} \cdot efffc_{r,cp,a,g} \right\}^{el^{inden}_{r,cp,a}}},$$

$$\forall r \in R, c \in ENE, a \in A$$

Energy input costs:

$$PENE_{r,a} \cdot QENE_{r,a} = \sum_{c \in C\_ENE} \left\{ PQD_{r,c,a} \cdot (1 + tqd_{r,c,a}) + PGHG_r \cdot \sum_{g \in G} gwp_g \cdot enur_{r,c,a} \cdot efffc_{r,c,a,g} \right\} \cdot QINT_{r,c,a},$$

$$\forall r \in R, a \in A$$

Energy consumption of energy transformation sector:

$$QINT_{r,c,a} = QENE_{r,a} \cdot iene_{r,c,a}, \qquad \forall r \in R, a \in ALEO\_ENE, c \in ENE$$

Freight transport generated by industrial activity:

$$QDTRS_{r,tr,a} = QINT_{r,"COM\_TRS",a} \cdot trscvf_{r,tr,a}, \qquad \forall r \in R, a \in A, tr \in TR\_FRT$$

Freight transport generated by household consumption:

$$QDTRS_{r,tr,h} = QH_{r,"COM\_TRS",h} \cdot trscvf_{r,tr,h}, \qquad \forall r \in R, h \in H, tr \in TR\_FRT$$

Total freight transport:

$$QDTRST_{r,tr} = \sum_{ac \in AC} QDTRS_{r,tr,ac}, \qquad \forall r \in R, tr \in TR\_FRT$$

Passenger transport (excluding household passenger car):

$$QDTRST_{r,tr} = trspss\_base_{r,tr} \cdot \left(\frac{GDP_r}{GDP\_base_r} \right)^{el^{pssincome}_{r,tr}}, \quad \forall r \in R, tr \in TR\_PSS$$

Transport energy demand:

$$TRS\_ENE_{r,tr} = QDTRST_{r,tr} \cdot trseneeffi_{r,tr} \cdot \left(\frac{PENE\_TR_{r,tr}}{pene\_tr\_base_{r,tr}} \right)^{el^{trspr}_{r,tr}}, \quad \forall r \in R, a \in A, tr \in TR$$

Transport energy price by mode:

$$PENE\_TR_{r,tr} = \frac{\sum\limits_{c \in C} TRS\_ENE\_FL_{r,tr,c} \cdot \left\{ PQD_{r,c} \cdot \left(1 + tqd_{r,c,"TRS"}\right) + PGHG_r \cdot \sum\limits_{g \in G} gwp_g \cdot enur_{r,c,"trs"} \cdot efffc_{r,c,"trs",g} \right\}}{\sum\limits_{c} TRS\_ENE\_FL_{r,tr,c}},$$

$$\forall r \in R, a \in A, tr \in TR$$

Transport energy source:

$$TRS\_ENE\_FL_{r,c,tr} = TRS\_ENE_{r,tr} \cdot \frac{\beta^{trsen}_{r,c,tr} \cdot \left\{ PQD_{r,c} \cdot \left(1 + tqd_{r,c,"TRS"}\right) + PGHG_r \cdot \sum\limits_{g \in G} gwp_g \cdot enur_{r,c,"trs"} \cdot efffc_{r,c,"trs",g} \right\}^{el^{trsen}_{r,c,tr}}}{\sum\limits_{cp \in ENE} \beta^{trsen}_{r,cp,tr} \cdot \left\{ PQD_{r,cp} \cdot \left(1 + tqd_{r,cp,"TRS"}\right) + PGHG_r \cdot \sum\limits_{g \in G} gwp_g \cdot enur_{r,cp,"trs"} \cdot efffc_{r,cp,"trs",g} \right\}^{el^{trsen}_{r,cp,tr}}},$$

$$\forall r \in R, c \in ENE, tr \in TR$$

Transport total energy consumption:

$$QENE_{r,a} = \sum_{tr \in TR} TRS\_ENE_{r,tr}, \quad \forall r \in R, a \in A\_TRS$$

Transport total energy consumption by fuel:

$$QINT_{r,c,a} = \sum_{tr \in TR} TRS\_ENE\_FL_{r,c,tr}, \quad \forall r \in R, c \in ENE, a \in A\_TRS$$

Commodity production and allocation:

$$QXAC_{r,a,c} = \theta_{r,a,c} \cdot QA_{r,a}, \quad \forall r \in R, a \in A, c \in CX$$

Commodity production and allocation (nonenergy commodities):

$$QX_{r,c} = \alpha^{ac}_{r,c} \cdot \left(\sum_{a \in A} \delta^{ac}_{r,a,c} \cdot QXAC_{r,a,c}{}^{-\rho^{ac}_{r,c}} \right)^{-\frac{1}{\rho^{ac}_{r,c}}}, \quad \forall r \in R, c \in (CX - ENE)$$

First-order condition for output aggregation function (nonenergy commodities):

$$PXAC_{r,a,c} = PX_{r,c} \cdot QX_{r,c} \cdot \left(\sum_{ap \in A} \delta^{ac}_{r,ap,c} \cdot QXAC_{r,ap,c}{}^{-\rho^{ac}_{r,ap}} \right)^{-1} \cdot \delta^{ac}_{r,a,c} \cdot QXAC_{r,a,c}{}^{-\rho^{ac}_{r,a}-1},$$
$$\forall r \in R, a \in A, c \in (CX - ENE)$$

Single sector commodity source:

$$PXAC_{r,a,c} = PX_{r,c}, \quad \forall r \in R, a \in A, c \in (CX - ENE)$$

Perfect substitution commodity sources:

$$\sum_{a \in A} QXAC_{r,a,c} = QX_{r,c}, \quad \forall r \in R, a \in A$$

Share of commodity production and allocation (energy commodities):

$$SHAC_{r,a,c} = \frac{\psi^{ac}_{r,a,c} PXAC_{r,a,c}{}^{\eta^{ac}_{r,a,c}}}{\sum_{ap \in A} \psi^{ac}_{r,ap,c} PXAC_{r,ap,c}{}^{\eta^{ac}_{r,ap,c}}}, \quad \forall r \in R, c \in (CX \cap ENE)$$

Commodity production and allocation (energy commodities):

$$QXAC_{r,a,c} = QX_{r,c} \cdot SHAC_{r,a,c}, \qquad \forall r \in R, a \in A, c \in (CX \cap ENE)$$

Balance of the output and commodity aggregate (energy commodities):

$$\sum_{a \in A} QXAC_{r,a,c} \cdot PXAC_{r,a,c} = QX_{r,c} PX_{r,c}, \qquad \forall r \in R, c \in (CX \cap ENE)$$

Output transformation (CET) function (nonenergy commodities):

$$QX2_{r,c} = \alpha^{t}_{r,c} \cdot \left(\delta^{t}_{r,c} \cdot DELTA\_AD2_{r,c} \cdot QE_{r,c}{}^{\rho^{t}_{r,c}} + \left(1 - \delta^{t}_{r,c} \cdot DELTA\_AD2_{r,c}\right) \cdot QD_{r,c}{}^{\rho^{t}_{r,c}} \right)^{\frac{1}{\rho^{t}_{r,c}}},$$
$$\forall r \in R, c \in (CE \cap CD - ENE)$$

Export constraint:

$$qe\_up_{r,c} \geq QE_{r,c} \perp DELTA\_AD2_{r,c} \geq 1, \quad \forall r \in R, c \in C\{qe\_up_{r,c} \geq 0\}$$

Export-domestic supply ratio (nonenergy commodities):

$$\frac{QE_{r,c}}{QD_{r,c}} = \left(\frac{PE_{r,c}}{PDS_{r,c}} \cdot \frac{1 - \delta^t_{r,c} \cdot DELTA\_AD2_{r,c}}{\delta^t_{r,c} \cdot DELTA\_AD2_{r,c}}\right)^{\frac{1}{\rho^t_{r,c}-1}}, \quad \forall r \in R, c \in (CE \cap CD - ENE)$$

Output transformation for domestically sold outputs without exports and for exports without domestic sales:

$$QX2_{r,c} = QD_{r,c} + QE_{r,c}, \quad \forall r \in R, c \in (CE \cap CEN) \cup (CD \cap CDN)$$

Share of the domestically sold and export (energy commodities):

$$SHQE_{r,c} = \frac{\psi^t_{r,c} PE_{r,c}{}^{\eta^t_{r,c}}}{\psi^t_{r,c} PE_{r,c}{}^{\eta^t_{r,c}} + \psi 2^t_{r,c} PDS_{r,c}{}^{\eta^t_{r,c}}}, \quad \forall r \in R, c \in (CE \cap CD \cap ENE)$$

Exported energy commodities:

$$QE_{r,c} = QX2_{r,c} \cdot SHQE_{r,c}, \quad \forall r \in R, c \in (CE \cap CD \cap ENE)$$

Domestically sold energy commodities:

$$QD_{r,c} - QX2_{r,c} \cdot (1 - SHQE_{r,c}), \quad \forall r \in R, c \in (CE \cap CD \cap ENE)$$

Composite supply (Armington) function (nonenergy commodities):

$$QQ_{r,c} = \alpha^q_{r,c} \cdot \left(\delta^q_{r,c} \cdot QM_{r,c}{}^{-\rho^q_{r,c}} + \left(1 - \delta^q_{r,c}\right) \cdot QD_{r,c}{}^{-\rho^q_{r,c}}\right)^{-\frac{1}{\rho^q_{r,c}}},$$
$$\forall r \in R, c \in (CM \cap CD - ENE)$$

Import-domestic demand ratio (nonenergy commodities):

$$\frac{QM_{r,c}}{QD_{r,c}} = \left(\frac{PDD_{r,c}}{PM_{r,c}} \cdot \frac{\delta^q_{r,c}}{1 - \delta^q_{r,c}}\right)^{\frac{1}{\rho^q_{r,c}+1}}, \quad \forall r \in R, c \in (CM \cap CD - ENE)$$

Composite supply for nonimported outputs and non-produced imports:

$$QQ_{r,c} = QD_{r,c} + QM_{r,c}, \qquad \forall r \in R, c \in (CD \cap CMN) \cup (CM \cap CDN)$$

Share of the domestically sold and imported (energy commodities):

$$SHQM_{r,c} = \frac{\psi^{m}_{r,c} PM_{r,c}{}^{\eta^{m}_{r,c}}}{\psi^{m}_{r,c} PM_{r,c}{}^{\eta^{m}_{r,c}} + \psi 2^{m}_{r,c} PDD_{r,c}{}^{\eta^{m}_{r,c}}}, \qquad \forall r \in R, c \in (CM \cap CD \cap ENE)$$

Imported energy commodities:

$$QM_{r,c} = QQ_{r,c} \cdot SHQM_{r,c}, \qquad \forall r \in R, c \in (CM \cap CD \cap ENE)$$

Domestically sold energy commodities:

$$QD_{r,c} = QQ_{r,c} \cdot (1 - SHQM_{r,c}), \qquad \forall r \in R, c \in (CM \cap CD \cap ENE)$$

Land input of activity:

$$QF_{r,f,a} = landeff_{r,f,a} \cdot QA_{r,a}, \qquad \forall r \in R, f \in FLND, a \in A$$

AEZ aggregation for activity a:

$$QF_{r,f,a} \cdot WF_{r,f} \cdot WFDIST_{r,f,a} = \sum_{fl \in FL} PLAND_{r,fl,a} \cdot QLAND_{r,fl,a}, \qquad \forall r \in R, f \in FLND, a \in A$$

AEZ selection for activity a:

$$QLAND_{r,fl,a} = QF_{r,\text{"lnd"},a} \cdot \frac{\delta^{ls}_{r,fl,a} \cdot PLAND_{r,fl,a}{}^{\rho^{ls}_{r,a}}}{\sum_{flp \in FL} \delta^{ls}_{r,flp,a} \cdot PLAND_{r,flp,a}{}^{\rho^{ls}_{r,a}}}, \qquad \forall r \in R, fl \in FL, a \in A$$

Crop land price:

$$QCROP_{r,fl} \cdot PCROP_{r,fl} = \sum_{a \in AAGR} PLAND_{r,fl,a} \cdot QLAND_{r,fl,a}, \qquad \forall r \in R, fl \in FL$$

AEZ land crop selection:

$$QLAND_{r,fl,a} = QCROP_{r,fl} \cdot \frac{\delta^{ltc}_{r,fl,a} \cdot PLAND_{r,fl,a}{}^{\rho^{ltc}_{r,fl}}}{\sum_{ap \in AAGR} \delta^{ltc}_{r,fl,ap} \cdot PLAND_{r,fl,ap}{}^{\rho^{ltc}_{r,fl}}}, \qquad \forall r \in R, fl \in FL, a \in AAGR$$

Grazing land price:

$$QGRZ_{r,fl} \cdot PGRZ_{r,fl} = \sum_{a \in ALIV} PLAND_{r,fl,a} \cdot QLAND_{r,fl,a}, \qquad \forall r \in R, fl \in FL$$

AEZ land grazing selection:

$$QLAND_{r,fl,a} = QGRZ_{r,fl} \cdot \frac{\delta^{ltc}_{r,fl,a} \cdot PLAND_{r,fl,a}{}^{\rho^{ltc}_{r,fl}}}{\sum\limits_{ap \in ALIV} \delta^{ltc}_{r,fl,ap} \cdot PLAND_{r,fl,ap}{}^{\rho^{ltc}_{r,fl}}}, \qquad \forall r \in R, fl \in FL, a \in ALIV$$

Crop and grazing selection for AEZ (crop):

$$QCROP_{r,fl} = QAGRT_{r,fl} \cdot \frac{\delta^{ltt}_{r,fl,"CROP"} \cdot PCROP_{r,fl}{}^{\rho^{ltt}_{r,fl}}}{\delta^{ltt}_{r,fl,"CROP"} \cdot PCROP_{r,fl}{}^{\rho^{ltt}_{r,fl}} + \delta^{ltt}_{r,fl,"GRZ"} \cdot PGRZ_{r,fl}{}^{\rho^{ltt}_{r,fl}}}, \quad \forall r \in R, fl \in FL$$

Crop and grazing selection for AEZ (grazing):

$$QGRZ_{r,fl} = QAGRT_{r,fl} \cdot \frac{\delta^{ltt}_{r,fl,"GRZ"} \cdot PGRZ_{r,fl}{}^{\rho^{ltt}_{r,fl}}}{\delta^{ltt}_{r,fl,"CROP"} \cdot PCROP_{r,fl}{}^{\rho^{ltt}_{r,fl}} + \delta^{ltt}_{r,fl,"GRZ"} \cdot PGRZ_{r,fl}{}^{\rho^{ltt}_{r,fl}}}, \quad \forall r \in R, fl \in FL$$

Aggregation of agriculture (in case missing either crop or graze):

$$QAGRT_{r,fl} = QCROP_{r,fl} + QGRZ_{r,fl}, \qquad \forall r \in R, fl \in FL$$

Aggregated agricultural price:

$$QAGR_{r,fl} \cdot PAGRT_{r,fl} = QCROP_{r,fl} \cdot PCROP_{r,fl} + QGRZ_{r,fl} \cdot PGRZ_{r,fl}, \qquad \forall r \in R, fl \in FL$$

Agriculture and forestry selection for AEZ (agriculture):

$$QAGRT_{r,fl} = QLANT_{r,fl} \cdot \frac{\delta^{ltf}_{r,fl,"AGR"} \cdot PAGRT_{r,fl}{}^{\rho^{ltf}_{r,fl}}}{\delta^{ltf}_{r,fl,"AGR"} \cdot PAGRT_{r,fl}{}^{\rho^{ltf}_{r,fl}} + \delta^{ltf}_{r,fl,"FRS"} \cdot PLAND_{r,fl,"FRS"}{}^{\rho^{ltf}_{r,fl}}}, \quad \forall r \in R, fl \in FL$$

Agriculture and forestry selection for AEZ (forestry):

$$QLAND_{r,fl,a} = QLANDT_{r,fl} \cdot \frac{\delta^{ltf}_{r,fl,''AGR''} \cdot PLAND_{r,fl,''FRS''}{}^{\rho^{ltf}_{r,fl}}}{\delta^{ltf}_{r,fl,''AGR''} \cdot PAGRT_{r,fl}{}^{\rho^{ltf}_{r,fl}} + \delta^{ltf}_{r,fl,''FRS''} \cdot PLAND_{r,fl,''FRS''}{}^{\rho^{ltf}_{r,fl}}},$$
$$\forall r \in R, fl \in FL, A \in AFRS$$

Aggregation of agriculture (in case missing agriculture):

$$QLANDT_{r,\,fl} = QLAND_{r,\,fl,''FRS''}, \qquad \forall r \in R, fl \in FL$$

Aggregation of agriculture (in case missing forestry):

$$QLANDT_{r,\,fl} = QAGRT_{r,\,fl}, \qquad \forall r \in R, fl \in FL$$

Aggregated total land price:

$$QLANDT_{r,fl} \cdot PLANDT_{r,fl} = QAGRT_{r,fl} \cdot PAGRT_{r,fl} + QLAND_{r,fl,''FRS''} \cdot PLAND_{r,fl,''FRS''},$$
$$\forall r \in R, fl \in FL$$

Total land:

$$QLANDT_{r,fl} = qlandtotara_{r,fl} \cdot \frac{\delta^{ara}_{r,fl,''USE''} \cdot PLANDT_{r,fl}{}^{\rho^{ara}_{r,fl}}}{\delta^{ara}_{r,fl,''USE''} \cdot PLANDT_{r,fl}{}^{\rho^{ara}_{r,fl}} + \delta^{ara}_{r,fl,''NUS''} \cdot plandt\_ini_{r,fl}{}^{\rho^{ara}_{r,fl}}},$$
$$\forall r \in R, fl \in FL$$

Positive land-use change emissions:

$$LUCHEM\_P_{r,fl} = \max\left[0, \left(qfrspre_{r,fl} - QPRMLANDT_{r,fl,''frs''}\right) \cdot ecoefluc_{r,fl,''frs''}\right],$$
$$\forall r \in R, fl \in FL$$

(EQ_LUCHEM_P)

Negative land use-change emissions:

$$prluch_{y,r,fl} \cdot cf\_v2t \cdot cdst\_wd \cdot 11/3 \cdot$$
$$LUCHEM\_N_{r,fl} = -\sum_{y \in Yold}\left[\exp\left[\alpha^{bms}_{fl} - \frac{30}{\{(Pyear - baseyear + 1) - Year_y\}} - \frac{30}{\{(Pyear - baseyear) - Year_y\}}\right]\right],$$
$$\forall r \in R, fl \in FL$$

(EQ_LUCHEM_N)

Factor income:

$$YF_{r,f} = \sum_{a \in A} WF_{r,f} \cdot WFDIST_{r,f,a} \cdot QF_{r,f,a} + transfr\_f_{r,f,"ROW"} \cdot EXR_r, \quad \forall r \in R, f \in F$$

Institutional factor incomes:

$$YIF_{r,i,f} = shif_{r,i,f} \cdot \left(\left(1 - tf_{r,f}\right) \cdot YF_{r,f} - transfr_{r,"ROW",f} \cdot EXR_r \right), \quad \forall r \in R, i \in INSD, f \in F$$

Income of nongovernmental domestic institution:

$$\begin{aligned} YI_{r,i} = & \sum_{f \in F} YIF_{r,i,f} + TRII\_Resource_{r,i} + shincome_{r,i} \cdot GHGTCOST_r \\ & + VRENCAPTOT_{r,i} - (PGHG\_G + PGHG\_IMP\_QUO_r - PGHG\_EXP\_QUO_r) \cdot \\ & GHG\_IMP_r \times EXR_r \cdot shincome_{r,i} + shres_{r,i} \cdot \sum_{a \in A} PRES_{r,a} \cdot QRES_{r,a} \\ & + shincome_{r,i} \cdot \sum_{a \in A} QENE_{r,a} \cdot PENE_{r,a} \cdot \left(\frac{1}{1 - ADEEI_{r,a}} - 1 \right) \\ & + shincome_{r,i} \cdot \sum_{a \in A} RQUOQA_{r,a} \cdot QA_{r,a} \cdot PA_{r,a} \qquad , \forall r \in R, i \in INSDNG \end{aligned}$$

Total rent of electricity capacity:

$$VRENCAPTOT_{r,"ent"} = \sum_{a \in A} VRENCAP_{r,a} \quad \forall r \in R$$

Household consumption expenditures:

$$\begin{aligned} EH_{r,h} = & (1 - shii\_use_{r,h})(1 - MPS_{r,h}) \cdot (1 - TINS_{r,h}) \cdot YI_{r,h} \\ & - PGHG_r \cdot \sum_{g \in G} gwp_{r,g} \cdot \left(EMALH_{r,h,g} + EMBIH_{r,h,g}\right) \\ & \forall r \in R, h \in H \end{aligned}$$

Passenger car by household:

$$QCARU_{r,h} = pcaru_{r,h} \cdot \left(\frac{EH_{r,h}/CPI_r}{EH\_base_{r,h}/CPI\_base_r} \right)^{pasch_{r,h}}, \quad \forall r \in R, h \in H$$

Energy consumption caused by passenger car in household[7]**:**

$$QCARUENET_{r,h} = QCARU_{r,h} \cdot careneeff_{r,h}, \qquad \forall r \in R, h \in H$$

Energy price for passenger car in household:

$$PENE\_H\_CAR_{r,h} \cdot QCARENET_{r,h} = \sum_{c \in C} QCARUENE_{r,h,c} \cdot \left(PQD_{r,c} \cdot dfpq_{r,c,h} \cdot (1 + tqd_{r,c,h}) + \sum_{g \in G} PGHG_r \cdot gwp_g \cdot effc_{r,c,h,g} \right), \quad \forall r \in R, h \in H$$

Energy fuel consumption caused by passenger car in household:

$$QCARUENE_{r,h,c} = QCARUENET_{r,h} \frac{\beta^{carh}_{r,c,h} \cdot \left\{ PQD_{r,c} \cdot (1 + tqd_{r,c,h}) + PGHG_r \cdot \sum_{g \in G} gwp_g \cdot enur_{r,c,h} \cdot effc_{r,c,h,g} \right\}^{el^{carh}_{r,c,h}}}{\sum_{cp} \beta^{carh}_{r,cp,h} \cdot \left\{ PQD_{r,cp} \cdot (1 + tqd_{r,cp,h}) + +PGHG_r \cdot \sum_{g \in G} gwp_g \cdot enur_{r,cp,h} \cdot effc_{r,cp,h,g} \right\}^{el^{carh}_{r,cp,h}}}, \quad \forall r \in R, h \in H, c \in ENE$$

Household consumption spending (LES):

$$PQH_{r,ch} \cdot QCH_{r,ch,h} = PQH_{r,ch} \cdot \gamma^m_{r,ch,h} + \beta^m_{r,ch,h} \cdot \left(EH_{r,h} - \sum_{c' \in C} PQH_{r,ch'} \cdot \gamma^m_{r,c'h,h} \right), \quad \forall r \in R, c \in C, h \in H$$

Household goods price:

$$QCH_{r,ch,h} \cdot PQH_{r,ch,h} = \sum_{c \in wch_{r,ch,c,h}} (QH_{r,c,h} - QCARENE_{r,c,h}) \cdot \left(PQD_{r,c} \cdot dfpq_{r,c,h} \cdot (1 + tqd_{r,c,h}) + \sum_{g \in G} PGHG_r \cdot gwp_g \cdot effc_{r,c,h,g} \right), \quad \forall r \in R, ch \in CH, h \in H$$

[7]This equation is currently not used.

Household goods consumption mapping:

$$QH_{r,c,h} = \sum_{ch \in wch_{r,ch,c,h}} QCH_{r,ch,h}, \quad \forall r \in R, c \in C, h \in H$$

Household energy source consumption:

$$\begin{aligned} QH_{r,h,c} = \; & QCARENE_{r,h,c} \\ & + QCH_{r,\text{"com\_ene"},h} \frac{\beta^{enec}_{r,c,h} \cdot \left\{ PQD_{r,c} \cdot \left(1 + tqd_{r,c,h}\right) + PGHG_r \cdot \sum_{g \in G} gwp_g \cdot enur_{r,c,h} \cdot efffc_{r,c,h,g} \right\}^{el^{enec}_{r,c,h}}}{\sum_{cp} \beta^{enec}_{r,cp,h} \cdot \left\{ PQD_{r,cp} \cdot \left(1 + tqd_{r,cp,h}\right) + +PGHG_r \cdot \sum_{g \in G} gwp_g \cdot enur_{r,cp,h} \cdot efffc_{r,cp,h,g} \right\}^{el^{enec}_{r,cp,h}}}, \\ & \forall r \in R, h \in H, c \in ENE \end{aligned}$$

Household utility (AIDADS):

$$Vu_{r,ch} = \sum_{ch} V\mu_{r,ch,h} \cdot \ln \left(\frac{QCH_{r,ch,c}}{poph_{r,h}} - \theta h_{r,ch,h} \right) - (\ln A + 1), \quad \forall r \in R, h \in H$$

Household marginal share (AIDADS):

$$V\mu_{r,ch,h} = \frac{\left[\alpha h_{r,ch,h} + \beta h_{r,ch,h} \cdot e^{uh_{r,h}}\right]}{\left[1 + e^{uh_{r,h}}\right]}, \quad \forall r \in R, ch \in CH, h \in H$$

Household consumption spending (AIDADS):

$$\begin{aligned} QCH_{r,ch,h} \cdot PQH_{r,ch,h} = \; & poph_{r,h} \cdot \\ & \left[\begin{aligned} & \theta h_{r,ch,h} \cdot PQH_{r,ch,h} + V\mu_{r,ch,h} \cdot \\ & \left(\begin{aligned} & EH_h - \sum_{c' \in C\_ENE} \left(PQD_{r,c'} \cdot dfpq_{r,c',h} \cdot \left(1 + tqd_{r,c',h}\right) + \sum_{g \in G} PGHG_r \cdot gwp_g \cdot efffc_{r,c',h,g} \right) \cdot QCARENE_{r,c',h} \\ & - \sum_{chp \in CH} PQH_{r,chp,h} \cdot \theta h_{r,chp,h} \end{aligned} \right) \end{aligned} \right], \\ & \forall r \in R, ch \in CH, h \in H \end{aligned}$$

Investment demand:

$$QINV_{r,c} = IADJ_r \cdot \overline{qinv_{r,c}}, \quad \forall r \in R, c \in C$$

Government consumption demand:

$$QG_{r,c} = \overline{GADJ_r} \cdot \overline{qg_{r,c}}, \quad \forall r \in R, c \in C$$

Government revenue:

$$\begin{aligned} YG_r = & \sum_{i \in INSDNG} TINS_{r,i} \cdot YI_{r,i} + \sum_{f \in F} tf_{r,f} \cdot YF_{r,f} \\ & + \sum_{a \in A} ta_{r,a} \cdot PA_{r,a} \cdot QA_{r,a} + \sum_{a \in A} tva_{r,a} \cdot PVA_{r,a} \cdot QVA_{r,a} \\ & + \sum_{c \in CM} tm_{r,c} \cdot PWM_c \cdot dis\_imp_{r,c} \cdot QM_{r,c} \cdot \overline{EXR_r} \\ & + \sum_{c \in CE} te_{r,c} \cdot PWE_c \cdot dis\_\exp_{r,c} \cdot QE_{r,c} \cdot \overline{EXR_r} \\ & + \sum_{c \in C} \sum_{a \in A} tqd_{r,c,a} \cdot dfpq_{r,c,a} \cdot PQD_{r,c} \cdot QINT_{r,c,a} \\ & + \sum_{c \in C} \sum_{h \in H} tqd_{r,c,h} \cdot dfpq_{r,c,h} \cdot PQD_{r,c} \cdot QH_{r,c,h} \\ & + \sum_{c \in C} tqd_{r,c,''gov''} \cdot dfpq_{r,c,''gov''} \cdot PQD_{r,c} \cdot QG_{r,c} \\ & + \sum_{c \in C} tqd_{r,c,''S-I''} \cdot dfpq_{r,c,''S-I''} \cdot PQD_{r,c} \cdot QINV_{r,c} \\ & + \sum_{f \in F} YIF_{r,''gov'',f} + TRII\_Resource_{r,''gov''} \\ & - TRII\_Use_{r,''gov''} + GHGTCOST_r \cdot shincome_{r,''gov''} \\ & + GHG\_IMP_r \cdot EXR_r \cdot (PGHG\_IMP\_QUO_r - PGHG\_EXP\_QUO_r) \\ & + shres_{r,''gov''} \cdot \sum_{a \in A} PRES_{r,a} \cdot QRES_{r,a} \qquad \forall r \in R \end{aligned}$$

Government expenditure:

$$EG_r = \sum_{c \in C} PQD_{r,c} \cdot dfpq_{r,c,''gov''} \cdot \left(1 + tqd_{r,c,''gov''}\right) \cdot QG_{r,c} \qquad \forall r \in R$$

Transfer use

$$TRII\_Use_{r,i} = shii\_use_{r,i} \cdot (1 - MPS_{r,i}) \cdot (1 - TINS_{r,i}) \cdot YI_{r,i}, \quad \forall r \in R, i \in INSDNG$$

Government transfer use:

$$TRII\_Use_{r,''gov''} = \overline{trnsfr\_CRT_{r,''gov''}} \cdot CPI_r \qquad \forall r \in R$$

Transfer resource:

$$TRII\_Resource_{r,i} = : shii\_resource_{r,i} \cdot \left(crt\_in_r \cdot \overline{EXR_r} - crt\_out_r \cdot \overline{EXR_r} + \sum_{i'} TRII\_Use_{r,i'} \right), \quad \forall r \in R, i \in INSD$$

Imported commodity:

$$QWM_{r,c} = QM_{r,c}, \quad \forall r \in R, c \in CM$$

Exported commodity:

$$QWE_{r,c} = QE_{r,c}, \quad \forall r \in R, c \in CE$$

World trade nominal balance:

$$\sum_{r \in R} (1 - tw_{r,c}) \cdot PWM_c \cdot QWM_{r,c} \cdot dis\_imp_{r,c} = \sum_{r \in R} PWE_c \cdot QWE_{r,c} \cdot dis\_\exp_{r,c} - \sum_{c \in C\_TRS} PTRS_c \cdot QTRS_c, \quad \forall c \in (CM \cap CE)$$

World trade volume balance:

$$\sum_{r \in R} QWM_{r,c} = \left(1 - \lambda_c^w\right) \cdot \sum_{r \in R} \left(QWE_{r,c} - QTRS_c\right), \quad \forall c \in (CM \cap CE)$$

Transport service demand:

$$QTRS_c = tsh_c \cdot \sum_r QWE_{r,c}, \qquad \forall c \in C\_TRS$$

CIF and FOB relationship:

$$PTRS_c \cdot QTRS_c = \sum_{r'} tw_{r',c} \cdot PWM_c \cdot QWM_{r',c} \cdot dis\_imp_{r',c}, \quad \forall c \in C\_TRS$$

Biomass consumption (household):

$$TBH_{r,h} = poph_{r,h} \cdot bioc_{r,h} \cdot biod_{r,h} \qquad \forall r \in R, h \in H$$

Biomass consumption (industry):

$$TBI_{r,a} = QA_{r,a} \cdot bioc_{r,a} \cdot biod_{r,a} \qquad \forall r \in R, a \in A$$

Emissions related to activity level (industrial activity):

$$EMALI_{r,a,g} = QA_{r,a} \cdot efacl_{r,a,g} \cdot \left(1 - NERED_{r,a,g}\right) \cdot \left(1 + \chi_{r,a,g}\right) - \sum_{emcm \in EMCM} \left(QRED_{r,emcm,a} / \eta_{emcm,a}\right), \quad \forall r \in R, a \in A, g \in G$$

Additional emission reductions related to activity level (industrial activity):

$$NERED_{r,a,g} = 1 - (PGHG_r + 1)^{-\sigma^{ghg}_{r,a,g}}, \qquad \forall r \in R, a \in A, g \in G$$

Emissions related to activity level (household):

$$EMALH_{r,h,g} = poph_{r,h} \cdot efacl_{r,h,g} \qquad \forall r \in R, h \in H, g \in G$$

Emissions related to fossil fuel combustion (industrial activity):

$$EMFFI_{r,c,a,g} = QINT_{r,c,a} \cdot enur_{r,c,a} \cdot effc_{r,c,a,g} \qquad \forall r \in R, c \in ENE, a \in A, g \in G$$

Emissions related to fossil fuel combustion (household):

$$EMFFH_{r,c,h,g} = QH_{r,c,h} \cdot effc_{r,c,h,g} \qquad \forall r \in R, c \in ENE, h \in H, g \in G$$

Emissions related to biomass combustion (industrial activity):

$$EMBII_{r,a,g} = TBI_{r,a} \cdot efbio_{r,a,g} \qquad \forall r \in R, a \in A, g \in G$$

Emissions related to biomass combustion (household activity):

$$EMBIH_{r,h,g} = TBH_{r,h} \cdot efbio_{r,h,g} \qquad \forall r \in R, h \in H, g \in G$$

GHG emission total in a region:

$$GHGT_r = \sum_{g \in G} gwp_g \cdot \left\{ \begin{array}{l} \left(\sum_{h \in H} \sum_{c \in C} QH_{r,c,h} \cdot enur_{r,c,h} \cdot effc_{r,c,h,g} + \sum_{a \in A} \sum_{c \in C} QINT_{r,c,a} \cdot enur_{r,c,a} \cdot effc_{r,c,a,g} \right) \\ + \left(\sum_{h \in H} EMALH_{r,h,g} + \sum_{a \in A} EMALI_{r,a,g} \right) \\ + \left(\sum_{h \in H} EMBIH_{r,h,g} + \sum_{a \in A} EMBII_{r,a,g} \right) \end{array} \right\}$$

$\forall r \in R$

GHG emission includes emission trading:

$$GHGT\_CT_r = GHGT_r - GHGT\_IMP_r \cdot \qquad \forall r \in R$$

GHG emission importing trading upper limit:

$$\overline{ghgt\_imp\_cap_r} - GHGT\_IMP_r \geq 0 \quad \perp PGHG\_IMP\_QUO_r > 0 \qquad \forall r \in R$$

GHG emission exporting trading upper limit:

$$GHGT\_IMP_r - \overline{ghgt\_\exp\_cap_r} \geq 0 \quad \perp PGHG\_EXP\_QUO_r \geq 0 \qquad \forall r \in R$$

GHG emission price and international price:

$$PGHG_r = EXR_r \cdot (PGHG\_G + PGHG\_IMP\_QUO_r - PGHG\_EXP\_QUO_r) \qquad \forall r \in R$$

GHG emission constraint:

$$\overline{ghgc_r} - GHGT\_CT_r \geq 0 \quad \perp PGHG_r \geq 0 \qquad \forall r \in R$$

GHG emission cost of nonenergy (industry):

$$GHGCA\_NENE_{r,a} = PGHG_r \cdot \sum_{g \in G} gwp_g \cdot (EMBII_{r,a,g} + EMALI_{r,a,g}) \qquad \forall r \in R, a \in A$$

GHG total cost:

$$GHGTCOST_r = \sum_{a \in A} \left(QINT_{r,c,a} \cdot enur_{r,c,a} \cdot efffc_{r,c,a,g} \right) + GHGCA\_NENE_{r,c,a} + \sum_{h \in H} PGHG_r \cdot \sum_{g \in G} gwp_g \cdot \left(\sum_{c \in C} QH_{r,c,h} \cdot efffc_{r,c,h,g} + EMALH_{r,h,g} + EMBIH_{r,h,g} \right) \quad \forall r \in R$$

Global GHG emission constraint:

$$\overline{ghgtot\_c} - \sum_{r \in R} GHGT\_CT_r \geq 0 \quad \perp PGHG\_G \geq 0$$

Global GHG emission constraint price:

$$PGHG\_G = PGHG_r, \qquad \forall r \in R$$

Global GHG emission trading total:

$$\sum_{r \in R} GHG\_IMP_r = 0$$

GHG emission related to the international transport:

$$EMFFINT_{tr,g} = \sum_{c \in C} INTTRSENE_{tr,c} \cdot efint\_trs_{tr,c,g} \qquad \text{(GAS\_FROM\_FF\_INT)}$$

International transport energy demand:

$$INTTRSENE_{tr,c} = inttrenecoef_{tr,c} \cdot \sum_{r} QDTRST_{r,tr} \cdot PGHG_r^{\sigma ghg\_int} \qquad \text{(INTTRSENEDEF)}$$

Nonenergy related gas emissions:

$$EMNEG_{r,g} = ecfneng_{r,g} \cdot GDP_r$$

Factor markets:

$$\sum_{a \in A} QF_{r,f,a} = \overline{QFS_{r,f}}, \qquad \forall r \in R, f \in F$$

Composite commodity markets:

$$QQ_{r,c} - \left(QX2_{r,c} + QM_{r,c}\right) \cdot loss_{r,c} + stch2_{r,c}$$
$$= \sum_{a \in A} QINT_{r,c,a} + \sum_{h \in H} QH_{r,c,h} + QG_{r,c} + QINV_{r,c}, \quad \forall r \in R, c \in C$$

Current-account balance for the rest of the world, in foreign currency:

$$\sum_{c \in CM} PWM_c \cdot dis\_imp_{r,c} \cdot QM_{r,c} + transfr\_crt\_out_r$$
$$+ GHG\_IMP_r \cdot PGHG_r + \sum_{f \in F} transfr_{r,"ROW",f} = \sum_{c \in CE} PWE_c \cdot dis\_\exp_{r,c} \cdot QE_{r,c}$$
$$+ transfr\_crt\_in_r + \overline{FSAV_r} + \sum_{f \in F} transfr\_f_{r,f,"ROW"}$$
$$\forall r \in R, f \in F$$

Government balance:

$$YG_r = EG_r + GSAV_r \qquad \forall r \in R$$

Direct tax rate:

$$TINS_{r,i} = \overline{tins_{r,i}} \cdot \left(1 + \overline{TINSADJ_r} \cdot tins01_{r,i}\right) + \overline{DTINS_{r,i}} \cdot tins01_{r,i}, \quad \forall r \in R, i \in INSDNG$$

Institutional savings rates:

$$MPS_{r,i} = \overline{mps_{r,i}} \cdot \left(1 + \overline{MPSADJ_r} \cdot mps01_{r,i}\right) + DMPS_r \cdot mps01_{r,i}, \quad \forall r \in R, i \in INSDNG$$

Savings-investment balance:

$$\sum_{i \in INSDNG} MPS_{r,i} \cdot (1 - TINS_{r,i}) \cdot YI_{r,i} + GSAV_r + \overline{FSAV_r} \cdot \overline{EXR_r}$$
$$= \sum_{c \in C} PQ_{r,c} \cdot dfpq_{r,c,"S-I"} \cdot \left(1 + tq_{r,c,"S-I"}\right) \cdot QINV_{r,i} + WALRAS_r \qquad \forall r \in R$$

Global investment balance:

$$\sum_{r \in R} FSAV_r = 0$$

Activity constraint (upper boundary):

$$renew\_up_{r,a} - QA_{r,a} \cdot \theta_{r,a,\text{"}COM\_ELY\text{"}} \geq 0 \quad VRENCAP_{r,a} \geq 0, \\ \forall r \in R, a \in A = \{renew\_up_{r,a} > 0\}$$

Activity constraint (QUOTA for aggregated region and activity):

$$quotaqa_{ragg,aagg} = \sum_{r \in Map\_Ragg(r,ragg)} \sum_{a \in Map\_aagg(a,aagg)} \sum_{c \in C} QA_{r,a} \cdot \theta_{r,a,c} \geq 0 \\ \perp RQUOQA\_agg_{ragg,aagg} \geq 0, \forall ragg \in Ragg, aagg \in Aagg$$

Activity constraint (QUOTA shadow price):

$$RQUOQA_{r,a} = \sum_{ragg \in Map\_Ragg(r,ragg)} \sum_{aagg \in Map\_aagg(a,aagg)} RQUOQA\_agg_{ragg,aagg} \geq 0, \\ \forall r \in R, a \in A$$

End-use device stock:

$$END\_STK_{r,ac,l} = \left(\left(1 - {}^{1}\!/_{\tau_l}\right) \cdot stka\_pre_{r,ac,l} \cdot \left(1 - \sum_{i \in LI(l,i)} END\_QEOR_{r,ac,i}\right) \right. \\ \left. + END\_QR_{r,ac,l} \right)$$

End-use device stock operation:

$$\left(1 + \lambda^d_{r,ac,l}\right) \cdot END\_STK_{r,ac,l} = END\_QXD_{r,ac,l}$$

End-use device service supply:

$$\left(1 + \varphi_{r,ac,i}\right) \cdot \sum_{l \in LI(l,i)} \left(ad_{r,ac,l,i} \cdot END\_QXD_{r,ac,l}\right) = QSD_{r,ac,i}$$

Industry service demand:

$$QSD_{r,a,i} = sd\_base_{r,a,i} \cdot \left(\frac{QA_{r,a}}{QA\_base_{r,a}}\right)^{indserincel_{r,a}} \quad \forall a \notin A\_TRS$$

Residential service demand:

$$QSD_{r,h,i} = sd\_base_{r,h,i} \cdot \left(\frac{EH_{r,h}}{EH\_base_{r,h}} \cdot \frac{CPI\_base_r}{CPI_r}\right)^{resserincel_{r,h}}$$

Transport service demand:

$$QSD_{r,a,i} = sd\_base_{r,a,i} \cdot \frac{\sum_{tr \in MTR(tr,i)} QDTRST_{r,tr}}{\sum_{tr \in MTR(tr,i)} QDTRST\_base_{r,tr}} \quad \forall a \in A\_TRS$$

End-use device new investment selection:

$$\sum_{i \in LI(l,i)} \left(\left(1 + \varphi_{r,ac,i}\right) \cdot ad_{r,ac,l,i} \cdot \left(1 + \lambda^{d}_{r,ac,l}\right) \cdot END\_QR_{r,ac,l}\right) = \sum_{i \in LI(l,i)} \left(\frac{\mu_{r,ac,l} \cdot (END\_PCOST_{r,ac,l})^{\eta^{end}_{r,ac,l}}}{\sum_{ll \in LI(ll,i)} \left(\mu_{r,ac,ll} \cdot (END\_PCOST_{r,ac,ll})^{\eta^{end}_{r,ac,ll}}\right)} \cdot END\_QRT_{r,ac,i} \right)$$

End-use device price:

$$END\_PCOST_{r,ac,l} = invc_{r,ac,l} + \left(oped_{r,ac,l} + \sum_{K\_END} \left(ied_{r,ac,l,K\_END} \cdot \left(END\_PED_{r,K\_END,ac} + \left(1 - \sum_{i \in LI(l,i)} BFRATIO_{r,ac,i}\right) \cdot PGHG_r \cdot \sum_{g \in G} \left(gwp_{r,g} \cdot emcoef\_end_{r,ac,K\_END,g}\right)\right)\right)\right) \cdot \left(1 + \lambda^{d}_{r,ac,l}\right)$$

End-use total new investment for service:

$$END\_QRT_{r,ac,i} \geq 0 \quad \perp \quad END\_QEOR_{r,ac,i} \geq 0$$

End-use energy price:

$$END\_PED_{r,K\_END,ac} = end\_ped\_base_{r,K\_END,ac} \cdot \sum_{c \in ENEENDMAP(c,K\_END)} \left(\frac{PQD_{r,c}}{PQD\_base_{r,c}} \right)$$

Energy use of service i in end-use energy:

$$END\_ENE_{r,ac,i,K\_END} = \sum_{l \in LI(l,i)} ied_{r,ac,l,K\_END} \cdot END\_QXD_{r,ac,l}$$

Energy use of service i in CGE energy:

$$QEND\_ENE_{r,c,ac,i} = \sum_{\substack{KAGG\_END \in MK\_AGGC \\ (KAGG\_END,c)}} \left(\sum_{\substack{K\_END \in MK\_AGGEND \\ (KAGG\_END,K\_END)}} (END\_ENE_{r,ac,i,K\_END}) \cdot \frac{\mu^{ea}_{r,c,ac,i} \cdot \left(dfpq_{r,c,ac} \cdot PQD_{r,c} \cdot (1 + tqd_{r,c,ac}) + PGHG_r \cdot \sum_{g \in G} \left(gwp_{r,g} \cdot enur_{r,c,ac} \cdot efffc_{r,c,ac,g} \right) \right)^{\eta^{ea}_{r,c,ac}}}{\sum_{\substack{cp \in MKAGGC \\ (KAGG\_END,CP)}} \left(\mu^{ea}_{r,cp,ac,i} \cdot \left(dfpq_{r,cp,ac} \cdot PQD_{r,cp} \cdot (1 + tqd_{r,cp,ac}) + PGHG_r \cdot \sum_{g \in G} \left(gwp_{r,g} \cdot enur_{r,cp,ac} \cdot efffc_{r,cp,ac,g} \right) \right)^{\eta^{ea}_{r,cp,ac}} \right)} \right)$$

Industry end-use energy mapping:

$$QINT_{r,c,a} = \sum_{i} QEND\_ENE_{r,c,a,i}$$

Industry end-use energy total:

$$QENE_{r,a} = \sum_{c \in C\_ENE(c)} QINT_{r,c,a}$$

Transport end-use energy mapping:

$$TRS\_ENE\_FL_{r,tr,c} = \sum_{i \in MTR\_I(tr,i)} QEND\_ENE_{r,c,\text{"TRS"},i} \quad \forall tr \in TR$$

Transport end-use energy total:

$$TRS\_ENE_{r,tr} = \sum_{c} TRS\_ENE\_FL_{r,tr,c}$$

Residential end-use energy mapping:

$$QH_{r,c,h} - QCARENE_{r,h,c} = \sum_{i} QEND\_ENE_{r,c,h,i}$$

Residential car end-use energy mapping:

$$QCARENE_{r,h,c} = \sum_{i \in MTR\_I("PC",i)} QEND\_ENE_{r,c,"TRS",i}$$

Residential car end-use energy total:

$$QCARENET_{r,h} = \sum_{c} QCARENE_{r,h,c}$$

End-use biofuel ratio:

$$BFRATIO_{r,ac,i} \cdot \sum_{c \in MKaggc("OIL",c)} QEND\_ENE_{r,c,ac,i} = QEND\_ENE_{r,"COM\_BIO",ac,i}$$

References

Bond TC, Bhardwaj E, Dong R, Jogani R, Jung S, Roden C, Streets DG, Fernandes S, Trautmann N (2007) Historical emissions of black and organic carbon aerosol from energy-related combustion, 1850–2000. Glob Biogeochem Cycles 21:GB2018. doi:10.1029/2006GB002840

EEA (Europe Environmental Agency) (2009) EMEP/EEA air pollutant emission inventory guidebook 2009, No. 9

Hyman RC, Reilly JM, Babiker MH, De Masin A, Jacoby HD (2003) Modeling non-CO_2 greenhouse gas abatement. Environ Model Assess 8:175–186

IPCC (2006) IPCC guidelines for national greenhouse gas inventories, prepared by the National Greenhouse Gas Inventories Programme. In: Eggleston HS, Buendia L, Miwa K, Ngara T, Tanabe K (eds). IGES, Japan

Lofgren H, Harris RL, Robinson S (2002) A standard computable general equilibrium (CGE) model in GAMS, IFPRI

Masui T, Matsuoka Y, Kainuma M (2006) Long-term CO_2 emission reduction scenarios in Japan. Environ Econ Policy Stud 7(3):347–366

Chapter 13
AIM/CGE V2.0: Basic Feature of the Model

Shinichiro Fujimori, Tomoko Hasegawa, and Toshihiko Masui

Abstract This chapter documents a main model structure and how to implement scenario assumptions for the analysis of long-term climate mitigation taken by AIM/CGE (Asia-Pacific Integrated Model/Computable General Equilibrium). There are six aspects which are going to be discussed. First, macroeconomy, labor, and population treatment are explained. Second, energy supply sector representation is described. Energy supply sectors are one of the key elements for decarbonizing economic systems. Third, energy demand sectors are discussed. Fourth, agriculture and land use are critically important for stringent climate mitigation policy since large bioenergy implementation combined with carbon capture and storage and afforestation would be thought as measures which enables so-called negative emissions. Fifth, nonenergy-related GHG reduction measures follow. They are mostly related to agricultural sectors. Sixth, we discuss how to add new sectors into the CGE system.

Keywords CGE • Documentation • Long-Term Scenario

13.1 Introduction

This chapter documents a main model structure and how to implement scenario assumptions for the analysis of long-term climate mitigation taken by AIM/CGE V2.0 (we call AIM/CGE from here). AIM/CGE is a multiregion and multi-sector recursive dynamic-type CGE model. There are several features to understand the behaviors of this model in conjunction with future scenarios. There are six topics as follows, and they will be explained by each section later:

S. Fujimori (✉) • T. Hasegawa
Center for Social and Environmental Systems Research, National Institute for Environmental Studies (NIES), 16–2 Onogawa, Tsukuba, Ibaraki 305–8506, Japan

International Institute for Applied Systems Analysis, Schlossplatz-1, Laxenburg 2361, Austria
e-mail: fujimori.shinichiro@nies.go.jp

T. Masui
Center for Social and Environmental Systems Research, National Institute for Environmental Studies (NIES), 16–2 Onogawa, Tsukuba, Ibaraki 305–8506, Japan

S. Fujimori et al. (eds.), *Post-2020 Climate Action*,
DOI 10.1007/978-981-10-3869-3_13

1. Macroeconomic, labor, and population
2. Energy supply
3. Energy demand
4. Agriculture and land use
5. GHG reduction measures other than changing energy system
6. How to implement new production sectors or goods (not accounted in the base year)

In this document, we put much more priority on showing how the scenario outcomes of this model are generated than just explaining either model structure or scenario assumptions. We keep a basic rule that each section provides following structure as much as possible:

- The importance of the factors
- The fundamental logics of the basic model structure
- How the exogenous assumptions are fed into the model for future simulations
- Actual parameter settings
- Limitation and remained issues that should be improved further version

13.2 Demographic and Macroeconomic Change

13.2.1 Population and Labor Participation

Future demographic change is one of the key drivers to change the future goods demand including energy and food. On the other hand, the production side is also affected by the demographic change through labor participation.

The population and labor forces are exogenous parameters for AIM/CGE. The macroeconomic growth is also the exogenous assumption.

The first-order effect of population change is the household consumption. The household consumption is formulated as LES (linear expenditure system) function (Eq. 13.1), and it consists of two factors for each commodity consumption, namely, subsistence consumption and marginal share of consumption. The population change directly effects on the former factor:

$$PQH_{r,ch} \cdot QCH_{r,ch,h} = PQH_{r,ch} \cdot \gamma^m_{r,ch,h} + \beta^m_{r,ch,h} \cdot \left(EH_{r,h} - \sum_{ch' \in CH} PQH_{r,ch'} \cdot \gamma^m_{r,ch',h} \right), \forall r \in R,\ ch \in CH,\ h \in H \quad (13.1)$$

$\gamma^m_{r,ch,h}$: subsistence consumption of marketed commodity c for household h

$\beta^m_{r,ch,h}$: marginal share of consumption spending on marketed commodity c for household h

$PQH_{r,ch,h}$: price of household category commodity *ch*
$QCH_{r,ch,h}$: household consumption of household commodity category *ch*
$EH_{r,h}$: household consumption expenditures
$r \in R$: a set of regions
$h \in H$: a set of households
$ch \in CH$: household commodity category *ch*

The population increase ratio relative to previous year is multiplied to the subsistence consumption parameter $\beta^{m}_{r,ch,h}$. The latter factor, marginal share of consumption, specifies how the household money excluding subsistence consumption is spent, and the expenditure share of each commodity is determined. Thus, it is not directly influenced by the population change. The intermediate inputs of agricultural commodity are also associated with population change, but it will be explained in Sect. 13.5.2.

There might be a discussion about the assumption in fixing population. The population and labor force could be affected by the future climate change impact through change of health condition including change of agricultural goods availability or migration induced by the environmental change. We ignore basically in the mitigation analysis. The exception can be seen in the study highlighting climate change impact (Hasegawa et al. 2016).

13.2.2 Macroeconomic Change

Future macroeconomic assumption also causes goods demand and supply. The macroeconomic growth is also the exogenous assumption. Usually the GDP change is used for the macroeconomic assumption for the integrated assessment models' (IAMs) future scenario simulation. However, the actual outcome from the model is not exactly same as the assumptions. Therefore, the GDP assumption is used for calculating total factor productivity (TFP), and TFP is totally exogenous parameter for the model.

The economic growth is realized by three factors: labor force change, accumulated capital, and total factor of productivity. The labor force is the one of the three main drivers of the GDP. The labor force change is exogenous variable for this model. The production sectors are basically[1] assumed to have multilevel nested CES (constant elasticity of substitution) function. Therefore, if the labor wage is constant, labor force change ratio directly affects to the GDP change.

The GDP assumptions for the future scenarios are used in the scenario analysis, while GDP is endogenously determined in the CGE model. As is mentioned previously, economic growth is realized by labor force, capital, and TFP. Thus,

[1]Power sectors have different structure as shown in Sect. 13.3.2.

there are four degrees of freedom. Labor is exogenous as indicated previously. Total capital accumulation is also determined a priori for a year since the total capital which is able to participate in production is previous year's capital formulation plus accumulation with certain of depreciation. Then, TFP is an unknown parameter. We use the GDP assumptions to calculate TFP change.

Let the total labor, capital, and GDP be given and the Hicks-neutral technical change is assumed. Then, the TFP annual change is determined as shown below:

$$GDP_r = tfp_r \cdot F(K_r, L_r), \quad \forall r \in R \tag{13.2}$$

where

tfp_r: TFP changes ratio to previous year in region r
gdp_r: expected GDP assumptions in region r
L_r: total labor participation in region r
K_r: total accumulated capital in region r

The CES function is used for the function F. The TFP change ratio derived from the above equation is multiplied to all sectors. This TFP change calculation is made only for the scenario excluding climate mitigation (what we call BaU scenario), and the scenarios with climate change mitigation adopt the BaU scenario's TFP change under same socioeconomic assumptions.

13.3 Energy Supply

The energy supply is classified into three parts. They are fossil fuel extraction, bioenergy supply, and power supply including renewable energy. These energy supply sectors play a key role to determine the future carbon price under climate mitigation scenarios. The current amount of CO_2 emission itself from power sectors is relatively larger than other sectors, and this sector has large potential to reduce the emissions.

13.3.1 Fossil Fuel Extraction

Fossil fuel extraction costs are projected to increase in the future in line with the current situation. This is mainly due to the limitation of cheap fossil fuel resource reserves and the increase in demand.

There are three fossil fuel extraction sectors within this CGE model, namely, coal, crude oil, and natural gas. The cost of fossil fuel mining is determined by the assumptions of those sectors' production function. The production structure is as

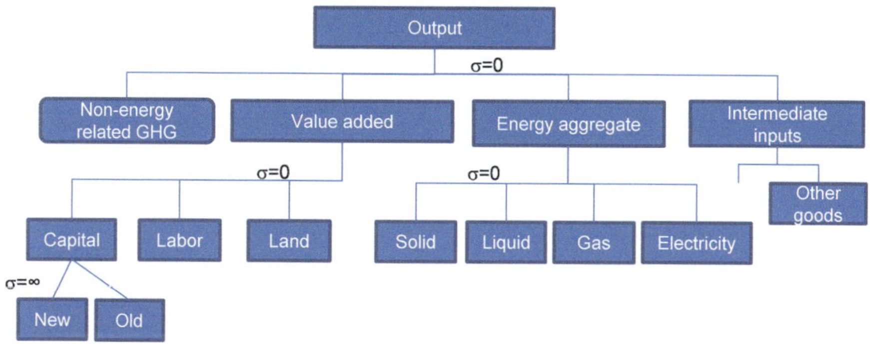

Fig. 13.1 Production structure (energy transformation sectors)

shown below. This structure is assumed for other sectors excluding energy transformation sectors.

Each producer (represented by an activity) is assumed to maximize profits, defined as the difference between revenue earned and the cost of factors and intermediate inputs. Profits are maximized subject to a production technology, the structure of which is shown in Fig. 13.1. At the top level, there is nonenergy-related GHG emission and conventional inputs. Conventional inputs technology is specified by a Leontief function of the quantities of energy, value-added, aggregate nonenergy intermediate input, and resource input.

The model calculates the extraction cost curve using Rogner (1997) and US Geological Survey (2000), and accordingly the production costs increase by region and fuel. We calculate cumulative fuel extraction every year in recursive iterations. The input of the fossil fuel sector is thought to increase based on the extraction cost curve. All input coefficients associated with production function are assumed to be increased according to the extraction cost curve.[2]

13.3.2 Power Generation

Given the electricity demand, the technological share (e.g., conventional fossil fuel fired, solar, wind, and so on) is determined by the power generation prices of each technology. A logit function is used for this selection. The power generation prices for each technology are determined by each production function.

The logit function which is used for determining power technological share is shown below:

[2]To avoid drastic change of the price, we put maximum annual change ratio as 5%.

$$SHAC_{r,a,c} = \frac{\psi^{ac}_{r,a,c} PXAC_{r,a,c}{}^{\eta^{ac}_{r,a,c}}}{\sum\limits_{ap \in A} \psi^{ac}_{r,ap,c} PXAC_{r,ap,c}{}^{\eta^{ac}_{r,ap,c}}}, \quad \forall r \in R, \ c \in C \tag{13.3}$$

where

$QXAC_{r,a,c}$: marketed output quantity of commodity c from activity a
$SHAC_{r,a,c}$: share of the commodity c produced by activity a
$\psi^{ac}_{r,a,c}$: share parameter of the commodity c produced by activity a
$\eta^{ac}_{r,a,c}$: exponent of domestic commodity aggregation
$c \in C$: a set of commodities (also referred to as c' and C')

There are two parameters in the above equation. The exponent of the logit is exogenously determined for not only future scenario simulation but also the calibration procedure, and we tentatively use the value 2.

The other parameter appeared in the equation is a share parameter for each power generation. That parameter is basically calibrated in the base year. The calibrated share parameter can be interpreted as the representation of preference, political decision, or the results of any factors other than cost. In the future simulation, we need to update this parameter since the currently unused or less used technologies such as renewable technologies are not appropriately represented in base year's share parameter. Therefore, we assumed that at a certain year, the sum of share parameters for the renewable technologies (wind, solar, and biomass) is supposed to have the same share as fossil-fired power, and the same values are shared by the renewable technologies (if the fired power share parameter is 1, the wind, solar, and biomass have 0.33 for each).[3] The year which has above treatment is tentatively 2050. To realize the objective parameters, we update the parameters with annual constant ratios for each technology.

The price of electricity generated by renewable energy source is assumed to be declined in the future. The future technological cost of renewable energy is uncertain. Hence, we follow the assumptions of Energy Technology Perspective (IEA 2012) for the reference. The input coefficients of intermediate goods and production factors in renewable energy sectors are changed overtime according to the IEA 2012s perspective. Power sector is disaggregated in detail as mentioned previously. Solar, wind, biomass, and geothermal energy power generation are included as renewable energy sources. The actual assumptions used in the model are shown in Table 13.1. The expected output price of each technology in 2050 is calculated based on these assumptions. The input coefficients of those sectors are assumed to be changed with constant ratio through 2050. This ratio is computed from the cost of base year and shown in Table 13.1. The conventional fossil-fired power is assumed to have an annual 0.5% improvement for the entire period.

[3]Thus, this approach has an advantage to foresee long-term future share of the technologies, but near-term forecast would not be relatively good at.

Table 13.1 Cost assumptions in 2050

	Solar	Wind (onshore)	Wind (offshore)
Investment cost (USD/KW)	1200	1500	2150
O&M cost (USD/KW/year)	12	30	65
Lifetime (year)	25	25	25
Technological capacity rate (–)	0.2	0.3	0.4
Interest rate (–)	0.1	0.1	0.1

13.3.3 Biomass Energy Supply

There are two types of biomass energy, what we call traditional and modern bioenergy. The former is mainly used in the low-income countries, and wood, charcoal, crop residue, and animal manure are the main energy sources. They are usually combusted directly. The latter biomass is used in two ways. One is refined for the bio-liquid fuel, and the other is combusted by power generation plants. The biofuel is used as transport sector. The demand of the biofuel will be addressed in Sect. 13.4.

The biofuel is supplied from first- and second-generation technologies. The former is made from cereals, sugars, and vegetable oils, which can easily be extracted using conventional technology. The latter is made from lignocellulosic biomass or woody crops and agricultural residues or waste, which makes it harder to extract the required fuel under current technology. This model assumes that the biomass supply is nested by logit function and there are three nodes, namely, first-generation biofuels, second-generation biofuel from energy purpose-grown crops (e.g., switch grass), and wastes (crop residue or wood residue). The three sources are individually treated as a single production sector (Fig. 13.2).

Each region has specific crops to be used for first-generation biofuel production. For instance, Brazil utilizes sugarcane while the USA produces it from maize. Therefore, we assume each region has such specific biofuel crops and only that crop can be used for biofuel production. The first-generation biofuel production sector is assumed to input such agricultural products, and the input coefficients are derived from biomass conversion energy efficiency 0.5. The other input coefficients are shown in Table 13.2.

The second-generation biofuels made from purpose-grown energy crops are assumed to input land. The yields of them for tropical, temperate, and boreal zones are assumed 15, 10, and 12 ton/hector, respectively. The other second-generation biofuels made from wastes do not input land or any agricultural commodities as production factors. Instead, this production sector is assumed to have available potential limitations which are calculated from the amount of crop residue, wood residue, and livestock manure generated in previous years. We assumed that half of the potential can be practically or economically available for the energy fuel production.

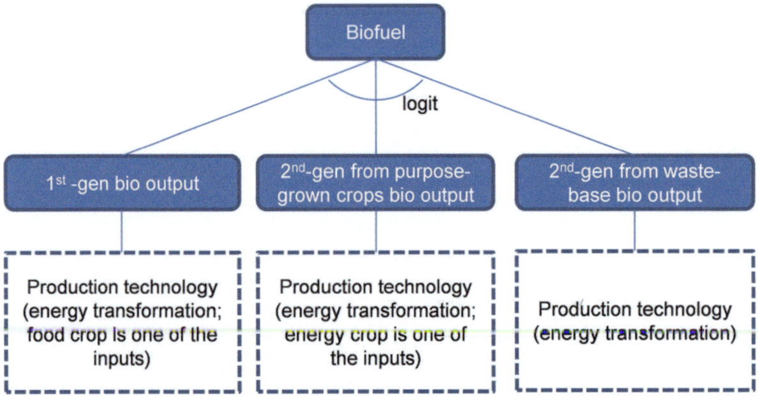

Fig. 13.2 Biomass supply

Table 13.2 Input coefficients of biofuel production (thousand $ per toe of biofuel production)

	1st gen	2nd gen from purpose-grown energy crops	2nd gen from waste
Agricultural products[a]	2.000		
Energy crop[a]		2.000	2.000
Chemistry	0.083	0.124	0.124
Transport	0.025	0.083	0.083
Other services	0.025	0.041	0.041
Labor	0.141	0.124	0.124
Capital	0.141	0.430	0.455

[a]Agricultural products and energy crops input are accounted as toe per toe of biofuel production

13.3.4 Other Energy Transformation Sectors

Other than those of energy transformation sectors such as petroleum refineries and town gas distribution sectors assumed to keep the same production structure as shown in power sector. Its input coefficients are also constant including energy transformation efficiency.

13.4 Energy Demand

The energy demand, in this document, is classified as final energy consumption defined in the energy balance table (International Energy Agency 2013). It generally consists of five or six sectors: industry, transportation, residential, commercial, nonenergy use, and other sectors.[4] Actual AIM/CGE has more disaggregated sectors, and, for example, iron and steel and food processing are classified as part of industrial sectors. The model has two options on how to determine these energy demands. One is using traditional functions such as CES function for production sectors (Fujimori et al. 2014a). The other option enables to consider bottom-up energy technological information and the energy demand explicitly determined by the detailed energy technologies (Fujimori et al. 2014b). In this section, these two options are separately explained.

13.4.1 Conventional Approach

Traditional approach to determine energy demand is using CES function for the production sectors and LES for residential sectors.

13.4.1.1 Production Sectors

1. Total energy

The general production structure of energy end-use sectors is illustrated in Fig. 13.3. The energy demand is associated with the bundle of value added and energy. The aggregated energy and value-added branch are assumed a fixed coefficient to output of that sector, while the energy and value added are assumed to have substitution relationship.

There are three main factors to determine energy demand. First is the production itself, and it is driven by the value-added growth which consists of labor, capital, and their productivitics. In other words, GDP growth drives energy demand. However, the GDP and energy demand are not linearly linked, and usually energy demand growth rate is slower than GDP; even the relationship of energy and value-added prices is constant. This is the second factor to change the energy demand and realized by technological progress (the effect of more efficient energy device) or energy service demand efficiency improvement (e.g., floor space of office is not so expand as the economic value of output), or mixture of them. We normally call this factor as AEEI (autonomous energy efficiency improvement) and multiply coefficients to the energy efficiency. The AEEI is changed as a function of GDP growth. In principle, the AEEI is high when a country has a high GDP growth rate, whereas it is low in low GDP growth areas (van Ruijven et al. 2010). If GDP growth is negative, AEEI is fixed as zero. If GDP growth ranges from 0–3%, 3–5%, and over 5%, annual AEEI is assumed to be 1%, 1.5%, and half of the GDP growth percentage, respectively.

---

[4]Residential and commercial are sometimes treated as single aggregated building sector.

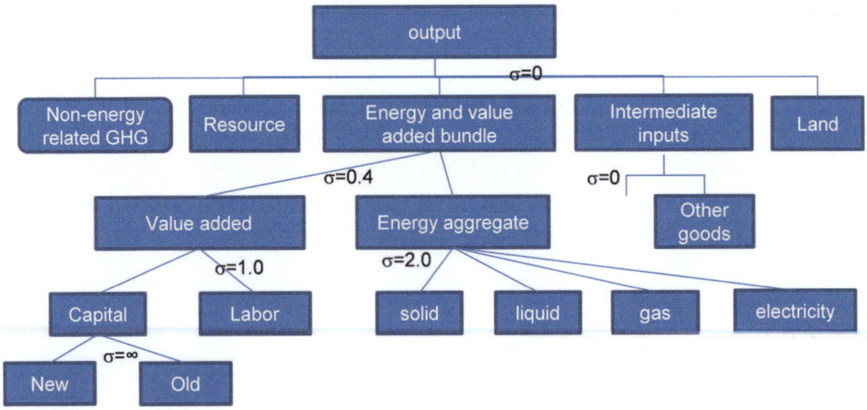

Fig. 13.3 Production structure of nonenergy transformation sectors

The third one is price factor. The ratio of energy and value-added inputs are determined by those of relative price relationship. The relative price and volume input relationship can be derived from the maximization profit condition subject to the CES production function. The ratio of value added and energy is determined by the share parameters (which are usually calibrated from base year's data) and the price ratio associated with substitution elasticity (Eq. 13.4):

$$\frac{QVA_{r,a}}{QENE_{r,a}} = \left(\frac{\delta^{vae}_{r,a}}{1-\delta^{vae}_{r,a}} \cdot \frac{PENE_{r,a}}{PVA_{r,a}}\right)^{\frac{1}{1+\rho^{vae}_{r,a}}}, \quad \forall r \in R, \quad a \in ACES \qquad (13.4)$$

where

$\alpha^{vae}_{r,a}$:	efficiency parameter in the CES energy and value-added function
$\delta^{vae}_{r,a}$:	CES energy and value-added function share parameter in activity a
$\rho^{vae}_{r,a}$:	CES energy and value-added function exponent
$PVA_{r,a}$:	price of (aggregate) value added
$QVA_{r,a}$:	quantity of (aggregate) value added
$PENE_{r,a}$:	price of (aggregate) energy input
$QENE_{r,a}$:	quantity of (aggregate) energy input
$a \in ACES$:	a set of activities which are nonenergy transformation sectors

2. Energy mix

The next story is how to determine the energy source mix to satisfy final energy demand. Energy has several forms such as solid, liquid, gaseous, and so on. The energy mix is associated with each sector's activity and its technologies. For instance, the boiler could use any types of energy to obtain certain amount of energy, while current car uses liquid (partly electricity and gas) but not solid-type fuel. Therefore, the energy source mix is not determined by simple cost minimization mechanism. However, the relatively cheaper energy sources may be attractive than others. To meet these mechanisms and to keep simplicity to solve the large system equilibrium model, the energy source mix is assumed to be determined by the CES function. The elasticity of substitution among the energy forms is 2.0 (Fig. 13.3). The AEEI should differ across energy sources to reflect the energy consumption composite switch from coal to oil, gas, or electricity. Therefore, coal, gas, and electricity are assumed to have 1%, −0.5%, and −1% annual changes. Because previous studies did not report these values, these numbers are arbitrarily assigned. The AEEI for the traditional biomass usage was assumed to be 1% annually.

13.4.1.2 Residential Sectors

The household consumption is determined by the LES as explained in Sect. 13.2.1. The energy demand is treated as two kinds of household goods. One is associated with the car usage energy demand, and the other is rest of the energy demand. There is a definition difference between the energy balance table and social accounting matrix in terms of how to account the energy consumption by private household cars. The social accounting matrix deals with the fuel consumption in order to drive household car is accounted as part of household consumption, while energy balance treats it as transport sector's energy consumption. To deal with the household energy demand consistently, we separately treat the car fuel and rest of energy.

The total car energy consumptions or other energy demand are determined goods prices and income of household under LES. We basically assumed 1.0 and 0.5 as income elasticity for the car energy usage and rest of household energy consumption. The income elasticity can be derived from the LES function, and it is fixed assumption for the future simulation. However, the derived income elasticity depends on the parameter $\beta^{m}_{r,ch,h}$ and goods prices (Eq. 13.5). In order to realize the constant income elasticity, the parameters are calibrated recursively.

Energy source mix is determined by logit function for each energy demand. The structure of household energy demand is shown in Fig. 13.4.[5] The exponent parameters for the car fuel are −1.0, and those of other energy demand are 0.2:

$$\frac{\partial QCH_{r,ch,h}}{\partial EH_{r,h}} = \frac{\beta^{m}_{r,ch,h}}{w_{r,ch,h}}, \quad \forall r \in R, \quad ch \in CH, \quad h \in H \tag{13.5}$$

$w_{r,ch,h}$: budget share of consumption spending on household commodity ch for household h

Currently the energy balance table does not account electricity consumption for road transport. However, electric vehicle could be one of the options in the future. Hence, the share parameter of the logit function is updated in the future. So in year

[5]Equation (13.1) describes upper part of this figure.

2015, the electric vehicle is assumed to start, and at that year, 0.1% of the share is assumed. This share is used for the logit functions' share parameter. Then, after that year the share parameter is linearly changed according to the electricity price. The biofuel for road transport is also similarly treaded, but the starting year is 2006.

13.4.2 Utilizing Detailed Energy Device Information

The other option to determine energy end-use demand is to utilize detailed energy device information, while the simplified function is used in the previous section. The overall structure is shown in Fig. 13.5.

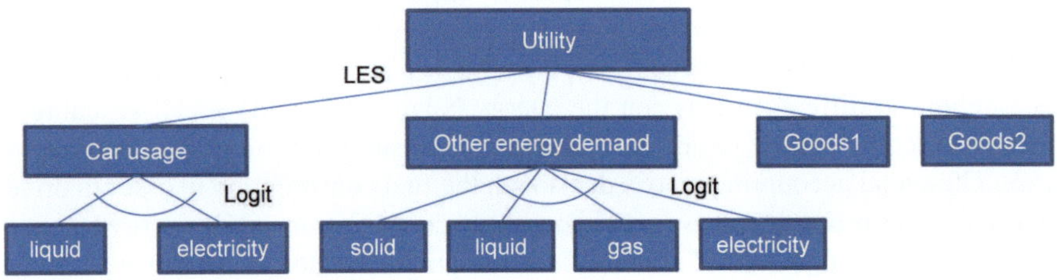

Fig. 13.4 Household demand structure

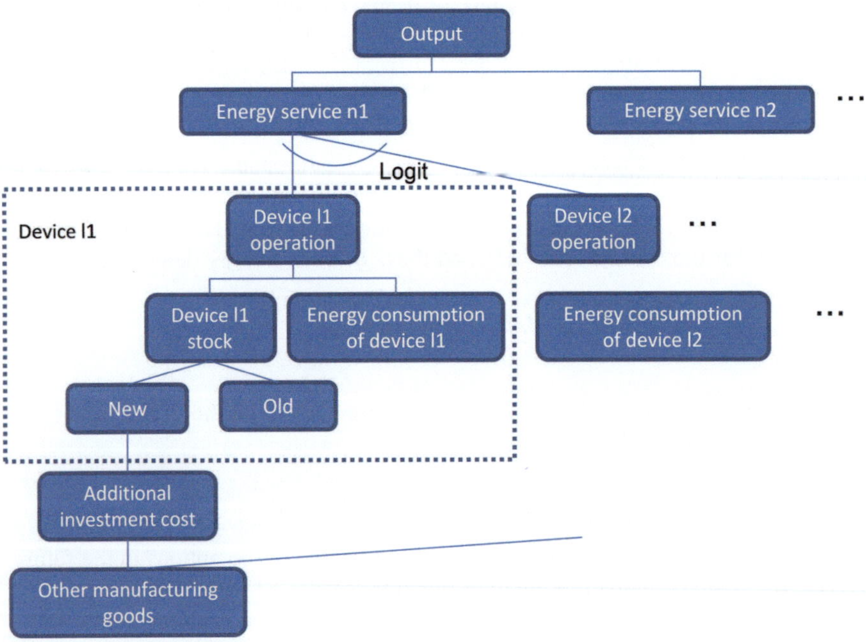

Fig. 13.5 Detailed energy device selection and energy demand structure

Each production sector inputs several types of energy service demand. The energy service demand is, for instance, energy requirement of boiler and the amount of operation of electric equipment in manufacturing sectors. To fulfill the energy service demand, the operation and stock of energy device are required. The energy device stocks consist of old and newly installed capital. Operating the devices requires energy consumptions and summation of the energy requirement caused by all of the energy devices, and energy services become a sector's energy consumption. The key determinant of the energy is how to distribute the share of the energy devices. Logit function is used for determining the share of energy device.

The device share determining the equation is expressed as Eq. (13.6). This equation means that the device which cost is expensive has relatively low share than cheap devices. The sensitivity to the cost is represented by the exponent:

$$SHDV_{r,a,l} = \frac{\beta_{r,a,l} CDV_{r,a,l}{}^{\rho_{r,a,l}}}{\sum_{ll \in LM} \beta_{r,a,ll} CDV_{r,a,ll}{}^{\rho_{r,a,ll}}} \tag{13.6}$$

where $SHDV_{r,a,l}$ is the share of sector a, device l; $CDV_{r,a,l}$ is annualized with a certain discount rates device cost of sector a, device l; $\beta_{r,a,l}$ is share parameter; and $\rho_{r,a,l}$ is an exponent. The device cost $CDV_{r,a,l}$ consists of components, namely, investment cost and operation and management (O&M) cost. Part of O&M cost is energy or related GHG emission tax cost.

The energy service demand is determined by multiplying the change ratio of the volume of a sector output from the previous year with elasticity. This assumption is obviously simplified and a source of uncertainty of the model analysis:

$$QSD_{r,a,i} = sdp_{r,a,i} \cdot \left(\frac{QA_{r,a}}{qap_{r,a}} \right)^{indserel_{r,a,i}}, \tag{13.7}$$

where $QSD_{r,a,l}$ is energy service demand quantity of energy service i in section a, $sdp_{r,a,l}$ is energy service demand quantity of energy service i in section a in previous year, $QA_{r,a}$ is a volume of output sector a, $qap_{r,a}$ is a volume of output sector a in previous year, and $indserel_{r,a,i}$ is an elasticity of energy service demand i to output volume change in sector a. $indserel_{r,a,i}$ is assumed to be 0.75 for all services.

The new investment cost is associated with the additional input of "other manufacturing." This is described as shown below:

$$QINT_{i,j} = ica_{i,j} \cdot QA_j + \sum_{l \in L} \{ dc_{l,j} \cdot (QXD_{j,l} - qxdb_{j,l}) \} \quad \forall i \in \text{Iothm} \tag{13.8}$$

where $QINT_{i,j}$ is intermediate inputs of sector j and goods i, $ica_{i,j}$ is intermediate input coefficient of sector j and goods i, $dc_{i,j}$ is annualized investment cost of device l in sector j, and $qsdb_j$ is base year's annualized device investment cost of device l in sector j. $i \in Iothm$; a set of other manufacturing goods.

If the carbon emission is priced and the low carbon or efficient technologies are selected. However, those are basically expensive technologies; the additional demand is put on the "other manufacturing" goods, and it eventually makes additional capital and labor to the "other manufacturing." Generally speaking, CGE models implicitly treat device investment costs as part of the capital cost, and our approach is different from that. One reason is that some energy devices which are not classified as capital by national account definition, e.g., lighting and heating, are accounted as intermediate inputs. Second, household consumption does not have the concept or definition of capital inputs, but they are treated as household goods consumptions. The other thing that should be noted here is that the cost relative to base year is taken into account as the additional cost and not direct absolute cost. This is due to the difficulties of extracting the energy device cost from social accounting matrix or input-output table. Base year's energy device cost is accounted as the part of capital, intermediate inputs, or household goods consumptions, and hence, the additional cost relative to base year is accounted in this model.

These structures and mechanism are applied to the household as well. In this case, the LES determines other than energy goods, and the energy goods are independently determined by the energy device information. The household energy service demand income elasticity is assumed 0.25 except for the private passenger car by household. The income elasticity of the passenger car service is assumed 0.75.

The list of energy service and devices are shown in Table 13.3. They are based on AIM/Enduse database (Akashi et al. 2012).

There are three parameters related to energy end-use technological assumptions depending on scenarios, namely, share parameters in the logit function, discount rates, and investment cost.

The share parameters are assumed to be changed by two aspects. First, the devices which have not introduced in base year are expected to be introduced in future. We assumed very small share in 2006, and we change the parameter that should be equalized among the options in 2050. Second, the coal and traditional biomass consumption has an externality generated by air pollutants in the residential and commercial sectors, and thus, these shares are expected to be decreased according to the income increase. We tentatively assume the share parameters of these devices in 2050 to be one tenth of base year since mostly the income is assumed to be increased particularly in developing countries.

The discount rate is also changed across scenarios. The basic assumption is based on Akashi and Hanaoka (2012). The device investment cost is expected to be lower in the future due to the technological progress. The default assumption is 0.5% annual improvement.

One thing that should be noted here is that steel industry is not treated as described in this section, but traditional CES function is used. It is because the activity of the steel industry has linkages with energy transformation. Considering such activities appropriately is one of the remained issues for our modeling.

Table 13.3 List of energy services and devices

Sector	Category	Technology options
Industry	Cement	Mill (e.g., ball mill, tube mill, vertical mill), kiln (e.g., wet kiln, semi-wet/dry kiln, dry long kiln, vertical shaft kiln, SP/NSP, advanced kiln with CCS)
	Other industries[a]	Boiler [level 1–5] [e.g., efficient boiler (coal, oil, gas), boiler with combustion control (coal, oil, gas), cogeneration (coal, oil, gas), regenerative gas boiler], process heat [e.g., efficient industrial furnace (oil, gas)], motors [level 1–2] (e.g., motor with inverter control, efficient motor)
Building	Cooling	Efficient air conditioner [level 1–3] (e.g., sold average in developed countries in 2005, top runner, highest performance)
	Warming	Efficient air conditioner [level 1–3] (e.g., sold average in developed countries in 2005, top runner, highest performance), thermal insulation (e.g., wall insulation for detached house, wall insulation, double-glazed glass with low-e)
	Hot water	Efficient water heater (kerosene, LPG, gas, coal), latent heat recovery water heater (kerosene, LPG, gas), CO_2 refrigerant heat pump water heater [level 1–3], solar thermal water heater, electric water heater, fuel cell
	Cooking	Efficient cooking stove (kerosene, LPG, gas, coal), electric IH cooker
	Lighting	Efficient fluorescent lamp [level 1–2], efficient compact fluorescent lamp [level 1–2], LED lamp
	Refrigerator	Efficient refrigerator [level 1–3] (sold average in developed countries in 2005, top runner, highest performance)
	TV	Efficiency TV [level 1–3] (sold average in developed countries in 2005, top runner, highest performance), liquid crystal display TV
Transport[b]	Passenger car	Efficient vehicle [level 1–3] (gasoline, diesel) [e.g., weight reduction, engine friction reduction, aerodynamic drag reduction, rolling resistance reduction, brake drag reduction, continuously variable transmission (CVT), variable valve life and time (VVLT) and cylinder reactivation, direct injection], hybrid electric vehicle (HEV) (gasoline, diesel), plug-in HEV (gasoline, diesel), electric passenger vehicle (EV), fuel cell passenger vehicle (FCV), biofuel
	Truck	Efficient small-sized truck (e.g., rolling resistance reduction, engine improvement, weight reduction, aerodynamic drag reduction), hybrid small-sized vehicle, small-sized electric vehicle, small-sized fuel cell vehicle, efficient large-sized truck, hybrid large-sized vehicle, natural gas vehicle, biofuel
	Passenger bus	Efficient bus (e.g., rolling resistance reduction, engine improvement), hybrid bus, biofuel
	Ship	Efficient ship, biofuel
	Aircraft	Efficient aircraft (e.g., engine improvement, weight reduction, drag reduction), biofuel
	Rail	Efficient train (electricity, diesel) (e.g., regenerative braking system with VVVF)

Taken from Akashi and Hanaoka (2012)

[a]Other industries have five energy services: boiler, process heat, motors, electrochemistry, and others

[b]Rail, navigation, and aviation have freight and passenger services

13.5 Agriculture and Land Use

Regarding agriculture and land use, there are three topics that should be discussed in this section. First is how the agricultural commodities' productions are determined. Second is the demand-side story. How the demand is modeled and traded is also included. Third, the production side has land completion, and how the land completion is realized within this model is explained.

13.5.1 Agricultural Commodities Production

Producers are assumed to maximize profits subject to technology (production functions) and prices of inputs. The first-order conditions for profit maximization essentially define the factor demands and output supply behavior of producers. The production structure is same as other industrial sectors except for land input treatment. The land input is assumed multiplying a coefficient to output. However, in some cases, this fixed coefficient approach makes it difficult to solve the program if the land constraint is substantially critical. Therefore, the term related to output price elasticity is multiplied to the fixed coefficient. The price elasticity is very small (say 0.05), and the model results can be interpreted as the land input is almost treated as Leontief-type input. Whether the treatment that land input is fixed technology has reality or not should be concerned carefully. The main technologies related to improving land efficiency are fertilizer inputs, advanced cultivar, optimizing planting date, and irrigation. When we consider the climate change impact, the constraints caused by climate change are water and temperature. These conditions can be partly controlled by the latter three technologies. The cultivars and planting date are what we call autonomous adaptation, and it would not change the production cost so much. GAEZ (Masutomi et al. 2009) results are usually fed into this CGE model to analyze climate change impact, and within GAEZ framework, the cultivars and planting date are already considered. Therefore, the last irrigation is the remained technology which could change the yield and production cost. However, the irrigation is not always accounted as production cost but as social infrastructure (reservoirs and canal). It makes it difficult to analyze the relationship of production cost and irrigation technologies. Tentatively, we don't touch about this issue in this version, but we need to consider this limitation to interpret the results of this model.

The future land input coefficient, which is yield, is exogenous assumption and based on IFPRI's IMPACT (Msangi et al. 2010), and Table 13.4 shows actual numbers used in the default scenario. It does not include climate change impact.

The livestock land productivity is assumed to be corresponding to the livestock goods demand change. For instance, if the meet demand is supposed to be 1.5-folds, then its land productivity is also assumed as well. This mechanism is based on the

Table 13.4 Future yield assumptions (annual growth percentage)

	Rice (%)	Wheat (%)	Other grains (%)	Oil crops (%)	Sugar crops (%)
USA	0.7	1.0	1.1	0.8	0.7
XE25	1.2	0.3	0.8	0.9	0.4
XER	1.0	1.3	1.3	1.4	1.1
TUR	0.2	1.0	0.7	0.5	1.1
XOC	0.1	1.0	1.3	0.6	0.4
CHN	0.8	0.9	1.5	1.0	0.9
IND	1.3	0.8	2.1	0.8	0.5
JPN	0.3	0.6	0.6	0.2	0.6
XSE	1.0	1.8	1.4	−0.4	0.8
XSA	0.9	1.4	1.4	0.9	1.4
CAN	–	1.8	1.6	0.1	1.0
BRA	0.8	1.7	1.8	0.9	1.1
XLM	0.6	1.3	1.4	0.6	1.1
CIS	1.7	2.0	2.1	1.4	1.4
XME	1.2	2.2	2.0	1.0	1.1
XNF	0.7	1.3	1.3	1.3	0.8
XAF	2.0	2.1	1.5	0.9	0.8

historical evidence that we could not find any relationship between grazing land productivity and other factors like economic development level, whereas the grazing area looked rather constant.

13.5.2 Demand of Agricultural Commodities

As Sect. 13.2.1 shows, the household expenditure is represented by LES function. The parameters dealt with LES are changed over time (written in Sect. 13.4.1.2).

At least food-related goods are crucial for this model; their parameters are carefully treated, since they are physically accounted in this model. We assumed food consumption income elasticity according to two sources. One is the historical observation. If we could find the relationship between the income level and food consumption, then that relationship is adopted. For the other goods we follow FAO's perspective (Food and Agriculture Organization of the United Nations 2013). The livestock goods are former cases, and one of the examples is plotted in Fig. 13.6. The latter goods income elasticity is shown in Table 13.5.

As for intermediate inputs not used as the feeding for livestock is determined by income and price elasticities derived from the household consumption function.

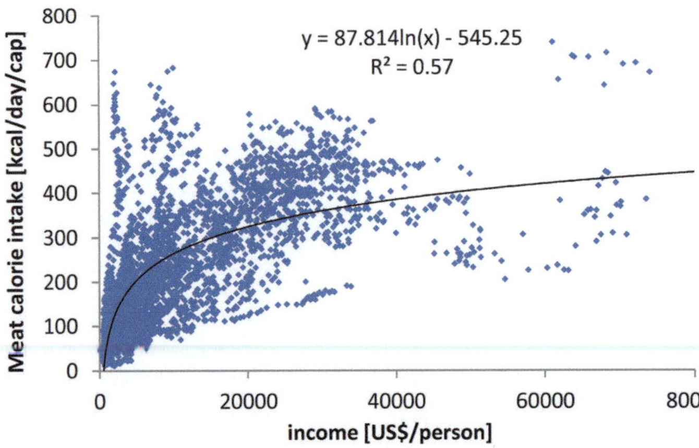

Fig. 13.6 Time series and cross-country data on meat calorie intake and income (1980–2009, source: World Bank (2013) for income and Food and Agriculture Organization of the United Nations (2013) for calorie intake)

Table 13.5 Income elasticity of crop consumption

	Cereal	Oil crops	Sugar crops	Other crops
USA	−0.04	0.10	−0.03	−0.14
XE25	−0.04	0.10	−0.03	−0.14
XER	−0.03	0.27	0.05	−0.04
TUR	−0.02	0.10	0.04	−0.01
XOC	−0.04	0.10	−0.03	−0.14
CHN	−0.06	0.19	0.22	−0.09
IND	0.03	0.26	0.10	0.18
JPN	−0.04	0.10	−0.03	−0.14
XSE	−0.01	0.23	0.12	0.08
XSA	0.03	0.26	0.10	0.18
CAN	−0.04	0.10	−0.03	−0.14
BRA	0.04	0.24	−0.02	−0.07
XLM	0.04	0.24	−0.02	−0.07
CIS	−0.03	0.27	0.05	−0.04
XME	−0.04	0.16	0.06	−0.02
XNF	−0.04	0.16	0.06	−0.02
XAF	0.22	0.37	0.39	0.07

13.5.3 Land Competition and Allocation Mechanism

A function whereby land is an input for the production of crops and livestock products, and landowners change its use in accordance with the prices of producer goods on cropland, pastureland, and forest. The model has a land nesting strategy, which is similar to the treatment in Sands and Edmonds (2005) and Wise and Calvin (2011). Land is categorized in one of three ecological zones, and there is a land

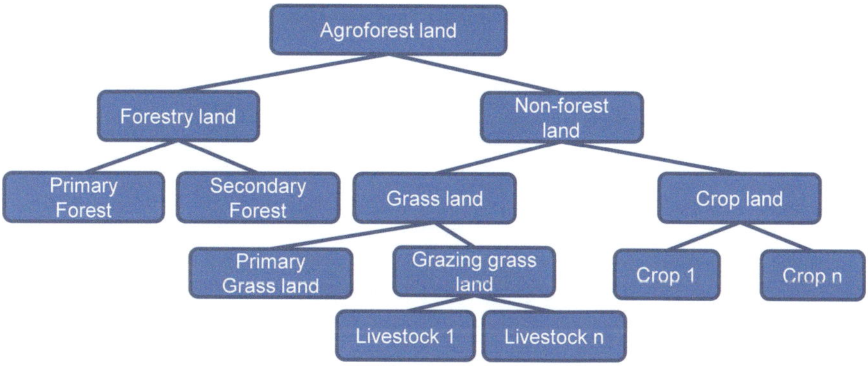

Fig. 13.7 Land allocation structure

market for each zone. Allocation of land by sector is formulated as a multi-nominal logit function to reflect differences in substitutability across land categories with land rent. In that, the function assumes that landowner of each region and Agro-ecological Zone (AEZ) decides land sharing among options with the land rent depending on production on each land (i.e., crops, livestock, and wood products). We deal with all land excluding desert, rock, ice, tundra, and built-up land.

Figure 13.7 shows the nesting diagram of land with an AEZ classification. We deal with all land, excluding desert, rock, ice, tundra, and built-up land. There are 18 AEZ classifications. At the top is all land, which is divided into two main types of nodes: forest land and non-forest land. The forest land node contains two competing uses: primary forest (unmanaged forest) and secondary forest (managed forest). The non-forest land could be divided, all grassland and cropland. The grassland could be divided into primary grassland (unmanaged pasture) and grazing grassland (managed pasture which feeds marketed livestock) which is divided further into each livestock (1–n). The cropland could be divided further into cropland for each crop (1–n) and fallow land. One approach of the nesting strategies is based on the assumption that the land regions are small enough that all competing options are equally substitutable. This assumption implies that it is easy to switch from forest to wheat as it is to switch from corn to wheat. However, this conversion would not happen unless wheat was more profitable than forest or corn. In that, the function assumes that landowner of each region and AEZ subregion decides land sharing among options depending on the land rent from production on each land (i.e., crops, livestock, and wood products). To calibrate the function for both of the managed and unmanaged lands in the base year, we used the mean base year land rent of the managed land as that of the unmanaged land because of unavailable data for the unmanaged land. Carbon stock on forest land is evaluated by price in the case of climate mitigation scenario. Land rent of forest includes both of revenue of wood product and the price of the carbon stock.

13.6 GHG Reduction Measures Other than Changing Energy System

13.6.1 CCS (Carbon Capture and Storage)

CCS technology is one of the key technologies for climate mitigation, and this subsection explains how it is treated in the model. CCS captures the CO_2 using chemical processes and stores the carbon underground or in the deep sea. It is mainly available for the large point CO_2 emission sources. Fired power plants, biomass power plants, oil refineries and coal transformation plants, nonmetal and mineral, chemical, and paper and pulp industries are supposed to be able to apply CCS technology in this study. These sectors input CCS services as intermediate inputs, and the CCS service is assumed to be provided by a CCS service sector which has independent production function.

The costs of the technology are different among sectors and are shown in Table 13.6. These costs have been taken from IEA (2008). Since IEA (2008) provides a range of the cost estimates, we took medium values. When the GHG emission price becomes higher than these costs, CCS technology is installed with a maximum increase ratio of 5% per year. CCS is assumed to be installable after 2020 since it is still under examination.

13.6.2 Non-CO_2 Reduction

As for the nonenergy-related GHG emissions reductions in areas such as agricultural CH4 and N2O emissions, the following equation is used. A similar method is already discussed by Hyman et al. (2003). The parameter is taken from Lucas et al. (2007):

Table 13.6 CCS technology cost

	Sectors	Price (US$/t$CO_2$)
Manufacturing	Petroleum refinery coal transformation	100
	Nonmetal and mineral	200
	Paper and pulp	150
	Chemical	150
Power	Coal fired	50
	Oil fired	70
	Gas fired	70
	Biomass fired	70

$$RD_{r,j,g} = (PGHG_r + 1)^{-\sigma^{ner}} \tag{13.9}$$

where

$RD_{r,j,g}$: nonenergy-related GHG emissions reductions ratio relative to the reference scenario for region r, sector j, and gas g ($/t$CO_2$)

$\sigma^{ner}_{r,j,g}$: the elasticity parameter of nonenergy reduction ratio relative to the reference scenario for region r, sector j, and gas g

13.6.3 Land Use-Related Countermeasures

Once carbon is priced, the stock of the forestry is potentially assumed to have value. This carbon value is assumed to be additional land rent for the forestry, and it makes forestry larger. The value itself is discounted carbon value. Lifetime of forestry that can absorb carbon is assumed as 60 years and interest rate is 5%. The amount of carbon absorption made by afforestation is depending on the age of the trees. The carbon stock is differentiated across AEZ.

13.7 How to Implement New Production Sectors or Goods (Not Accounted in the Base Year)

The sectors or goods which are not recorded in the base year (e.g., solar power, biofuel, and so on) need to be assumed explicitly in some future years. Advanced renewable energy or CCS technologies are under this treatment. The list of the sectors and goods and the starting years are shown in Table 13.7.

Table 13.7 List of sectors and goods which are not accounted base year but newly introduced in the future scenarios and the starting years

	Goods or sectors	Starting year
Goods	Biofuel	2006
	CCS service for cement	2021
	CCS service for furnaces	2021
	CCS service for other manufacturing	2021
	CCS service for coal-fired power plant	2021
	CCS service for other fired power plant and biomass refinery	2021
Sectors	1st-gen biofuel	2006
	2nd-gen biofuel from purpose-grown energy crops	2015
	2nd-gen biofuel from waste base	2015
	Purpose-grown energy crops	2015
	Nuclear	2006
	Wind	2006
	Solar	2006
	Biomass-fired power	2006

To implement those sectors or goods, the production input coefficients or demand of the goods should be formulated. The same production structures as those of the USA are assumed for the nuclear, renewable power energies. The coefficients of the production are shown in Table 13.8 for power generation. The demand of the generated power is assumed very small portion in the starting year (0.1% of the total generation), and this assumption is made on the share parameter discussed in Sect. 13.3.2.

The CCS technologies input five factors as labor, capital, chemical products, transport, and other services. The cost shares are assumed 0.1, 0.4, 0.1, 0.3, and 0.1, respectively. The assumed input coefficients are based on (IEA 2008).

13.8 List of Regions and Industries

The list of regions for global model and industries is shown as in Tables 13.9 and 13.10. For the national assessment, we use a single country model which fixes international price and uses exactly same assumption and model structure as the global model.

Table 13.8 Input coefficients for power generation (thousand $ per toe of electricity production)

	Nuclear	Solar	Wind	Biomass fired
Energy crops[a]				2.5
Other manufacturing	0.022	0.027	0.027	0.058
Electricity[a]	0.067			
Construction	0.041	0.049	0.049	0.099
Transport	0.068	0.081	0.080	0.161
Other services	0.106	0.126	0.125	0.250
Labor	0.095	0.113	0.113	0.226
Capital	0.545	0.537	0.538	0.206

[a]Electricity and energy crop inputs are accounted as toe per toe of electricity production

Table 13.9 List of regions

Code	Region	Code	Region
JPN	Japan	TUR	Turkey
CHN	China	CAN	Canada
IND	India	USA	United States
XSE	Southeast Asia	BRA	Brazil
XSA	Rest of Asia	XLM	Rest of South America
XOC	Oceania	XME	Middle East
XE25	EU25	XNF	North Africa
XER	Rest of Europe	XAF	Rest of Africa
CIS	Former Soviet Union		

Table 13.10 List of industries

Agricultural sectors	Energy supply sectors	Other production sectors
Rice	Coal mining	Mineral mining and other quarrying
Wheat	Oil mining	Mineral mining and other quarrying
Other grains	Gas mining	Food products
Oil seed crops	Petroleum refinery	Textiles and apparel and leather
Sugar crops	Coal transformation	Wood products
Other crops	Biomass transformation (1st generation)	Paper, paper products, and pulp
Ruminant livestock	Biomass transformation (2nd generation with energy crop)	Chemical, plastic, and rubber products
Raw milk	Biomass transformation (2nd generation with residue)	Iron and steel
Other livestock and fishery	Gas manufacture distribution	Nonferrous products
Forestry	Coal-fired power	Other manufacturing
	Oil-fired power	Construction
	Gas-fired power	Transport and communications
	Nuclear power	Other service sectors
	Hydroelectric power	CCS service
	Geothermal power	
	Photovoltaic power	
	Wind power	
	Waste biomass power	
	Other renewable energy power generation	
	Advanced biomass power generation	

References

Akashi O, Hanaoka T (2012) Technological feasibility and costs of achieving a 50% reduction of global GHG emissions by 2050: mid- and long-term perspectives. Sustain Sci 7(2):139–156. doi:10.1007/s11625-012-0166-4

Akashi O, Hijioka Y, Masui T, Hanaoka T, Kainuma M (2012) GHG emission scenarios in Asia and the world: the key technologies for significant reduction. Energy Econ 34:S346–S358. doi:10.1016/j.eneco.2012.04.011

Food and Agriculture Organization of the United Nations (FAO) (2013) FAOSTAT. http://faostat.fao.org/default.aspx, Rome, Italy

Fujimori S, Kainuma M, Masui T, Hasegawa T, Dai H (2014a) The effectiveness of energy service demand reduction: a scenario analysis of global climate change mitigation. Energy Policy 75(0):379–391. doi:http://dx.doi.org/10.1016/j.enpol.2014.09.015

Fujimori S, Masui T, Matsuoka Y (2014b) Development of a global computable general equilibrium model coupled with detailed energy end-use technology. Appl Energy 128(0):296–306. doi:http://dx.doi.org/10.1016/j.apenergy.2014.04.074

Hasegawa T, Fujimori S, Takahashi K, Yokohata T, Masui T (2016) Economic implications of climate change impacts on human health through undernourishment. Clim Chang 136 (2):189–202

Hyman R, Reilly J, Babiker M, De Masin A, Jacoby H (2003) Modeling non-CO_2 Greenhouse Gas Abatement. Environ Model Assess 8(3):175–186. doi:10.1023/A:1025576926029

IEA (2008) CO_2 capture and storage – a key carbon abatement pption. OECD/IEA

IEA (2012) Energy Technology Perspectives 2012

International Energy Agency (IEA) (2013) Energy balances for OECD countries. International Energy Agency, Paris

Lucas PL, van Vuuren DP, Olivier JGJ, den Elzen MGJ (2007) Long-term reduction potential of non-CO_2 greenhouse gases. Environ Sci Pol 10(2):85–103. doi:10.1016/j.envsci.2006.10.007

Masutomi Y, Takahashi K, Harasawa H, Matsuoka Y (2009) Impact assessment of climate change on rice production in Asia in comprehensive consideration of process/parameter uncertainty in general circulation models. Agric Ecosyst Environ 131(3–4):281–291. doi:10.1016/j.agee.2009.02.004

Msangi S, Ewing M, Rosegrant MW, Zhu T (2010) Biofuels, food security, and the environment: a 2020/2050 perspective, pp 65–94. doi:10.1007/978-3-642-04615-5_4

Rogner H-H (1997) An assessment of world hydrocarbon resources. Annu Rev Energy Environ 22(1):217–262. doi:10.1146/annurev.energy.22.1.217

Sands R, Edmonds J (2005) Climate change impacts for the conterminous USA: an integrated assessment. Clim Chang 69(1):127–150. doi:10.1007/s10584-005-3616-5

U.S. Geological Survey (2000) U.S. geological survey world petroleum assessment 2000 description and results

van Ruijven B, de Vries B, van Vuuren DP, van der Sluijs JP (2010) A global model for residential energy use: uncertainty in calibration to regional data. Energy 35(1):269–282. doi:10.1016/j.energy.2009.09.019

Wise M, Calvin K (2011) GCAM 3.0 Agriculture and land use; technical description of modeling approach

World Bank (2013) World development indicators. http://data.worldbank.org/6data-catalog/world-development-indicators

MIX
Papier aus verantwortungsvollen Quellen
Paper from responsible sources
FSC® C105338

If you have any concerns about our products,
you can contact us on
ProductSafety@springernature.com

In case Publisher is established outside the EU,
the EU authorized representative is:
Springer Nature Customer Service Center GmbH
Europaplatz 3, 69115 Heidelberg, Germany

Printed by Libri Plureos GmbH
in Hamburg, Germany